D1358008

Genetic Programming ~

An Introduction

On the Automatic Evolution of Computer Programs and Its Applications

Genetic Programming ~

An Introduction

On the Automatic Evolution of Computer Programs and Its Applications

Wolfgang Banzhaf

Peter Nordin

Robert E. Keller

Frank D. Francone

Morgan Kaufmann Publishers, Inc.
San Francisco, California

dpunkt
Verlag für digitale Technologie GmbH
Heidelberg

Copublished by dpunkt.verlag and Morgan Kaufmann Publishers, Inc.

Morgan Kaufmann Publishers, Inc.

Sponsoring Editor	Michael B. Morgan
Production Manager	Yonie Overton
Production Editor	Elisabeth Beller
Cover Design	Ross Carron Design
Cover Photo	Chris Howes/Masterfile
Cover Illustration	Cherie Plumlee
Proofreader	Robert Fiske
Printer	Courier Corporation

dpunkt.verlag

Sponsoring Editor	Michael Barabas
Production Manager	Josef Hegele
Copyeditor	Andrew Ross

This book has been author-typeset using LaTEX.

Available in Germany, Austria, and Switzerland from
dpunkt —Verlag für digitale Technologie GmbH
Ringstrasse 19
D-69115 Heidelberg
Germany
Telephone +49/6221/1483-12
Facsimile +49/6221/1483-99
Email hallo@dpunkt.de
WWW www.dpunkt.de

Available in all other countries from
Morgan Kaufmann Publishers, Inc.
Editorial and Sales Office
340 Pine Street, Sixth Floor
San Francisco, CA 94104-3205
USA
Telephone 415/392-2665
Facsimile 415/982-2665
Email mkp@mkp.com
WWW www.mkp.com

Order toll free 800/745-7323

Printed in the United States of America

02 01 5 4 3

Library of Congress Cataloging-in-Publication Data
Genetic programming—an introduction : on the automatic evolution of
computer programs and its applications / Wolfgang Banzhaf ... [et al.].
p. cm.
Includes bibliographical references and index.
ISBN 1-55860-510-X
1. Genetic programming (Computer science) I. Banzhaf, Wolfgang, date.
QA76.623.G46 1998
006.3'1—dc21 97-51603
CIP

MKP ISBN: 1-55860-510-X
dpunkt ISBN: 3-920993-58-6

To Pia, Teresa, Benedikt and Judith, who had to sacrifice many evenings and weekends for this book to become possible.

Wolfgang Banzhaf

To my family and friends, who were there when times got tough. To the lovely Miss Christina Espanto, who makes even the tough times good.

Frank D. Francone

To those who missed me while I was working on this book.

Robert E. Keller

To my parents Set and Inga.

Peter Nordin

Foreword by John R. Koza

Genetic programming addresses the problem of automatic programming, namely, the problem of how to enable a computer to do useful things without instructing it, step by step, on how to do it.

Banzhaf, Nordin, Keller, and Francone have performed a remarkable double service with this excellent book on genetic programming.

- First, they have written a book with an up-to-date overview of the automatic creation of computer programs by means of evolution. This effort is especially welcome because of the rapid growth of this field over the past few years (as evidenced by factors such as the more than 800 papers published by some 200 authors since 1992).

- Second, they have brought together and presented their own innovative and formidable work on the evolution of linear genomes and machine code in particular. Their work is especially important because it can greatly accelerate genetic programming.

The rapid growth of the field of genetic programming reflects the growing recognition that, after half a century of research in the fields of artificial intelligence, machine learing, adaptive systems, automated logic, expert systems, and neural networks, we may finally have a way to achieve automatic programming. When we use the term *automatic programming*, we mean a system that

1. produces an entity that runs on a computer (i.e., either a computer program or something that is easily convertible into a program),

2. solves a broad variety of problems,

3. requires a minimum of user-supplied problem-specific information,

4. in particular, doesn't require the user to prespecify the size and shape of the ultimate solution,

5. implements, in some way, all the familiar and useful programming constructs (such as memory, iteration, parameterizable subroutines, hierarchically callable subroutines, data structures, and recursion),

6. doesn't require the user to decompose the problem in advance, to identify subgoals, to handcraft operators, or to tailor the system anew for each problem,

7. scales to ever-larger problems,

8. is capable of producing results that are competitive with those produced by human programmers, mathematicians, and specialist designers or of producing results that are publishable in their own right or commercially usable, and

9. is well-defined, is replicable, has no hidden steps, and requires no human intervention during the run.

Genetic programming is fundamentally different from other approaches to artificial intelligence, machine learning, adaptive systems, automated logic, expert systems, and neural networks in terms of (i) its representation (namely, programs), (ii) the role of knowledge (none), (iii) the role of logic (none), and (iv) its mechanism (gleaned from nature) for getting to a solution within the space of possible solutions.

Among these four differences, representation is perhaps the most important distinguishing feature of genetic programming. Computers are programmed with computer programs – and genetic programming creates computer programs.

Computer programs offer the flexibility to perform computations on variables of many different types, perform iterations and recursions, store intermediate results in data structures of various types (indexed memory, matrices, stacks, lists, rings, queues), perform alternative calculations based on the outcome of complex calculations, perform operations in a hierarchical way, and, most important, employ parameterizable, reusable, hierarchically callable subprograms (subroutines) in order to achieve scalability.

In attacking the problem of automatic programming, genetic programming does not temporize or compromise with surrogate structures such as Horn clauses, propositional logic, production rules, frames, decision trees, formal grammars, concept sets, conceptual clusters, polynomial coefficients, weight vectors, or binary strings. Significantly, human programmers do not commonly regard any of the above surrogates as being suitable for programming computers. Indeed, we do not see computers being ordinarily programmed in the language of any of them.

My view is that if we are really interested in getting computers to solve problems without explicitly programming them, the structures that we need are computer programs.

This book will be coming out almost exactly ten years since my first run of genetic programming in October 1987 (solving a pair of linear equations and inducing the Fibonacci sequence). Certainly I could not have anticipated that this field would have grown the way it has when I thought of the idea of genetic programming while flying over Greenland on my return from the 1987 meeting of the International Joint Conference on Artificial Intelligence in Italy.

We know from Yogi Berra that predictions are risky, particularly when they involve the future. But, it is a good guess that genetic programming will, in the future, be successfully expanded to greater levels of generality and practicality.

In trying to identify future areas for practical application, the presence of some or all of the following characteristics should provide a good indication:

1. areas where the interrelationships among the relevant variables are poorly understood (or where it is suspected that the current understanding may well be wrong),

2. areas where finding the size and shape of the ultimate solution to the problem is a major part of the problem,

3. areas where conventional mathematical analysis does not, or cannot, provide analytic solutions,

4. areas where an approximate solution is acceptable (or is the only result that is ever likely to be obtained),

5. areas where small improvements in performance are routinely measured (or easily measurable) and highly prized,

6. areas where there is a large amount of data, in computer readable form, that requires examination, classification, and integration (such as molecular biology for protein and DNA sequences, astronomical data, satellite observation data, financial data, marketing transaction data, or data on the World Wide Web).

The four authors are to be congratulated on producing a fine book and the reader will be rewarded by reading it.

John R. Koza
Stanford University
July 8, 1997

Contents

Preface

When we first conceived of this text, we were concerned it might be too early for such an undertaking. After all, genetic programming (GP) had grown very rapidly in the few years since 1992. Would not such a young discipline be much too fluid for being pinned down in a text book? The published literature in the field is diverse. Different approaches to genetic programming manifest varying degrees of complexity. Finally, there are no firmly established paradigms that could serve as guide posts.

At the same time, however, we could not escape the impression that genetic programming had accumulated enough real substance for a systematic overview to be of use to the student, to the researcher, and to engineers interested in real-world applications of this field. So we proceeded despite the relative newness of genetic programming. The results of our effort can be seen here.

We have written this text for more than one audience. Accordingly, the book has many entry points, depending on the level of knowledge the reader brings to the table (see Figure 1).

The text's core is divided into three parts with four chapters each. Where appropriate, a chapter ends with exercises and recommendations for further reading.

- Part I describes fundamentals we regard as prerequisites to a deeper understanding of genetic programming. This part is intended to set the groundwork for our main theme without actually describing genetic programming in detail. While the chapters in this part are recommended for newcomers to genetic programming, they should also contain useful information for readers who are more familiar with the field.

  As a general introduction we recommend Chapter 1. This chapter should also serve well for those readers who seek a better understanding of how genetic programming fits into the overall discipline of machine learning. Chapter 2 should be of importance to readers interested in the connection between genetic

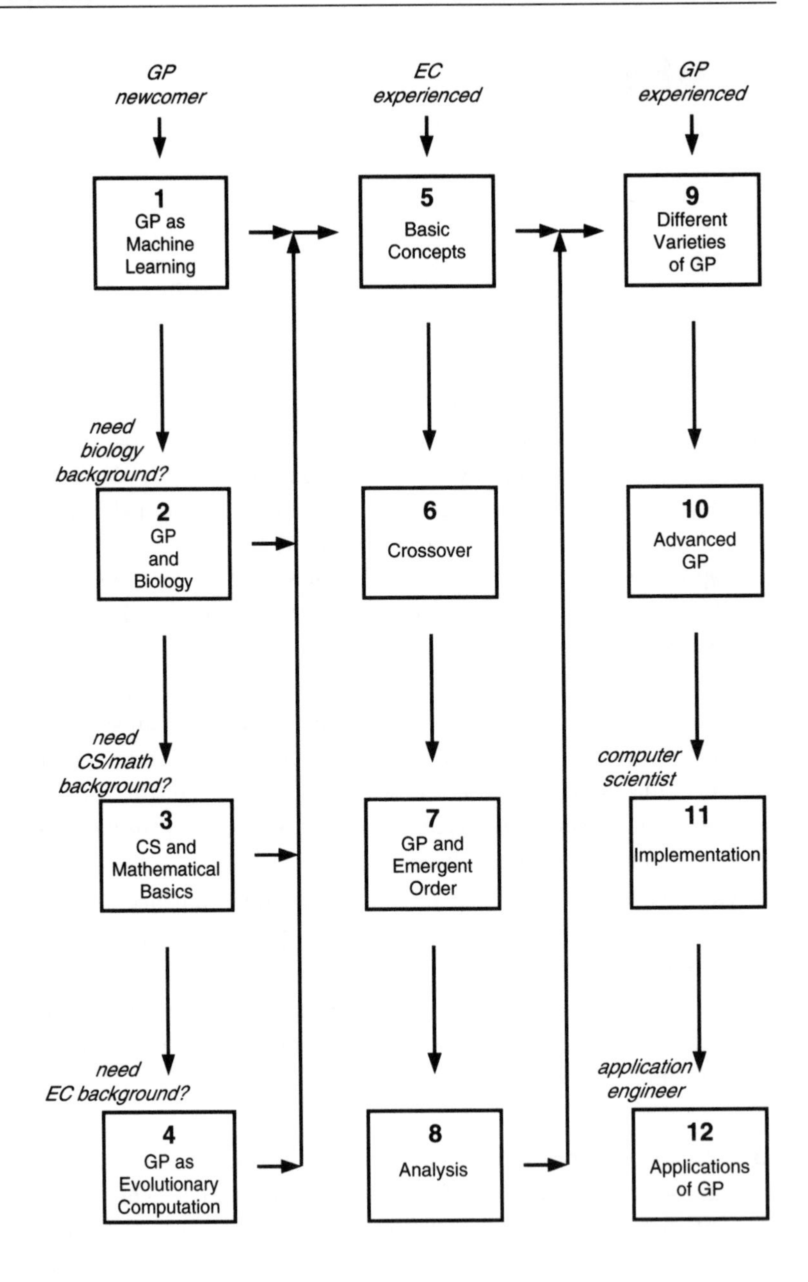

Figure 1
Navigating the book. Italic text indicates entry points.

programming, evolution, development, and molecular biology. Chapter 3 offers background in aspects of mathematics and computer science that are important in genetic programming, while Chapter 4 provides a general description of the field of evolutionary computation, of which genetic programming is a part. Each of these chapters may be read separately, depending on the background and interests of the reader.

- Part II is central to this book. Chapter 5 introduces three basic paradigms of genetic programming – tree, linear and graph based systems – while the remaining chapters introduce and analyze the important problems in the field. Readers who already have basic knowledge might want to start with Chapter 5 directly.

  Chapter 6, 7 and 8 focus on prominent unsolved issues in genetic programming, such as the effect and power of the crossover operator, introns, genetic programming as an emergent system, and many others. These three chapters should be read sequentially and only by those with a firm understanding of the basics of genetic programming described in Chapter 5. In addition, we emphasize that the materials in Chapter 6 to Chapter 8 would best be understood by a reader well versed in the principles of evolution and molecular biology described in Chapter 2. It is, however, possible to skip Chapter 6 to Chapter 8 in the first pass through the book and nevertheless have the background to move on to Part III.

- Part III offers material for readers familiar with genetic programming and comprises a more subjective selection of topics. Chapter 9 provides a detailed look at various genetic programming systems, followed in Chapter 10 by a discussion of advanced techniques to improve the basic algorithm. Chapter 11 discusses important implementation issues. As such, it is a possible starting point for computer scientists who want to explore genetic programming from a technical perspective. Chapter 12 describes an – admittedly subjective and incomplete – spectrum of applications to which genetic programming has already been applied with *at least* some degree of success. Engineers familiar with genetic programming might want start with this chapter and then digest other parts of the book at their leisure.

- Four appendices summarize valuable resources available for the reader: Appendix A contains printed and recorded resources, Appendix B suggests web-related resources, Appendix C discusses GP software tools, including Discipulus™, the GP software developed by the authors, and Appendix D mentions events most closely related to the field of genetic programming. URLs can be found online at `http://www.mkp.com/GP-Intro`.

It took us approximately two years to complete this project. Considering that two of the authors live in Germany, one in Sweden, and one in the United States of America, our collaboration was remarkably smooth. During that time, the four of us met only twice in person. Instead, we made heavy use of the Internet. Aside from our co-authors, our most frequent "professional" companions during the past two years were e-mail, FTP, TELNET and the World Wide Web. It would be fair to say that this book would probably not exist had it not been catalyzed by modern communications technology.

Cooperating this way is not atypical of what is happening in many newly emerging fields in science and technology. We are convinced that electronic means for collaboration greatly accelerate the pace of progress in the particular fields involved. As for this book, we hope that readers from all over the world will find it both useful and enjoyable. May genetic programming continue to thrive on international cooperation.

Wolfgang Banzhaf
Dortmund, Germany

Peter Nordin
Goeteborg, Sweden

Robert E. Keller
Dortmund, Germany

Frank D. Francone
Oakland, California

August 1997

Acknowledgments

First and foremost, we would like to express our gratitude to John R. Koza from Stanford University, California, for investing a significant amount of his time reviewing the text and for writing his foreword to this book.

We are further indebted to Peter J. Angeline, Markus Brameier, Markus Conrads, Jason M. Daida, Brian Dilkes, Jürgen Friedrich, David E. Goldberg, Larry Gritz, Frederic Gruau, James K. Hahn, Ulrich Hermes, August Ludvikson, Holger Kemper, Thorsten Michaelis, Stefan Mintert, Kathy Nelton, William F. Punch, Hilmar Rauhe, Michael L. Raymer, Justinian P. Rosca, Lee Spector, Astro Teller, and Michael Walper for their help with various stages of producing this book.

We are very grateful to the friendly team at our German publisher *dpunkt*, Michael Barabas, Ulrike Meuer, Christa Preisendanz, Gerhard Rossbach, and Maren Seyfahrt, to Andrew Ross, our copy editor, and to Michael Morgan and the wonderful staff at our U.S. publisher *Morgan Kaufmann* in San Francisco, especially Marilyn Alan, Elisabeth Beller, Cyd Harrell, Patricia Kim, and Yonie Overton, for making this book possible.

While the above-mentioned persons have much contributed to this book, only the authors claim responsibility for the mistakes that may have sneaked into the final product. Any suggestions for improvement are most welcome.

We also would like to acknowledge that we made use of LaTeX for typesetting under the *dpunkt* style. It is amazing what LaTeX can do if one keeps talking to it in a friendly manner; it is equally amazing what one can do after some time if LaTeX keeps responding in unfriendly ways. Figures were incorporated by using the EPSFIG package. The contents tables at the beginning of each chapter were generated by the MINITOC package. Producing many indexes, like this book's person and subject index, is easily done by using the MULTIND package. Some of the tree figures were designed with the DAVINCI visualization tool. Most figures were done with ISLANDDRAW, XFIG, and GNUPLOT. Postprocessing of certain figures was accomplished using XV and ISLANDDRAW.

The Deutsche Forschungsgemeinschaft (DFG) provided support under grant Ba 1042/5-1 and within the Sonderforschungsbereich *Computational Intelligence* under grant SFB 531 - B2. A sabbatical stay of W.B. at the International Computer Science Institute, UC Berkeley, CA was very helpful in the last stage of this project.

Part I

Prerequisites of Genetic Programming

1 Genetic Programming as Machine Learning

Contents

> Evolution is Nature's mistake. Intelligence is its insistence on making the same mistake.
>
> S.LEM, GOLEM XIV, 1981

1.1 Motivation

Automatic programming will be one of the most important areas of computer science research over the next twenty years. Hardware speed and capability has leapt forward exponentially. Yet software consistently lags years behind the capabilities of the hardware. The gap appears to be ever increasing. Demand for computer code keeps growing but the process of writing code is still mired in the modern day equivalent of the medieval "guild" days. Like swords in the 15th century, muskets before the early 19th century and books before the printing press, each piece of computer code is, today, handmade by a craftsman for a particular purpose.

The history of computer programming is a history of attempts to move away from the "craftsman" approach – structured programming, object-oriented programming, object libraries, rapid prototyping. But each of these advances leaves the code that does the real work firmly in the hands of a craftsman, the programmer. The ability to enable computers to learn to program themselves is of the utmost importance in freeing the computer industry and the computer user from code that is obsolete before it is released.

Since the 1950s, computer scientists have tried, with varying degrees of success, to give computers the ability to learn. The umbrella term for this field of study is "machine learning," a phrase coined in 1959 by the first person who made a computer perform a serious learning task, Samuel.

Machine Learning and Genetic Programming

Originally, Samuel used "machine learning" to mean computers programming themselves [Samuel, 1963]. That goal has, for many years, proven too difficult. So the machine learning community has pursued more modest goals. A good contemporary definition of machine learning is due to Mitchell: "[machine learning] is the study of computer algorithms that improve automatically through experience" [Mitchell, 1996].

Genetic programming, GP for short, aspires to do precisely that – to induce a population of computer programs that improve automatically as they experience the data on which they are trained. Accordingly, GP is part of the very large body of research called machine learning (ML).

Within the machine learning community, it is common to use "genetic programming" as a shorthand for any machine learning system that evolves tree structures. The focus on tree structures is really a tribute to the immense influence of Koza. In 1992, he wrote a treatise entitled "Genetic Programming. On the Programming of Computers by Means of Natural Selection." Before this work, a number of researchers had used genetic or evolutionary operators to induce computer programs. But these earlier works went largely unrecognized because they were, in effect, buried in the mass of genetic algorithm research. In his seminal book, Koza was the first to recognize that GP was something new and different – he even gave the new discipline its name. Koza's results were achieved by evolving tree structures. It is not surprising, therefore, that many use the term "genetic programming" to mean the evolution of tree structures, nor is it surprising that most of the work in this discipline is with various tree-based systems.

Our Definition of GP

The definition of GP used in this book will be less restrictive than the definition referred to above.

1. First and foremost we will consider the induction of computer programs by evolutionary means. Accordingly, in this book, the term "genetic programming" shall include systems that constitute or contain explicit references to programs (executable code) or to programming language expressions. So, for example, evolving LISP lists are clearly GP because LISP lists constitute programming language structures and elements of those lists constitute programming language expressions. Similarly, the common practice among GP researchers of evolving C data structures that contain information explicitly referring to programs or program language tokens would also be GP.

2. It is already clear from the GP literature that programs or programming language structures may be represented in ways other than as trees. Research has already established the efficacy of both linear and graph-based genetic programming systems. Therefore, we do not limit our definition of GP to include only systems that use (expression) trees to represent programs. Instead, all means of representing programs will be included.

3. Not all algorithms running on computers are primarily programs. For example, neural networks are (learning) algorithms, but their implementation is usually of secondary concern. Nevertheless, we shall not exclude these algorithms from being legitimate members of the GP family. There already exist numerous applications in the algorithmic domain, and excluding them would unnecessarily deprive GP of an important source of inspiration.

4. We do not limit our definition of GP to include only systems that use certain operators, such as crossover. As long as there is a population of programs or algorithms used for the benefit of the search, and as long as some kind of indeterminism is applied to generate new variants, we think we can legitimately call a system a genetic programming system.

Genetic Programming is a Kind of Program Induction

With the above discussion in mind, it is possible to define genetic programming as the direct evolution of programs or algorithms for the purpose of inductive learning. Thus, in a very real sense, GP returns to Samuel's original goal for machine learning in 1959 – teaching computers to program themselves.

Today, we program by telling the computer exactly how to do every possible thing that we think it might need to do – how to respond to every possible request from a user or a network. Not only is this cumbersome, it is impossible for software packages that routinely occupy fifty megabytes of hard disk space. The great goal of machine learning, and especially GP, is to be able to tell the computer what task we want it to perform and to have it learn to perform that task. GP would do so by letting the computer program itself or other computers.

Is GP capable of such a feat today? In a general sense, no. That is, there is no GP system that will generically accept any problem and then automatically generate computer code that addresses that problem. It is clearly not a human programmer. Notwithstanding this limitation, in only a few short years since the publication of Koza's book, GP has already changed the wisdom on the range of problems machine learning can solve and has equaled or exceeded the performance of other machine learning systems in various studies. In fact, GP has already evolved programs that are *better* than the best programs written by people to solve a number of difficult engineering problems. Finally, GP has introduced a level of freedom of representation into the machine learning world that did not previously exist. That is why we urge the reader to look carefully at this exciting and dynamic new branch of computer science.

Chapter Overview

This chapter will describe some of the central issues in machine learning and will show where genetic programming fits in. Our purpose is not to describe the entire field of ML exhaustively – this chapter will paint with a very broad brush. Rather, we intend to place GP in the context of the overall field.[1]

[1] A good general discussion of artificial intelligence, machine learning, and genetic programming's place in machine learning is contained in [Angeline, 1994].

We will begin with a brief history of machine learning from the 1950s until the present. After that, we will look at machine learning *as a process.* This *process* is remarkably similar from one machine learning paradigm to another. Moreover, understanding this process will be essential to understanding GP itself. Finally, we will examine some of the details of the machine learning process. It is in the details that machine learning paradigms diverge and genetic programming becomes quite distinctive.

1.2 A Brief History of Machine Learning

Although genetic programming is a relative newcomer to the world of machine learning, some of the earliest machine learning research bore a distinct resemblance to today's GP. In 1958 and 1959, Friedberg attempted to solve fairly simple problems by teaching a computer to write computer programs [Friedberg, 1958] [Friedberg et al., 1959].

Early Attempts at Program Induction

Friedberg's programs were 64 instructions long and were able to manipulate, bitwise, a 64-bit data vector. Each instruction had a virtual "opcode" and two operands, which could reference either the data vector or the instructions. An instruction could jump to any other instruction or it could manipulate any bit of the data vector. Friedberg's system learned by using what looks a lot like a modern mutation operator – random initialization of the individual solutions and random changes in the instructions.

Friedberg's results were limited. But his thinking and vision were not. Here is how Friedberg framed the central issue of machine learning:

> If we are ever to make a machine that will speak, understand or translate human languages, solve mathematical problems with imagination, practice a profession or direct an organization, either we must reduce these activities to a science so exact that we can tell a machine precisely how to go about doing them or we must develop a machine that can do things without being told precisely how... . The machine might be designed to gravitate toward those procedures which most often elicit from us a favorable response. We could teach this machine to perform a task even though we could not describe a precise method for performing it, provided only that we understood the task well enough to be able to ascertain whether or not it had been done successfully. ...In short, although it might learn to perform a task without being told precisely how to perform it, it would still have to be told precisely how to learn.
>
> R.M.FRIEDBERG, 1958

Artificial Intelligence Rules the Day

Friedberg's analysis anticipated the coming split between the artificial intelligence community (with its emphasis on expert *knowledge*) and machine learning (with its emphasis on *learning*). Just a few years after Friedberg's work, ML took a back seat to expert knowledge systems. In fact, artificial intelligence (AI) research, the study of domain-knowledge and knowledge systems, was the dominant form of computational intelligence during the 1960s and 1970s. Expert system domain-knowledge in this era was generally human knowledge encoded into a system. For example, an expert system might be developed by polling human experts about how they make particular kinds of decisions. Then, the results of that polling would be encoded into the expert system for use in making real-world decisions.

The type of intelligence represented by such expert systems was quite different from machine learning because it did not learn from experience. In paraphrasing Friedberg's terms, AI expert systems attempt to reduce performing specific tasks "...to a science so exact that we can tell a machine precisely how to go about doing them" [Friedberg, 1958].

The expert system approach, in the 1960s and thereafter, has had many successes, including:

- MYCIN – Diagnosis of Infectious Diseases
- MOLE – Disease Diagnosis
- PROSPECTOR – Mineral Exploration Advice
- DESIGN ADVISOR – Silicon Chip Design Advice
- R1 – Computer Configuration

The Reemergence of Learning

Notwithstanding this success, expert systems have turned out to be brittle and to have difficulty handling inputs that are novel or noisy. As a result, in the 1970s, interest in machine learning reemerged. Attention shifted from the static question of how to represent knowledge to the dynamic quest for how to acquire it. In short, the search began in earnest to find a way, in Friedberg's words, to tell a computer "precisely how to learn."

By the early 1980s, machine learning was recognized as a distinct scientific discipline. Since then, the field has grown tremendously. Systems now exist that can, in narrow domains, learn from experience and make useful predictions about the world. Today, machine learning is frequently an important part of real-world applications such as industrial process control, robotics control, time series prediction, prediction of creditworthiness, and pattern recognition problems such as optical character recognition and voice recognition, to name but a few examples [White and Sofge, 1992] [Biethahn and Nissen, 1995].

High-Level Commonalities among ML Systems

At the highest level, any machine learning system faces a similar task – how to learn from its experience of the environment. The *process* of machine learning, that is, the defining of the environment and the techniques for letting the machine learning system experience the environment for both training and evaluation, are surprisingly similar from system to system. In the next section of this chapter, we shall, therefore, focus on machine learning *as a high-level process.* In doing so, we will see what many ML paradigms have in common.

Implementation Differences among Machine Learning Systems

On the one hand, many successful machine learning paradigms seem radically dissimilar in how they learn from the environment. For example, given the same environment some machine learning systems learn by inducing conjunctive or disjunctive Boolean networks (see Section 1.5.2). The implicit assumption of such systems is that the world may be modeled in formal Aristotelian and Boolean terms. On the other hand, connectionist systems such as fuzzy adaptive or neural networks create models of the same environment based (loosely) on biological nervous systems. They regard the world as non-linear, highly complex, and decidedly non-Aristotelian (see Section 1.5.3). The variety does not end there because various systems also search through possible solutions in different ways. For example, blind search, beam search, and hill climbing are principal search paradigms (see Section 1.6). Each may be broken into many subdisciplines and each has grown out of different philosophies about how learning works, and indeed, what learning is.

Accordingly, later in this chapter we shall overview the ways in which machine learning systems are distinct from each other. In other words, we will look at the details of how different machine learning systems attack the problem of learning.

1.3 Machine Learning as a Process

Machine learning is a process that begins with the identification of the learning domain and ends with testing and using the results of the learning. It will be useful to start with an overview of how a machine learning system is developed, trained, and tested. The key parts of this process are the "learning domain," the "training set," the "learning system," and "testing" the results of the learning process. This overall *process* of machine learning is very important for the reader to understand and we urge special attention in this section if the reader is not already familiar with the subject matter.

The Learning Domain

Machine learning systems are usually applied to a "learning domain." A learning domain is any problem or set of facts where the researcher is able to identify "features" of the domain that may be measured, and a result or results (frequently organized as "classes") the researcher would like to predict. For example, the stock market may be the chosen domain, the closing S&P index[2] for the past 30 days may be the features of the domain selected by the researcher, and the closing S&P index tomorrow may be the result that the researcher wants to predict. Of course, the features (past index values) ought to be related in some manner to the desired result (the future index value). Otherwise, a machine learning system based on these features will have little predictive power.

In the GP world, a "feature" would more likely be referred to as an "input" and the "class" would more likely be referred to as the "output." These are mostly differences of terminology.[3] Regardless of terminology, once the features (inputs) are chosen from the learning domain, they define the overall dimensions of the environment that the ML system will experience and from which it will (hopefully) learn.

Training Sets, Training Data

But the selection of features (inputs) does not completely define the environment from which the system will learn. The researcher must also choose specific past examples from the learning domain. Each example should contain data that represent one instance of the relationship between the chosen features (inputs) and the classes (outputs). These examples are often referred to as "training cases" or "training instances." In GP, they are called "fitness cases." Collectively, all of the training instances are referred to as the "training set." Once the training set is selected, the learning environment of the system has been defined.

Training

Machine learning occurs by training. An ML system goes through the training set and attempts to learn from the examples. In GP, this means that the system must learn a computer program that is able to predict the outputs of the training set from the inputs. In more traditional machine learning terminology, GP must find a computer program that can predict the class from the features of the learning domain.

Generalization and the Testing Set

Finally, the researcher must appraise the quality of the learning that has taken place. One way to appraise the quality of learning is

[2] A leading stock market indicator in the United States.

[3] The use of the term "class" is actually due to the historic focus of mainstream machine learning on classification problems. We will maintain that terminology here for simplicity. Both GP and many other ML paradigms are also capable of dealing with domains that require numeric output for problems that are not classification problems. In this case, the terminology would be problem specific.

to test the ability of the best solution of the ML system to predict outputs from a "test set." A test set is comprised of inputs and outputs from the same domain the system trained upon. Although from the same domain, the test set contains different examples than the training set. The ability of a system to predict the outputs of the test set is often referred to as "generalization," that is, can the learned solution *generalize* to new data or has it just *memorized* the existing training set? Much of Chapter 8 is devoted to this very important issue. There, we shall also see that using a training set and a test set only oversimplifies the problem of generalization.

An Example: Iris Classification

An example might be useful here: the "Iris data set."[4] The Iris data set presents a "classification" problem – that is, the challenge is to learn to identify the *class* of Iris to which a photograph of a particular iris plant belongs. The set itself is based on a sample of 150 different photographs of irises. The photos represent irises from three different classes – class 0 is *Iris Setosa*, class 1 is *Iris Versicolour*, and class 2 is *Iris Virginica*. The data set itself is comprised of measurements made from these 150 photos.

The learning domain is, in this example, all photos of irises of these three types. The 150 photos are not the learning domain – they are just specific examples drawn from the domain. When the researcher chose what measurements to make off the photos, he or she identified the features of the domain. Here are the inputs (or features) that were chosen:

- Input 1. Sepal length in cm.
- Input 2. Sepal width in cm.
- Input 3. Petal length in cm.
- Input 4. Petal width in cm.

There is, of course, a value for each of these inputs in each of the 150 training instances.

The 150 instances are then divided into two groups, a training set and a test set. At this point, a machine learning system is given access to the training data and its training algorithm is executed.

[4] A word of caution: the Iris data set is often referred to in the machine learning literature as a "classic." This may imply to the reader that it might be a good idea actually to use the Iris data to test an ML system. However, the Iris domain is trivially simple, at least for GP, and its use as a test of ML systems is discouraged [Francone et al., 1996]. We use the example in the text only because it is a simple example of a learning domain.

The goal in training is to take the sepal and petal measurements (the features) in the training set and to learn to predict which of the three classes a particular iris belongs to. Not only must the system predict the class of Iris for the training set, it should also be able to do so for the test set in a manner that is statistically significant.

With this overview of the process of machine learning in place, we can now look at some of the details of learning for various machine learning paradigms.

1.4 Major Issues in Machine Learning

Until this time, the manner in which learning occurs has been ignored so that we could focus on issues common to all ML systems. But the choices made in designing a learning algorithm are crucial. The learning algorithm defines the system in which it is used more than any other single factor. Not only that, the learning algorithm is where ML systems diverge. GP systems use a learning algorithm based on an analogy with natural evolution. "Multilayer feedforward neural networks" are based on an analogy with biological nervous systems. Bayes/Parzen systems are based on statistics.

There are many ways to classify learning algorithms. Here, we will classify them by how they answer the following four questions about the "how to" of learning:

1. How are solutions represented in the algorithm?
2. What search operators does the learning algorithm use to move in the solution space?
3. What type of search is conducted?
4. Is the learning supervised or unsupervised?

Each of these four questions raises important and sometimes contentious issues for learning algorithm design. In looking at some of the different ways in which different ML systems have answered these questions, the place of GP in machine learning will become clearer.

1.5 Representing the Problem

1.5.1 What Is the Problem Representation?

An ML system's problem representation is its definition of what possible solutions to the problem look like – what kinds of inputs do the solutions accept, how do they transform the inputs, how do they produce an output? In short, the problem representation defines the set of all possible solutions to a problem that a particular ML system can find. We will frequently refer to *possible* solutions to a problem as "candidate solutions." In other words, the representation of the problem defines the space of candidate solutions an ML system can find for a particular problem.

Example: A Polynomial

A simple example illustrates this point. Suppose we wanted to predict the value of variable y from values of variable x. In the terms of the previous section, y is the output and x is the input. A very simple representation of this problem might take the form of a second order polynomial such as:

$$y = ax^2 + bx + c \tag{1.1}$$

The types of solutions this system could find would be very limited – all the system could do would be to optimize the parameters a, b, and c. One possible candidate solution in this representation would be:

$$y = 2.01x^2 + 6.4x + 7 \tag{1.2}$$

The representation could be made more complex by allowing the system to change the order of the polynomial. Then, it could explore a solution space that included both higher and lower order polynomials. In other words, the representation of the problem defines and limits the space of possible solutions the system is capable of finding.

There are actually three different levels on which a problem may be represented [Langley, 1996].

1. **Representation of the input and output set**
   In the polynomial example above, the training set would be pairs of numbers, one value for the input x and another for the output y. The representation of the inputs and outputs would, therefore, be an integer or real number representation. It is also common to represent inputs as Boolean values, real numbers between 0 and 1, enumerated sets, or in many other ways.

2. **Representation of the set of concepts the computer may learn**
   This may be referred to as the "concept description language." The manner in which learned concepts can be expressed in machine learning is diverse. Likewise, the complexity of the organization of the learned concepts varies widely. Different systems use, among other things, simple conjunctive Boolean expressions, disjunctive lists of features, class summaries, case-based descriptions, inference trees, threshold units, multilayer

feed forward networks, decision trees, and in GP, computer programs.

3. **Interpretation of the learned concepts as outputs**
   Concepts are important, but they need to be converted to reality. The interpreter does just that. For instance, a medical diagnosis system may take as input: whether the patient has chest pain, numbness in the extremities, and is more than 20 pounds overweight. The three inputs may be held in simple concepts such as:

```
if chest_pain = TRUE then
    high_heart_attack_risk := TRUE
    else
       high_heart_attack_risk := FALSE;
```

   It is not clear from such concepts how to generate an output when the concepts are combined. What should be done, for example, where a patient has chest pain and is thirty pounds overweight but no numbness is occurring? That is what the interpreter would do. An interpreter could predict risk of heart attack (that is, generate an output) by requiring that all three concepts be true (a Boolean interpretation). On the other hand, it could require only that two of the three be true (a threshold interpretation).

It would be impossible to survey even a substantial portion of the types of representational schemes that have been implemented in various machine learning systems.[5] A survey would be made even more complex by the fact that many systems mix and match types of representations. For example, a system could represent the inputs and outputs as Boolean while the concepts could be stored as case-based instances and the interpretation of the concepts could use threshold units.

We will try to follow the above three threads through the examples below.

1.5.2 Boolean Representations

Some machine learning systems represent problems in Boolean terms. By Boolean, we mean that each training instance contains an indication whether a feature (or input) of the system is true or false. In a

[5] An excellent discussion of the details of these and many other machine learning systems and other issues may be found in [Langley, 1996], from which this chapter draws heavily.

pure Boolean system the inputs (or features) are expressed in Boolean terms and the system describes the concepts that it has learned as Boolean conjunctions or disjunctions of the input features (the concept description language). We will examine how Boolean systems might represent features of the comic strip world of Dick Tracy as a machine learning problem.

Conjunctive Boolean Representations

We begin by describing a conjunctive Boolean system and how it might describe the features of Dick Tracy's world. By conjunctive, we mean that the system uses the Boolean AND to join features (or inputs) together into concepts and outputs.

Assume that a researcher wants to be able to predict whether a cartoon character in the Dick Tracy comic strip is a "bad guy." The researcher carefully examines years of old comic pages and determines that the following features might be useful in distinguishing the class of "bad guys" from everyone else:

Table 1.1
Inputs for classification

Feature	Value
Shifty eyes	True or False
Scarred face	True or False
Skull tattoo	True or False
Slouches while walking	True or False
Hooked nose	True or False
Wears two-way wrist radio	True or False

All of these features (inputs for classification) are Boolean (true or false) values. A completely Boolean system would also express the concepts that could be learned as Boolean values. A conjunctive Boolean system might learn the following concepts in classifying good guys from bad guys:

Table 1.2
Conjunctive concepts

Concept 1	Shifty eyes AND Scarred face AND Has skull tattoo
Concept 2	Hooked nose AND Wears two-way wrist radio

But the descriptions themselves do not suffice; the concepts have to be interpreted into classifications and the interpretation may be represented in different ways.

Here is an example of how a Boolean Dick Tracy might go about classifying the above concepts. Dick Tracy himself would immediately recognize that a value of TRUE for Concept 1 was indicative of criminality from his "crime watchers" guide. On the other hand, Concept 2 (hooked nose and two-way wrist radio) is consistent with a good guy – Dick Tracy himself. So here is how Dick Tracy would use the concepts to classify suspects as bad guys:

Table 1.3
Classification concepts

Concept	Value	Bad guy?
1	`True`	`True`
2	`True`	`False`

In Concept 1, there are three factors that indicate criminality – shifty eyes, scarred face, and skull tattoo. Must all three be true before we declare someone a bad guy? In a conjunctive Boolean system, the answer is yes. Therefore, without a scarred face, a man with shifty eyes and a skull tattoo would not be classified as a bad guy.

Disjunctive Boolean Representations

Now we may look briefly at a disjunctive Boolean system. By *disjunctive*, we mean that the interpreter joins the simpler concepts with the Boolean `OR` function. In a disjunctive Boolean system, if any of the simple learned concepts evaluate as true in a particular training instance, the interpreter evaluates the training instance as having an output of true also. Three simple concepts from Dick Tracy's world might be represented as follows:

Table 1.4
Disjunctive concepts

Concept	Description	Value
Concept 1	Shifty eyes	`True` or `False`
Concept 2	Scarred face	`True` or `False`
Concept 3	Has skull tattoo	`True` or `False`

In this example, if *any* one of the three concepts in the list evaluated as true, the system would evaluate the entire training instance as true. Of course, disjunctive Boolean systems would ordinarily be applied to a list of concepts considerably more complex than those in the above table. In that event, the list holds the concepts and it dictates the order in which the concepts are evaluated.

Disjunctive systems can describe more complex learning domains using conjunctive concepts and vice versa. For example, in Figure 1.1, the two classes A and B are linearly separable – that is, one could draw a line (or in three or more dimensions, a plane or a hyperplane) that separates all instances of one class from those of the other.

Both conjunctive and disjunctive systems can fully describe a domain that is linearly separable. On the other hand, in Figure 1.2, the two classes are not linearly separable. Although this is a more difficult task, Boolean systems can completely describe the domain in Figure 1.2.

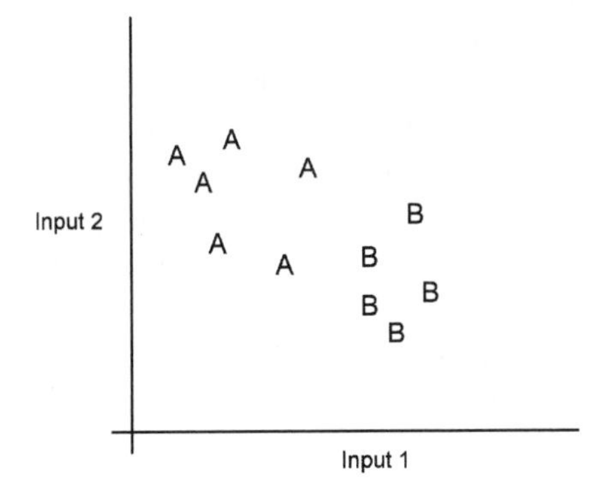

Figure 1.1
Two classes (A and B) that are linearly separable

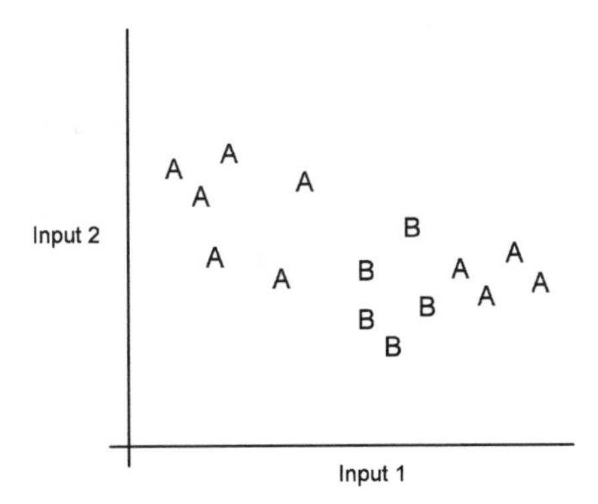

Figure 1.2
Two classes (A and B) that are not linearly separable

1.5.3 Threshold Representations

Numeric threshold representations are more powerful than Boolean representations. A threshold unit produces a value only if its input exceeds some threshold. When its outputs are expressed in Boolean terms, a threshold unit has a value of `TRUE` only if the input to that unit exceeds a particular value. Note that the use of the term "unit" makes the threshold approach very general. A threshold unit may appear as an input, a concept, or an interpreter in a machine learning system.

Dick Tracy Revisited

The Dick Tracy example illustrates how a threshold unit may be used as an interpreter in the problem representation. In the discussion above, the conjunctive Boolean interpreter of concept 1 (shifty eyes `AND` skull tatoo `AND` scarred face) required that *all* of the features of concept 1 be `true` before the interpreter could evaluate the concept itself as being `true`. In short, if there are n features in a conjunctive Boolean concept, n of n features must be `true` for the expression to evaluate as `true`. In a disjunctive Boolean system, 1 out of n features needed to be `true`.

A simple threshold interpreter unit produces quite a different result. Where $1 \leq m < n$, one type of threshold unit requires that only m of the n features in a concept be `true` for the interpreter to evaluate the entire concept as being `true`. Such an m of n interpreter would apply to concept 1 from our Dick Tracy example as follows. Suppose that $m = 2$. A threshold unit would assess a suspect with shifty eyes `AND` a skull tattoo as a bad guy, even though the suspect did not have a scarred face. In other words, only two of the three elements in concept 1 above would have to be `true` for the threshold interpreter to evaluate the concept as `true`.

Multilayer Feedforward Neural Network

The previous paragraph discusses a simple threshold interpreter for a Boolean concept description language. But as noted above, threshold units may be used in any part of a representation, not just for the interpreter. In multilayer feedforward neural network, threshold concepts are used in all parts of the problem representation. A multilayer feedforward neural network uses "neurons" as its threshold units. Figure 1.3 shows a simple example with three input neurons, two "hidden" neurons, and one output neuron.

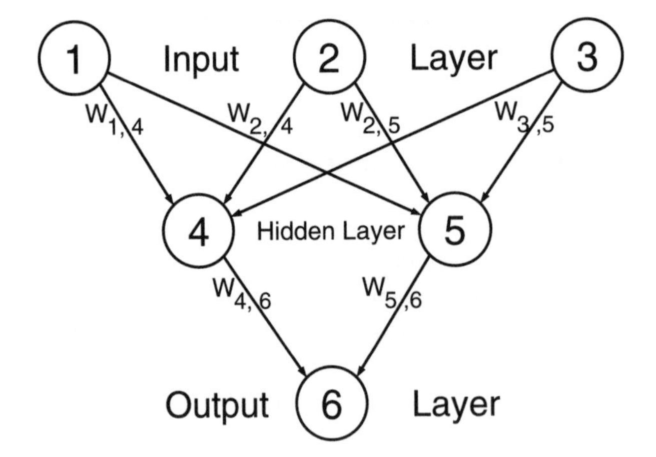

Figure 1.3
A multilayer feedforward neural network with nodes 1 ... 6 and weights $W_{1,4}$ ... $W_{5,6}$

Each neuron sums all of its inputs together and then determines whether its total input exceeds a certain threshold. If it does not, the output value of the neuron is typically 0. But if the inputs do exceed the threshold, the neuron "fires," thereby passing a positive value on to another neuron. Multilayer feedforward neural networks can be very powerful and can express very complex relationships between inputs and outputs.

1.5.4 Case-Based Representations

Another type of machine learning stores training instances as representations of classes or stores general descriptions of classes by averaging training instances in some way. A very simple instance averaging system for two classes, represented by A and B, is illustrated in Figure 1.4.

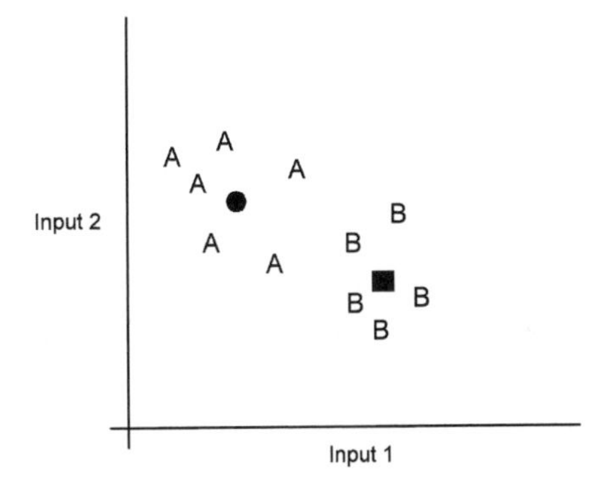

Figure 1.4
Classification based on instance averaging

Each class has two inputs, the x and y values on the two axes. The average of the inputs for the A class is the circle. The average of the inputs for the B class is the square. To classify a new set of inputs, the system simply calculates how close the new inputs are to each of the two averages. The closer of the two averages determines the class of the new input.

While this simple averaging system may be fine for simple learning domains as shown in Figure 1.4, it would clearly have a difficult time dealing with the linearly non-separable classes shown in Figure 1.2. The problem with using instance averaging on the learning domain demonstrated in Figure 1.2 is how to determine a value for class A. A simple average is obviously unacceptable for class A.

K-Nearest Neighbor Method

Other case-based learning systems handle linear non-separability much more gracefully. For example, the K-nearest neighbor approach does so by storing the training instances themselves as part of the problem representation. A new input is classified by finding the class of the input's K nearest neighbors in the stored training instances. Then the new input is classified as a member of the class most often represented among its K nearest neighbors. For example, suppose a particular K-nearest neighbor system looks at the 3 nearest neighbors of a new input; if 2 of those 3 neighbors were from class A, then the system would classify the new input as being in class A also (this is a threshold interpreter combined with a case-based system).

In Figure 1.5 the training instances for the classes A and B are linearly non-separable. A K-nearest neighbor classifier with $K = 3$

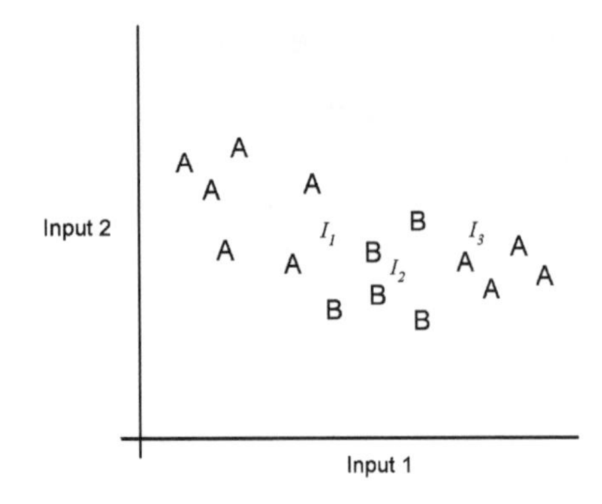

Figure 1.5
Using K-nearest neighbors to classify inputs

would classify the three new inputs I_1, I_2, and I_3 as being in classes A, B, and A, respectively. However, if $K = 5$, I_1 would be classified as a member of class B.

Bayes/Parzen Classification

Bayes/Parzen classification is treated at length in many statistics textbooks and is loosely related to the K-nearest neighbor system [Masters, 1995a]. Bayes proved that if we know the true probability density function for each class, then an optimal decision rule for classification may be formulated.

Of course, the problem is approximating that probability density function. In 1962, Parzen determined an excellent method for estimating such functions from random samples. In fact, as the sample size increases, Parzen's method converges on the true density function [Parzen, 1962]. Parzen uses a potential function for each training instance [Meisel, 1972]. The function is centered on the training instance and decreases in value rapidly as it moves away from the training instance. Parzen's estimator is simply the scaled sum of the potential function for all sample cases from a class. So in a sense, Parzen's method is a very sophisticated relative of the K-nearest neighbor systems. Parzen's estimator has been extended to multiple variable situations [Cacoullos, 1966] and is the backbone of two neural network paradigms, the "probabilistic neural network" and the "general regression neural network" [Specht, 1990] [Specht, 1991].

1.5.5 Tree Representations

Many problem space representations are based on decision trees. Consequently some of the most popular and successful machine learning systems use tree representations, including Quinlan's ID3 algorithm [Quinlan, 1979] and its variants.[6] In the ID3 algorithm, the

[6]The most recent version of the ID3 Algorithm is called C4.5 [Quinlan, 1993].

concepts are represented as a decision tree – a type of directional graph. Each internal node in the tree is a feature of the domain. Each edge in the graph represents a possible value of an attribute of the node above it. Each leaf node of the tree is a classification.

Let us look again at the Dick Tracy comic book example from above. Internal nodes in a decision tree are features of the system. So a node could be labeled `Shifty eyes`. The edges below that node would be the attributes of the feature – `TRUE` or `FALSE`. The tree below each of the attribute edges would represent the path in the decision tree consistent with `Shifty eyes` being true or false.

The ID3 learning algorithm chooses the best feature for each new node by sorting the training set by the attributes of each potential feature. It measures how much extra information about the training set each feature adds. The feature that adds the most useful information about the training set is added to the decision tree as the next node.

1.5.6 Genetic Representations

Genetic or evolutionary representations have been applied in a number of ways. Here we shall mention only one of them. We shall go into more detail about other possible representations in Chapter 4.

The GA Representation

A genetic algorithm (GA) has fixed length binary strings. Each bit is assigned a meaning by the researcher. Bits may be freely assigned any meaning and this lends great freedom to the representation. For example, the GA need not have any of the inputs represented in the bit string – the bit string could also represent a series of transformations to perform on the inputs. The bit string can represent the weights, biases, and structure of a neural network or it can represent transformations to be performed for a Boolean multiplexer type problem. The bit string is the concept definition language for the GA and the meaning assigned to the bits would be analogous to the interpreter.

There are good theoretical reasons for supposing that the low cardinality of the bit string representation is optimal for GA search. In practice, however, many researchers have used higher cardinalities in their GA representations with great success [Goldberg, 1989].

The GP Representation

Genetic programming, on the other hand, represents its concepts and its interpreter as a computer program or as data that may be interpreted as a computer program.[7] GP systems are capable of rep-

[7] GP sometimes interprets the output of its evolved solutions. That occurs in a wrapper that takes the output of the program and transforms it in some manner [Koza, 1992d]. In this case, the interpreter resides, in a sense, outside of the program.

resenting the solution to a problem with any possible computer program. In fact, at least two GP systems evolve programs in languages provably Turing complete [Teller, 1994c] [Nordin and Banzhaf, 1995b].

All machine learning systems other than GP are or may be run on computers. This means that all other systems of machine learning may be represented as a computer program. We could say, therefore, that GP is theoretically capable of evolving *any* solution that may be evolved by any of the above machine learning representations.

This is more than a theoretical consideration. For example:

- GP systems may (and often do) include Boolean operators. Boolean representations are, therefore, easy to evolve in GP.

- A threshold function is no more than a particular instantiation of an `IF/THEN` structure. Given the proper form of inputs, GP could evolve threshold functions or the researcher could make threshold functions an explicit part of the function set (see Chapter 5).

- Conditional branching structures such as `IF/THEN` or `SWITCH` statements make it possible for GP to evolve decision trees, or the researcher could constrain the GP search space to permit only program structures that permit decision trees to evolve.

- GP systems may be implemented so that memory is available for a solution to store aspects of the training set, as would a case-based system.

GP Is a Superset of Other ML Representations

The point here is that the GP representation is a *superset* of all other machine learning representations. Therefore, it is theoretically possible for a properly designed GP system to evolve *any* solution that *any* other machine learning system could produce. The advantage of this is almost complete freedom of representation – the only limits are what a Turing-complete computer can do, and the speed of the computer. On the other hand, there are advantages to constrained representations when it comes to conducting the search, as we will see in the next section.

One other key aspect of the GP representation is that, unlike many other machine learning systems, the programs that GP evolves are *variable* in length. For example, a neural network, while training, usually has a fixed size. This feature may be the source of much of GP's power. We will spend some time addressing this issue in Chapter 7.

1.6 Transforming Solutions with Search Operators

The representation issue discussed in the previous section defines the set of all candidate solutions that may be evaluated as possible solutions to the problem. It should be clear by now that the number of candidate solutions that can be evaluated by most ML systems is huge for non-trivial problems. Evaluating the entire space of *all* candidate solutions is, however, usually completely impractical. Therefore, each system must define how it will search through a limited portion of such large solution spaces. That is, which candidate solution will it evaluate first, which next, and next, and when will it stop?

Search operators define *how* an ML system chooses solutions to test and in what order. Assume that an ML system starts at step 0 and chooses a candidate solution to evaluate for step 1. It evaluates that solution. It then repeats that process n times for n steps or until some termination criterion is met. So where $0 \leq i < n$, the search operators define what solution will be chosen for each step $i+1$ from each step i. Search or transformation operators, therefore, define and limit the area of the representation space that actually will be searched. It should be obvious that a good machine learning system would use search operators which take a path through solution spaces that tends to encounter good solutions and to bypass bad ones. Some of the different types of search operators used in machine learning are discussed below.

1.6.1 Generality/Specificity Operators

In both Boolean and threshold representations, it is possible to conduct a search from the most general possible solution to the most specific.[8] For example, in a conjunctive Boolean system, every time a new conjunctive term is added to a concept, the concept is more specific than previously. Likewise, in a threshold system, increasing the threshold makes the concept described by the threshold unit more specific and vice versa [Langley, 1996]. It is on observations such as these that general to specific transformation operators were devised.

[8] Searches may also be conducted from the most specific to the most general. However, such searches are of limited interest because they tend to overfit with solutions that are far too specific. See Chapter 8.

In a small problem domain, one could start a general to specific search with all possible concept expressions that contained one feature. The next level of search could add one feature, expressed conjunctively, to each expression. Those expressions that were inconsistent with the training set could be discarded and then yet another term could be added at a third level of search. Effectively, what this search is doing is starting very general and becoming more and more specific.

1.6.2 Gradient Descent Operators

Many neural network systems use gradient descent operators to transform the networks. The weights of the network are adjusted according to a gradient descent algorithm such as back propagation or cascade correlation until a termination criterion is met [Rao and Rao, 1995].

Recall Figure 1.3, the multilayer feedforward neural network with three inputs, two hidden neurons, and one output. During training of this network, the values of the weights (w_{ij}) between the neurons would be adjusted in small amounts by a deterministic hill climber until improvement stops. To apply this procedure in multilayer feedforward neural networks, the error on a training instance is "backpropagated" through the weights in the network, starting at the output. Because of this, the effect of each adjustment in the weights is usually small – the system is taking little steps up the local hill.

1.6.3 Genetic Programming Operators

In GP, the primary transformation operators are "crossover" and "mutation" (see Chapter 5). Mutation works by changing one program; crossover by changing two (or more) programs by combining them in some manner. Both are, to a large extent, controlled by pseudo-random number generators.

Crossover and the Building Block Hypothesis

The predominant operator used in GP is crossover. In fact, the crossover operator is the basis of the GP *building block hypothesis.* That hypothesis is an important part of the basis upon which GP makes its claim to be more than a massively parallel hill climbing search. In crossover, two parents are chosen and a portion from each parent is exchanged to form two children. The idea is that useful building blocks for the solution of a problem are accumulated in the population and that crossover permits the aggregation of good building blocks into ever better solutions to the problem [Koza, 1992d]. If the building block hypothesis is correct, then GP search should be more efficient than other machine learning search techniques (see Chapter 6).

Mutation

Mutation is the other of the two main transformation operators in GP. Although not as popular as crossover, mutation is actually quite important. For our purposes here, we shall define mutation as being any sort of (random) manipulation that can be performed on one program alone. We shall argue later that it is actually mutation that brings innovation to GP.

Other Operators

Elements of general/specific search operators do appear in GP. For example, the operators devised by Koza in creating and modifying automatically defined functions are expressly based on the generality/specificity approach [Koza, 1994a]. In addition, many of the genetic operators in GP could be viewed as having generality/specificity effects. For example, adding an `if`-conditional at the bottom of a subtree is very likely to make that subtree more specific. Removing it has the opposite effect. Aside from the automatically defined functions, however, all such exploration on the basis of generality and specificity happens only as a side-effect of the other genetic operators. Such exploration has not been deliberately designed into the system.

1.7 The Strategy of the Search

While the search operators define what *types* of jumps a system can make through the search space, the *extent* of the search conducted is quite a different matter. There are different types of search used in machine learning systems. We will look at only three here:

- Blind search,
- Hill climbing,
- Beam search.

1.7.1 Blind Search

Blind search means searching through the solution space and picking a solution using no information about the structure of the problem or results from previous steps in the search. In other words, blind search proceeds without knowledge of the search space or the benefit of heuristics (rules of thumb) to direct the search. Often, blind search moves through a tree representing the search space. In that tree, each node represents a candidate solution and the edges represent permissible jumps through the search space among nodes. The edges, therefore, represent the effect of the search operators.

Blind Search is Based on Structure Only

Blind search proceeds through a tree by applying a specific strategy for movement based only on the tree structure and what nodes have been previously visited in the search. Two such strategies are breadth-first and depth-first tree search. The former searches each level of the tree until a good solution is found. The latter goes to the maximum depth of the tree down the first path dictated by the tree. If it reaches a dead end (a branch without an acceptable solution), depth-first search backtracks up the tree until it finds a branch not yet taken. It takes that branch. The process continues until the search space is completely explored or the algorithm finds an acceptable solution. In this form, both depth-first search and breadth-first search represent a type of "exhaustive search" because they search the tree until it is finished or an acceptable solution has been found.[9]

Needless to say, exhaustive search works only where the solution space is very small. For genetic programming, exhaustive search would be completely impractical. GP works in a combinatorial space suffering from the so-called curse of dimensionality. That is, the volume of the solution space increases so quickly with the addition of new dimensions that here is no practical way to do an exhaustive search of that space. This problem exists for most machine learning systems. For nearly all interesting learning domains, the search space of possible solutions is far too large for exhaustive search to be completed in reasonable time frames.

1.7.2 Hill Climbing

Hill climbing starts in one spot in the search space, transforms that solution, and keeps the new solution if it is a better solution. Otherwise the new solution is often (although not always) discarded and the original solution is again transformed, evaluated, and discarded or kept until a termination criterion is met. No record is kept of the path that has been traversed already.

Simulated annealing (SA) and many neural network training algorithms are typical of this approach. Only one solution is considered at a time and only one path through the solution space is explored.

Simulated Annealing

Simulated annealing [Kirkpatrick et al., 1983] is based on an analogy with the cooling of metal in a process called annealing. Early in the cooling, the molecular structure of the metal changes in large steps. By analogy, early in an SA run, the changes are "large"

[9] Blind search techniques need not be exhaustive. For example, iterative deepening is a hybrid version of depth-first search and breadth-first search that combines the strengths of both search strategies by avoiding the extensive storage requirements for breadth-first search and the extensive time requirements for the depth-first search. In a nutshell, it is a constrained depth-first search. As such, it is an example of an entire class of search strategies, namely, partial searches that deal with the trade-off between search time and search space.

steps. As the metal cools, the changes in the metal settle down. SA makes random changes to an existing solution, retains the transformed solution if it is better than the original solution, and sometimes retains the transformed solution if it is worse. As an SA run continues, the temperature parameter is decreased to make the likelihood of retaining negative transformations less and less. SA has been applied with real success to program induction by O'Reilly [O'Reilly and Oppacher, 1994a]. Because simulated annealing does not use a crossover-type operator or maintain a population of candidate solutions, however, it is usually not classified as genetic programming, even though it attacks the same problem as does GP – program induction from experience of the problem domain.

Back Propagation and Cascade Correlation

Neural networks are frequently trained with search algorithms such as back propagation or cascade correlation. Although these algorithms seek the bottoms of valleys instead of the tops of hills in the fitness landscape, they are properly categorized as hill climbing algorithms. Unlike SA, which uses random steps, these algorithms use deterministic steps. That is, as soon as the parameters and the starting point are chosen, the paths the algorithms take through the search space are already determined.

Back propagation and cascade correlation train the network in many dimensions simultaneously by varying the values of the weights (w_{ij}) of the network (see Figure 1.3) in fixed step sizes. The direction of each step is chosen using derivatives to find the optimal direction for the step. When a neural network is trained using such an algorithm, the run will start at one point in the solution space and proceed along one path through the space to the bottom of the nearest valley.

1.7.3 Beam Search

All of the foregoing algorithms are single point-to-point searches in the search space. GA, GP, and beam search maintain a *population* of search points. Beam search is a compromise between exhaustive search and hill climbing. In a beam search, some "evaluation metric" is used to select out a certain number of the most promising solutions for further transformation. All others are discarded. The solutions that are retained are the "beam." In other words, a beam search limits the points it can find in search space to all possible transformations that result from applying the search operators to the individuals in the beam. Beam search has been a well-established technique in heuristic machine learning systems for many years [Langley, 1996].

Angeline recognized in 1993 that GP is a form of beam search because it retains a population of candidate solutions that is smaller than the set of all possible solutions [Angeline, 1993] [Tackett, 1994] [Altenberg, 1994b]. His insight revealed a fundamental similarity between GP and machine learning that had previously been concealed because of GP's roots in evolutionary algorithms. That is, the evolutionary nomenclature used in GP tended to conceal that GP is a flavor of this very popular machine learning method. In particular:

- The machine learning evaluation metric for the beam is called the "fitness function" in GP.
- The beam of machine learning is referred to as the "population" in GP.

Machine learning systems have operators that regulate the size, contents, and ordering of the beam. GP of course regulates both the contents and ordering of the beam also. The contents are regulated by the genetic operators and the ordering is, for the most part, regulated by fitness-based selection. Simply put, the more fit an individual, the more likely it will be used as a jumping-off point for future exploration of the search space.[10]

1.8 Learning

It may suffice here to quickly name three major approaches to learning that can be used with genetic programming.

1. **Supervised learning**
   Supervised learning takes place when each training instance is an input accompanied by the correct output. The output of a candidate solution is evaluated against that correct answer.

   Many GP applications use supervised learning – the fitness function compares the output of the program with the desired result.

2. **Unsupervised learning**
   Unsupervised learning takes place when the ML system is *not*

[10] GP incorporates a "reproduction" operator, in which an individual is allowed to duplicate itself unchanged – so that, after reproduction, there would be two copies of the individual in the population. Reproduction is not a search operator – it makes no change to the individual. It is best viewed as a way of regulating the ordering and contents of the beam. It regulates the contents because reproduction doubles the number of a particular individual in the beam. It also increases the likelihood that the individual will be chosen for future genetic transformations, thus it also regulates the ordering of the individuals in the beam.

told what the correct output is. Rather, the system itself looks for patterns in the input data.

The operation of a Kohonen neural network [Kohonen, 1989] is a good example of unsupervised learning. Given a set of training instances to be classified and a specification of the number of classes to be found (note no outputs are given to the network), a Kohonen network will devise its own classes and assign the training instances to those classes.

GP is not normally used for unsupervised training. However, it would be possible to use it for that purpose.

3. **Reinforcement learning**
Reinforcement learning [Barto et al., 1983] falls between supervised and unsupervised learning. Although correct outputs are not specified as in supervised learning, a general signal for quality of an output is fed back to the learning algorithm. Thus, there is more information than in unsupervised learning, although it is rather unspecific.

Many of the fitness functions in GP are more complex than just comparing the program output to the desired output. These systems could be considered as reinforcement learning systems.

1.9 Conclusion

From the above it is apparent that, viewed as just another machine learning system, GP may be described as follows:

- GP represents a problem as the set of all possible computer programs or a subset thereof that are less than a designated length. This is one of its great strengths. The GP representation is a superset of all other possible machine learning representations;

- GP uses crossover and mutation as the transformation operators to change candidate solutions into new candidate solutions. Some GP operators explicitly increase or decrease the generality of a solution;

- GP uses a beam search, where the population size constitutes the size of the beam and where the fitness function serves as the evaluation metric to choose which candidate solutions are kept in the beam and not discarded;

- GP typically is implemented as a form of supervised machine learning. However, this is no more than convention. It is perfectly possible to use GP as a reinforcement or an unsupervised learning system.

GP is therefore most distinctive in its use of variable length programs to represent candidate solutions during training and in its explicit use of analogies to evolutionary genetics in its search operators. Chapter 2 will look closely at the analogy with biology. Chapter 7 will discuss the power of variable length problem representation.

Exercises

1. Is GP concept representation a superset of conjunctive/disjunctive Boolean and threshold concept representations?

2. Why is GP concept representation a superset of a decision tree concept representation?

3. What is the interpreter in a GP system?

4. Devise two rules, other than GP populations with natural selection, for maintaining a beam during a search. List the advantages and disadvantages of each relative to GP.

5. Would you expect crossover or gradient descent to produce bigger jumps (on average) in the quality of a solution. Why?

6. Design a gradient descent training system for GP. In doing so, consider what sort of operators could work on a program along a gradient.

7. Design a constrained syntax for programs so that a GP system could only evolve conjunctive Boolean solutions.

8. Would a `switch`-statement be a helpful program structure to use to constrain GP so that it evolved only threshold representations? Would it be helpful for evolving a disjunctive Boolean system? Why?

9. Design a generality/specificity mutation operator for genetic programming. When would you use it and why?

Further Reading

E.A. Bender,
MATHEMATICAL METHODS IN ARTIFICIAL INTELLIGENCE.
IEEE Computer Society Press, Los Alamitos, CA, 1996.

L. Pack Kaelbling, M. Littman, A. Moore,
REINFORCEMENT LEARNING: A SURVEY.
J. Artificial Intelligence Research **4** 237-283, 1996.

P. Langley,
ELEMENTS OF MACHINE LEARNING.
Morgan Kaufmann, New York, NY, 1996.

T. Masters,
ADVANCED ALGORITHMS FOR NEURAL NETWORKS.
Wiley, New York, NY, 1995.

T. Mitchell,
MACHINE LEARNING.
McGraw-Hill, New York, NY, 1996.

V.B. Rao and H.V. Rao,
C++ NEURAL NETWORKS & FUZZY LOGIC, 2ND ED.
MIT Press, New York, NY, 1995.

S. Shapiro (ed.),
ENCYCLOPEDIA OF ARTIFICIAL INTELLIGENCE.
Wiley, New York, NY, 1996.

D. White and D. Sofge (eds.),
HANDBOOK OF INTELLIGENT CONTROL.
NEURAL, FUZZY AND ADAPTIVE APPROACHES.
Van Norstrand Reinhold, New York, NY, 1992.

2 Genetic Programming and Biology

Contents

> Looking back into the history of biology, it appears that wherever a phenomenon resembles learning, an instructive theory was first proposed to account for the underlying mechanisms. In every case, this was later replaced by a selective theory. Thus the species were thought to have developed by learning or by adaptation of individuals to the environment, until Darwin showed this to have been a selective process. Resistance of bacteria to antibacterial agents was thought to be acquired by adaptation, until Luria and Delbrück showed the mechanism to be a selective one. Adaptive enzymes were shown by Monod and his school to be inducible enzymes arising through the selection of preexisting genes. Finally, antibody formation that was thought to be based on instruction by the antigen is now found to result from the selection of already existing patterns. It thus remains to be asked if learning by the central nervous system might not also be a selective process; i.e., perhaps learning is not learning either.
>
> N.K. JERNE, 1967

Genetic programming is the automated learning of computer programs. GP's learning algorithm is inspired by the theory of evolution and our contemporary understanding of biology and natural evolution. Viewed as a learning process, natural evolution results in very long-term learning from the collective experience of generations of populations of organisms. In other words, every living creature is the result of millions of years of learning by its ancestors about how to survive on Earth long enough to reproduce.

Information learned through biological evolution is regularly stored in DNA base pairs. Sequences of DNA base pairs act like instructions or partial instructions in computer programs, mediating the manufacture of proteins and the sequence of manufacture [Eigen, 1992]. This program-like nature of DNA, together with the variable length structure of DNA, explains the appeal of biological evolution as a model for computer program induction.

The Loose Connection

Our choice of words above was deliberate – GP's learning algorithm was *inspired* by the theory of evolution and molecular biology. No claim is made here or in the GP community that the GP learning algorithm *duplicates* biological evolution or is even closely modeled on it. At most we can say that GP learning algorithms have been loosely based on biological models of evolution and sexual reproduction. This chapter touches on some aspects of evolution and biology that may help the reader understand GP.

2.1 Minimal Requirements for Evolution to Occur

Darwin argued that

> ...if variations useful to any organic being do occur, assuredly individuals thus characterized will have the best chance of being preserved in the struggle for life; and from the strong principle of inheritance they will tend to produce offspring similarly characterized. This principle of preservation, I have called, for the sake of brevity, Natural Selection.
>
> C. DARWIN, 1859

In other words, there are four essential preconditions for the occurrence of evolution by natural selection:

1. Reproduction of individuals in the population;
2. Variation that affects the likelihood of survival of individuals;
3. Heredity in reproduction (that is, like begets like);
4. Finite resources causing competition.

Those factors, Darwin [Darwin, 1859] [Maynard-Smith, 1994] argued, result in natural selection which changes (evolves) the characteristics of the population over time. To some extent, the remainder of this chapter will be about one or more of these factors.

We begin by looking at evolution at work in a very simple environment – laboratory test tubes.

2.2 Test Tube Evolution – A Study in Minimalist Evolution

Evolution occurs even in simple non-living systems, such as *in vitro* (test tube) environments. For example, evolution may be observed in simple experiments using the enzyme $Q\beta$ replicase and RNA. Orgel has done a series of such experiments [Orgel, 1979] which are an excellent starting point for our considerations because they bear many similarities to GP systems. The significance of the $Q\beta$ replicase experiments may only be understood with some knowledge of the experimental setup, hence we shall discuss them here in more detail.

Experimental Design

$Q\beta$ replicase will make an adequate copy of any strand of RNA (as long as it has a supply of the monomers from which the new RNA may be made). Imagine a series of test tubes. Each tube contains a solution of $Q\beta$ replicase and the proper monomer mix. Although there are many different types of RNA, an initial RNA template containing only one type of RNA is introduced to test tube 1. The $Q\beta$ replicase immediately begins making copies of the RNA template. After 30 minutes, take a drop out of test tube 1 and place it into test tube 2. Thirty minutes later, repeat that process from test tube 2 to test tube 3 and so on.

Experimental Results

Four features of the $Q\beta$ replicase experiments are noteworthy here [Orgel, 1979] [Maynard-Smith, 1994] because genetic programming runs exhibit much the same behavior.

1. The *structure and function* of the RNA in the test tubes evolves, often dramatically. For example, in one experiment, the size of the RNA molecules in test tube 75 were only *one-tenth* as long as the molecules in the original RNA template and the $Q\beta$ replicase was making new RNA molecules at more than *twenty times* the rate in test tube 1. Clearly, the RNA population had, in less than a day, evolved to be *much* shorter and to replicate *much* faster.

2. The mix of RNA in the *last* test tube varies. However, each experiment evolves to a stable and repeatable final state that depends on the initial conditions of the experiment. Evolution ceases when that state has been reached.

3. Different initial conditions result in a final mix specifically adapted to those conditions. For example:

   - By reducing the amount of solution transferred from one test tube to the next, the experiments isolated a strain of RNA that could reproduce successfully even if only one RNA molecule was transferred to a new tube.
   - When the amount of CTP[1] in the monomer mix was reduced, the final mix contained an RNA strain that reproduced rapidly but had relatively low cytosine content.
   - When an antibiotic was included in the test tubes, the final mix contained RNA that was resistent to the antibiotic.

4. Finally, the RNA that evolves in these test tube experiments would have been extremely unlikely to evolve by random chance.

Fast Replicating RNA

One common end product in the $Q\beta$ replicase experiments illus-

[1]CTP (for cytosine triphosphate) is an energy-rich monomer containing the nucleotide cytosine. Its energy is consumed when cytosine is added to a string of RNA. Similarly, ATP, GTP, and UTP are used to elongate RNA strings.

trates these points. This end product (call it "fast RNA") is copied very quickly by the $Q\beta$ replicase enzyme for two reasons:

- Fast RNA is only 218 bases long and is, therefore, very short compared to the original RNA template (> 4000 bases). The shorter the RNA, the faster it replicates.
- The three-dimensional structure of fast RNA makes it especially easy for $Q\beta$ replicase to copy quickly.

In many of the $Q\beta$ replicase experiments, fast RNA evolves from the initial RNA template until it completely dominates the mix, even though it was not in the initial RNA template [Maynard-Smith, 1994].

RNA is not alive. It cannot copy itself. More important, there is little or no variation in the initial RNA template population. If multiplication and variation are supposedly two necessary preconditions for the occurrence of evolution, how is it that fast RNA evolves?

The key lies in the fact that copies of RNA produced by $Q\beta$ replicase are not always perfect copies. About one in every ten thousand replications results in errors – bases are inadvertently added, deleted, or improperly transcribed. It is these errors that introduce variability into the population of RNA. Variants of RNA that reproduce faster in a $Q\beta$ replicase solution have an evolutionary advantage – over any period of time they will, on average, produce more copies than other types of RNA. Although it may take many replication errors to move from the initial RNA template (over 4000 base pairs in length) to fast RNA (only 218 base pairs in length), natural selection operating on tiny variants is able to accomplish such a transition in fairly short order.

The Power of Simple Evolutionary Search

Inducing the fast RNA structure in less than geologic time scales is, by itself, a remarkable accomplishment. Consider the magnitude of the task. There are more than 10^{128} possible RNA molecules of 218 base pairs in size. To sample them one by one would take longer than the age of the universe. Accordingly, finding the form of fast RNA would be very unlikely using random search, even if we knew the correct number of base pairs (218) in advance. We may safely conclude that evolutionary search can, therefore, learn good solutions much more rapidly than random search and with no knowledge about what the final product should look like.

Eigen writes that the power of evolutionary search resides in the population. In exploring the search space from many points in parallel, evolutionary search can allocate more trials to superior mutants, with the result that:

> The (quantitative) acceleration of evolution that this brings about is so great that it appears to the biologist as a surprisingly new

> *quality*, an apparent ability of selection to 'see ahead', something that would be viewed by classical Darwinians as the purest heresy!
>
> M. EIGEN, 1992

"Test Tube" Evolution with an Explicit Fitness Function

Orgel's evolution of fast RNA was an early demonstration of the power of simple evolutionary search. Orgel's mechanism was different from GP in that Orgel did not use a defined fitness function for selection. Rather, Orgel's selection mechanism was inherent in the experimental setup, which selected for fast replicating RNA molecules. By way of contrast, Tuerk and Gold [Tuerk and Gold, 1990] have devised techniques to evolve RNA and protein translation complexes using an expressly designed selection mechanism – in GP terms, a "fitness function." Tuerk and Gold call their procedure SELEX, which stands for "systematic evolution of ligands by exponential enrichment."

SELEX starts with a diverse population of RNA molecules. This brew of RNA molecules is then passed through an "affinity column," in which RNA molecules that bind (at least weakly) to a *target molecule* are recovered and then replicated. This procedure is repeated by passing the replicated RNA through the affinity column again and again. Note: all of the elements for evolution are present in Tuerk and Gold's SELEX algorithm – variation and selection, replication and heredity. Their results were, in retrospect, not surprising. After four rounds of selection and replication Tuerk and Gold had evolved a population of RNA molecules with strong, selective binding to the target molecule in the affinity column.

Bartel and Szostak have used a SELEX approach to evolve ribozymes customized to catalyze a particular chemical reaction. Bartel and Szostak characterize their approach as "iterative *in vitro* selection." The customized ribozymes were evolved from a random population of ribozymes. The evolved ribozymes were very effective at catalyzing the chosen reaction – two to three orders of magnitude more effective than the most effective ribozyme located by random search [Bartel and Szostak, 1993].

Lessons for Genetic Programming

There are a number of important lessons for genetic programmers in these simple but elegant experiments:

- A simple system may evolve as long as the elements of multiplication, variance, and heredity exist.
- Evolutionary learning may occur in the absence of life or of *self*-replicating entities.
- Evolutionary learning may be a very efficient way to explore learning landscapes.

- Evolution may stagnate unless the system retains the ability to continue to evolve.
- The selection mechanism for evolutionary learning may be implicit in the experimental setup (Orgel) or may be explicitly defined by the experimenter (SELEX).

The evolution of a simple population of RNA in test tubes is, of course, a "toy" model in comparison to the complexity of actual evolution occurring in populations of living organisms. But so too is genetic programming. In fact, in many ways, these simple RNA models are good starting points for studying genetic programming because they ignore many of the complexities encountered in studying evolution in living organisms – the separation of genotype and phenotype, the apparatus for reproduction, sexual recombination, and ontogeny to name but a few.

Occam's Evolutionary Razor

We will, of course, move on and discuss some aspects of the evolution of living organisms. But it is important to keep in mind that the complexity that accompanies natural evolution is *not* a condition for the occurrence of evolution. Biological evolution as we have come to understand it is the manner in which evolution expresses itself given the complex set of constraints imposed by organic chemistry, DNA synthesis and replication, protein manufacture and functionality, and the fitness landscape encountered by living organisms. Evolution is *not* the complexity itself. Evolution is a process, an algorithm if you will, that occurs spontaneously both in complex populations of living organisms and in much simpler systems such as the *in vitro* RNA experiments and genetic programming runs as long as certain conditions are met.

Nevertheless, biological evolution is the single best example we have of evolution at work. GP has deliberately imitated its mechanism in a number of ways. So we will now take a look at important aspects of natural evolution. Our trip through molecular biology and population genetics will necessarily be brief and greatly oversimplified.

2.3 The Genetic Code – DNA as a Computer Program

DNA, the principal constituent of the genome, may be regarded as a complex set of instructions for creating an organism. Human DNA is comprised of approximately three billion base pairs. Many species have DNA many times longer than human DNA. While the number of instructions contained in one strand of DNA probably dwarfs the number of instructions in all software ever written by humans, the mechanism by which DNA stores instructions for the creation of organisms is surprisingly simple.

The Genetic Code in Brief

In brief overview, the basic unit of the genetic code is the DNA base pair. Three DNA base pairs combine to form a codon, which codes for the production of an amino acid. Sequences of codons code for the assembly of amino acids into RNA, polypeptides (protein fragments), proteins, or functional RNA. The products so formed mediate the growth and development of organisms.

2.3.1 The DNA Alphabet

Base pairs are the low-level alphabet of DNA instructions – each pair representing, say, part of an instruction for the creation of a particular amino acid. A base pair is comprised of two nucleic acid bases that are chemically bonded.

Only four different bases appear in DNA, adenine, guanine, cytosine, and thymine, abbreviated A, G, C, and T, respectively. The rules for base pairings are simply A pairs with T; G pairs with C.[2] Thus, any one of the following four base pair configurations comprises a single piece of information in the DNA molecule:

$$A \longleftrightarrow T \; T \longleftrightarrow A$$
$$G \longleftrightarrow C \; C \longleftrightarrow G$$

The base pairs then bond to each other, forming a ladder of base pairs that, because of the three-dimensional properties of the strands, forms a double helix. A section of DNA of six base pairs in length is shown in Figure 2.1.

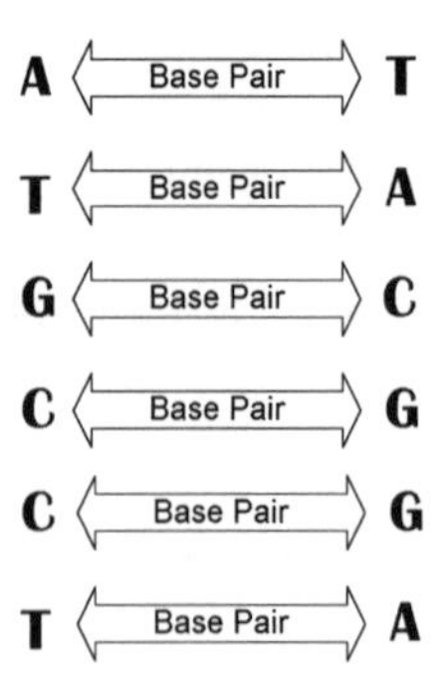

Figure 2.1
Base pairs in a DNA segment

Each base (pair) could thus be regarded as the equivalent of a bit in the DNA computer. Each DNA information element has a cardinality of four because there are four possible base pairs. Note the

[2] Actually, other pairings are possible, but much less stable.

elegance of this structure. Each of the strands of DNA is redundant – for example, if a G appears in one strand, a C must appear in the same position in the other strand. The entire DNA molecule could be reconstructed from just one of the two strands. In fact, DNA has many repair mechanisms that exploit this redundancy [Watson et al., 1987, pages 347–348].

2.3.2 Codons and Amino Acid Synthesis

Codons

In a computer program, a bit is only part of an instruction to the CPU. The entire instruction is comprised of a sequence of bits. The same is true in DNA. Each low-level instruction in DNA is actually comprised of a sequence of three base pairs. Three consecutive RNA bases are a "codon." Using the abbreviations for the base pairs above, a typical codon would be represented by biologists as "AGA," which would symbolize a codon comprised of an adenine, guanine, adenine sequence.

A codon is a template for the production of a particular amino acid or a sequence termination codon. A few examples of codons and the amino acids for which they code are: ATG which codes for methionine; CAA which codes for glutamine; CAG which also codes for glutamine.

Codon Redundancy

In all, there are sixty-four different codons – that is, there are sixty-four different ways to order four different bases in three different locations. But, there are only twenty amino acids for which DNA codes. This means that all amino acids and a stop order codon (a codon that says "quit here") could be specified with only twenty codons. What happens to the rest of the codons? The answer is that there are often several different codons that produce the same amino acid. An example is shown above: the codons CAG and CAA both code for the production of the amino acid glutamine.

Small or Neutral Mutations

This redundancy in DNA's coding for the same amino acids may be of some importance in GP for two reasons:

- The efficiency of different codons in producing the *same* amino acid can vary widely from codon to codon. Multiple codons that transcribe for the same amino acid at different rates may be of some importance in mutation. Simply put, this effect allows random mutation – normally a highly destructive event – to accomplish a relatively small change in phenotypic behavior. When one codon that produces glutamine is mutated to another codon that also produces glutamine, the most that changes is the rate of production of the protein in which the glutamine is included. The protein itself is not changed.

- Kimura has noted the possible importance that neutral mutation plays in evolution [Kimura, 1983]. Where the translation rate of two different codons that produce the same amino acid is roughly balanced, redundancy in coding for amino acids provides one important route for such neutral mutations. That is, if the segment of base pairs that produces a codon for serine is mutated to produce another one of the serine producing codons, the functionality of the DNA is unchanged although the structure of the DNA has been changed.[3]

DNA Instruction Sequencing

If DNA were no more that a collection of codons that amount to "instructions," the analogy between DNA and a computer program would be weak indeed. A computer program has an additional element – the instructions have a sequence of execution. So, too, does DNA. This sequence arises from the fact that adjacent base pairs bind from a 5 prime site to a 3 prime site. That bond is directional from the 5 prime to the 3 prime site. This directional vector is used by the organism in protein synthesis. In short, DNA not only has instructions with specific meanings, the instructions have an implicit order of execution also.

2.3.3 Polypeptide, Protein, and RNA Synthesis

Codons produce amino acids. But the end product of DNA instructions is not to produce amino acids. Rather, DNA acts on the rest of the world by *providing the information* necessary to manufacture polypeptides, proteins, and non-translated RNA (tRNA and rRNA) molecules, each of which carry out various tasks in the development of the organism. Proteins are complex organic molecules that are made up of many amino acids. Polypeptides are protein fragments. Non-transcribed RNA is an RNA molecule (similar to DNA but having only a single strand) that is not merely an intermediate step in the production of a protein or a polypeptide.

DNA Transcriptional Segments

The intricacies of how DNA causes the production of proteins, polypeptides, and RNA is far beyond the scope of this chapter and this discussion glosses over many complexities such as the synthesis of messenger RNA, tRNA, and the like. Generally speaking, DNA transcribes RNA molecules, which then translate into one or more proteins or polypeptides. The point here is that there are segments

[3] In many circumstances, further redundancy lies in the protein structures that are produced by the sequences of codons. Some amino acids in a protein may be replaced by another with little or no change in protein functionality. In this case, a mutation that switched for coding from amino acid 1 to amino acid 2 would be functionally neutral [Watson et al., 1987].

of DNA that engage in transcriptional activity. Transcriptional activity is necessary for the direct or indirect synthesis of the brew of polypeptides, proteins, and RNA molecules produced by DNA. As Watson et al. put it:

> DNA molecules should thus be functionally regarded as linear collections of discrete transcriptional units, each designed for the synthesis of a specific RNA molecule.
>
> J.D. WATSON ET AL., 1987

2.3.4 Genes and Alleles

This brings us to the difficult task of defining genes and alleles. In popular literature, a gene is a location on the DNA that decides what color your eyes will be. Slight variations in that location make your eyes green or brown. The variations are called "alleles." This explanation greatly oversimplifies what we know about DNA and the manner in which it mediates the creation of an organism. Such a notion of a gene is misleading, particularly when it comes to designing a GP system around such an explanation. What biologists try to express with the concepts of "gene" and "alleles" is more complex:

- Adjacent sequences of DNA do act together to affect specific traits. But a single gene can affect more than one trait. Moreover, DNA at widely scattered places on the DNA molecule may affect the same trait. For example, the synthesis of the amino acid arginine is mediated by eight separate enzymes (proteins). Each of those enzymes is created by a different gene located in a different place on the DNA [Watson et al., 1987, pages 218–219].

- The portions of the DNA that engage in transcriptional activity (the genes) are separated by long sequences of DNA that have no apparent function called "junk DNA." Junk DNA is, for all we know, just inert. It does not ever activate to transcribe or affect the transcription process. Clearly, junk DNA does not qualify as a gene.[4]

- The portions of DNA that engage in transcriptional activity are located in the regions between the long junk DNA sequences referred to above. Genes are, therefore, located in these regions also. However, not all of the intragenic DNA sequence engages

[4] There is discussion in the biological community whether junk DNA is really junk or not; see later sections of this chapter.

in transcription. These intragenic sequences are comprised in part of alternating sections of "exons" and "introns."

Put simply, exons transcribe for proteins, polypeptides or mRNA. Introns, on the other hand, are removed from the RNA before translation to proteins. Junk DNA has no apparent effect on the organism. It is, apparently, inert. Although it would seem that introns should be equally inert (they are not transcribed to mRNA and are therefore not expressed as proteins), the issue is not so clear. The presence of and contents of introns frequently have a measurable effect on the amount and biological effect of proteins produced by the gene in which they occur [Watson et al., 1987, 867-868] [Maniatis, 1991] [Rose, 1994] [McKnight et al., 1994]. It may also be that introns play a role in preventing damage to exons during recombination and in enabling the evolutionary process to experiment with shuffling and combining slightly different variations of the functional parts that make up a protein [Watson et al., 1987, 645-646]. We will deal with this issue at somewhat greater length in Chapter 7.

- DNA's functions – even those we know about – are much more complex than the translation of polypeptides and proteins. The intragenic sequences referred to above also contain "control sequences" that turn the translation of proteins and polypeptides in other DNA sequences on and off [Watson et al., 1987, page 233], almost a "wet" form of conditional branching or of GOTO statements. Whether to include these control sequences as "genes" is not at all clear.

The One Gene–One Protein Rule

Needless to say, these ambiguities have lead to more than one definition of a gene among biologists. For example, biologists have at various times defined a gene as the segment or segments of DNA that produced a single protein – the "one gene–one protein" rule. More recently, others have suggested that, in reality, DNA sequences often code for less than a protein. They code for polypeptide sequences, which may be combined into proteins. This has lead to the "one gene–one polypeptide" rule [Watson et al., 1987, page 220].

Watson et al. conclude that the best working definition would:

> ...restrict the term *gene* to those DNA sequences that code for amino acids (and polypeptide chain termination) or that code for functional RNA chains (e.g. tRNA and rRNA), treating all transcriptional control regions as extragenic elements. DNA chromosomes, so defined, would be linear collections of genes interspersed with promoters, operators, and RNA chain termination signals.
>
> J.D. WATSON ET AL., 1987

The lesson here, with respect to genes, alleles, and other DNA structures is to be *very careful* when applying such concepts to another medium such as digital computers. Biologists are hard pressed to agree on a definition. The first question in applying such concepts to new media is to define clearly what is meant in the first medium and then to determine whether the term has any meaning at all when applied to the other medium. As we will see in later chapters, these terms may have little or no meaning in a GP population made up of randomly generated computer programs.

2.3.5 DNA without Function – Junk DNA and Introns

We have been speaking above about sequences of DNA that have a function – they create proteins or polypeptides or control their creation. But much of the DNA of many organisms apparently just sits there and does nothing. This is referred to as junk DNA and as introns. The word *apparently* is important here. We know that junk DNA and introns are not "transcriptional segments" of DNA – they do not code for proteins. On the other hand, proving that they do not code for proteins does not prove that junk DNA and introns do not have functions of which we are unaware.

All but a few percent of the DNA of eukaryotic organisms (all higher organisms are eukaryotic) consist of non-coding DNA comprising junk DNA, control regions, and introns. On the other hand, procaryotes (many of the bacteria species) have no introns at all.

The reason for the existence of junk DNA and introns is the subject of debate in the biological community. We have noted above that introns have some role in increasing the efficiency of protein translation. In any event, junk DNA and introns will have some importance later as we examine apparent GP analogs of the intron structures that emerge spontaneously during GP evolution, in Chapter 7.

We have spent some time in the molecular world of DNA, polypeptides and the like. It is now time to discuss the relationship between the tiny-scale world of DNA and the much larger world of the organisms DNA creates.

2.4 Genomes, Phenomes, and Ontogeny

In 1909, Johannsen realized that it was important to distinguish between the appearance of an organism and its genetic constitution. He coined the words *phenotype* and *genotype* to label the two concepts [Johannsen, 1911]. The distinction is still an important one. Evolution interacts differently with the genotype and the phenotype in biological evolution. In GP, the distinction is more elusive. Some GP systems explicitly distinguish between the genotype and phenotype [Gruau, 1992a] [Banzhaf, 1993a] [Keller and Banzhaf, 1996], whereas others [Koza, 1992d] [Nordin et al., 1995] do not.

Genotype

The genome or genotype of an organism is the DNA of that organism.[5] Half of the genome (DNA) is passed from parent to child. Thus, heredity is passed through the genome. The genome is also the principal mechanism for variance within a population because genetic changes caused by mutation and recombination are passed with the genome.

Phenotype

The phenome or phenotype is the set of observable properties of an organism. In a colloquial sense, the phenotype is the body and the behavior of the organism. Natural selection acts on the phenotype (not on the genotype) because the phenotype (the body) is necessary for biological reproduction. In other words, the organism (the phenotype) must survive to reproduce.

Ontogeny

Ontogeny is the development of the organism from fertilization to maturity. Ontogeny is the link between the genotype (DNA), the phenotype (the organism's body and behavior), and the environment in which the organism's development takes place. The organism's DNA mediates the growth and development of the organism from birth to death. The environment of the organism is frequently an important element in determining the path of development dictated by the DNA.

In biological evolution, ontogeny is a one-way street. That is, changes in an organism's DNA can change the organism. However, except in rare instances [Maynard-Smith, 1994], changes in the organism do not affect the organism's DNA. Thus, the village blacksmith may have a large son. The son is large because his father passed DNA for a large body to the son, not because the father built up his own muscles through a lot of physical labor before siring his son. The blacksmith's acquired trait of large muscles is, emphatically, not passed on to the blacksmith's son. Ignoring the complication of sexual reproduction, Figure 2.2 diagrams this mechanism.

All of the mechanisms of biological heredity (copying the parent's DNA and passing it to the child, mutation of the parent's DNA, and recombination of the parent's DNA) take place at the genotype (**G**) level. On the other hand, natural selection acts only at the level of the

---

[5]This simple definition glosses over the difference between procaryotic and eukaryotic organisms. For eukaryotes, it would be more accurate to define the genome as the genes contained in a single representative of all chromosome pairs. Because the most important constituent in the chromosome pairs is DNA, the definition in the text is sufficient for our purposes.

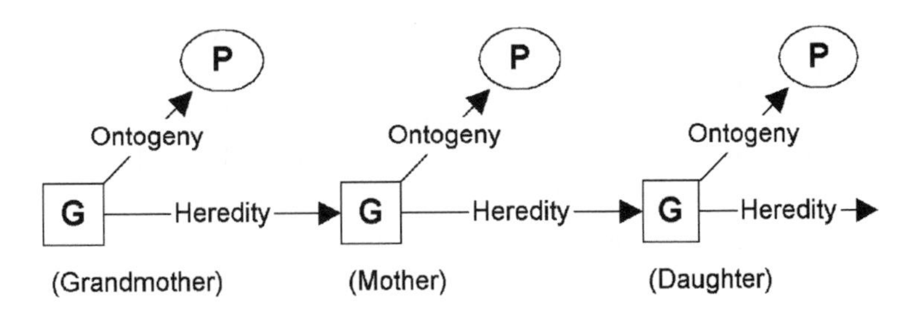

Figure 2.2
The mechanism of Darwinian heredity

phenotype (**P**). Therefore, natural selection acts upon the genotype only indirectly.

The RNA *in vitro* experiments discussed above have much to say about the design of GP systems in regard to separation of the genotype and the phenotype and in regard to ontogeny:

- **Evolution is possible even where there is no physical difference between the genotype and the phenotype.**
  In biological evolution, the DNA (genotype) is quite distinct from the organism's body and behavior (phenotype). By way of contrast, the RNA in Orgel's experiments has no phenotype that is separate from the genotype (the RNA itself).[6] The RNA does, however, engage in *behavior* that could be regarded as phenotypical. Similarly, in GP, the genotype is the evolved program and, in many systems, there is no separate phenotype. But the *behavior* of the GP program when it executes is, like the RNA's behavior after folding, phenotypical behavior.

  While some phenotypical behavior on which natural selection can act appears to be a necessary component of evolution, a separate structure that may be labeled the "phenotype" is not.

- **Evolution is possible with or without ontogeny.**
  The RNA experiments also suggest that ontogeny is not a requirement of evolution. In the Orgel experiments, the RNA did not undergo any development process between reproduction events.[7]

[6]One could argue that the folded RNA strand is physically different from the unfolded strand and thus constitutes a phenotype.

[7]One could argue that the temperature conditions required for folding of RNA are the appropriate environment and the folding process itself is ontogeny.

The lesson here for GP is that it is not necessary to create a separate phenotype structure from a genotype by some sort of ontological process. However, ontogeny and a separate phenome are *tools* that may be used by the GP researcher to improve the quality of GP search. One example of GP's use of such tools is the cellular encoding technique, which explicitly uses the genotype (a tree GP structure) to define a process of ontogeny by which a single "primordial" neuron develops into an elegantly complex neural network structure (the phenotype) [Gruau, 1992a]. More recently, finite automata have been treated in a similar way [Brave, 1996].

2.5 Stability and Variability of Genetic Transmission

The way genetic material is passed from the parent to the child is the most important factor in determining genetic stability (like begets like) and the genetic variability in a population. For evolution to occur, genetic transmission must be simultaneously stable and variable. For example, the transmission of genetic material from parent to child must, with high probability, pass parental traits to the child. Why? Without such stability of inheritance, natural selection could select a parent with good traits but the children of that parent would not be likely to inherit that good trait. At the same time, there must also be a chance of passing useful new or different traits to the child. Without the possibility of new or different traits, there would be no variability for natural selection to act upon.

This section explores the mechanisms of stability and variation.

2.5.1 Stability in the Transmission of Genetic Material

In natural evolution, reproduction involves copying the parent's DNA and transmitting all or part of that DNA to the child. We have already pointed to some of the principal mechanisms of stability in heredity:

- **Redundancy**
  The elegant redundancy of the base pair structure of DNA.

- **Repair**
  Complex repair mechanisms reside in the DNA molecule for repairing damage to itself and for the correction of copying errors. This function is so important that several percent of some bacteria's DNA is devoted to instructions for DNA repair [Watson et al., 1987, page 339 onward].

- **Homologous Sexual Recombination**
  Sexual recombination is often regarded as a source of genetic variability, and it is. But recombination is also a major factor in the genetic stability of a species. It tends to prevent the fixing of negative mutations in the population (thereby reducing variability) [Maynard-Smith, 1994]. In addition, Watson et al. argue that *recombination's most vital function is probably the repair of damaged DNA* [Watson et al., 1987, page 327].

2.5.2 Genetic Variability

Genetic variation in a population is the result of three principal forces.

Mutation

Entropy-driven variation, such as mutation, is the principal source of variability in evolution. There are many types of mutation, including [Watson et al., 1987, pages 340–342]:

- Changes from one base pair to another. These often produce neutral or useful variations. Although a base pair switch occurs about once every ten million replications or less, there are hot spots where base pair switching is up to twenty-five times the normal rate.

- Additions or deletions of one or more base pairs. This is called a frameshift mutation and often has drastic consequences on the functioning of the gene.

- Large DNA sequence rearrangements. These may occur for any number of reasons and are almost always lethal to the organism.

Homologous and Non-Homologous Genetic Transfer in Bacteria

The exchange of genetic material among bacteria through mechanisms such as phages, plasmid conjugation, and transposons is also a source of bacterial genetic variability. Of particular interest are:

- **Hfr Conjugation**
  A bacterium in the Hfr state actually injects a copy of part of its genetic material into another bacterium, where homologous recombination occurs. The bacteria need not be of the same species.

- **Transposons**
  A transposon is able to insert entire genes into the genetic sequence of the recipient bacterium. Transposons are thought to be responsible for conveying the genes for antibiotic resistance among bacteria of different species.

These mechanisms have been mostly ignored by GP programmers because of the GP focus on sexual reproduction. The authors believe that this is an area ripe for research because these are the methods of genetic exchange in the simpler evolutionary pattern of asexual production and may be more appropriate to the complexity level of a GP run.

Homologous Sexual Reproduction

Exchange of genetic material in sexual reproduction happens through recombination. The DNA from both parents is recombined to produce an entirely new DNA molecule for the child. GP crossover models sexual recombination in the sense that there are two parents and that portions of the genetic material of each parent are exchanged with portions of the other. On the other hand, GP does not model the *homologous* aspect of natural recombination, as will be discussed in detail below.

2.5.3 Homologous Recombination

The concept of *homologous* genetic transfers is clearly an important one. Most recombination events seem to be homologous. The reason for that will become clear in this section. Homology appears throughout procaryotic and eukaryotic genetic exchanges and is an important element in the stability/variability mix of natural evolution.

Homologous Variation versus Mutation

Homologous genetic exchange occurs during "crossover" in sexual recombination. It also occurs in bacteria during events such as plasmid conjugation. Homology is the reason that genetic exchange is a source of both genetic variation *and* genetic stability in a population.

We have seen that mutation causes random changes in DNA – normally quite damaging changes. Homologous exchange is completely different – it encourages changes in DNA of a very narrow and specified type. Although we will go into more detail later, in homologous exchange, genetic material is exchanged in a manner that preserves the function of the all-important DNA transcription segments (genes) and the length of both DNA molecules. The result of this highly controlled exchange is that sexual recombination has a success rate in generating *successful* variations that is remarkable. Most children of sexual recombination are viable.[8]

The Requirements of Homologous Exchange

Homologous exchange will not work unless two requirements are met:

1. Homologous exchange can only occur between two identical or almost identical DNA segments. In higher organisms, this means that horses mate with horses, not dogs. Why? One horse's DNA can be matched up closely with another's so that recombination preserves gene functionality. In a nutshell, this preservation of gene functionality is why clearly defined species evolve for organisms that reproduce sexually.

2. Homologous exchange can occur only if the two DNA segments to be exchanged can be matched up so that the swap point is at functionally identical points on each strand. In fact, DNA strands are able to align themselves where their base pairs are identical or almost identical before recombination. [Watson et al., 1987] [Maynard-Smith, 1994].

Figure 2.3 demonstrates the first point – the DNA strands that engage in genetic exchange must be identical or nearly identical. Figure 2.3 shows what happens when the DNA segments are dissimilar. Suppose we recombined the DNA from two species, a horse and a finch. Short DNA sequences from each species are shown in Figure 2.3. Before recombination, the horse (species 1) has genes that transcribe for: protein A (Pr A), polypeptide B (Pp B), and protein C (Pr C). The finch (species 2) has genes that transcribe for: polypeptide D (Pp D), polypeptide E (Pp E), and an rRNA molecule of type F (rRNA F). The swap point for recombination is shown in the figure.

After crossover, the horse's foal would be able to make finch polypeptide E and finch mRNA F. But it would likely have lost the ability to make horse polypeptide B or protein C. If polypeptide B or protein C are at all important to the horse, this non-homologous recombination would likely be a disaster for both the horse and the finch and is probably a good reason for a horse not to mate with a finch.

In fact, non-homologous recombination is nothing more than a massive mutation. It affects not just one or two base pairs but the entire strand of DNA from the crossover point on. Thus, it falls into

[8] With three exceptions (transposons, integrative viruses and agrobacterium tDNA) the known forms of genetic exchange among individuals involve *homologous* exchange. Even these three mechanisms appear to involve homologous-like mechanisms.

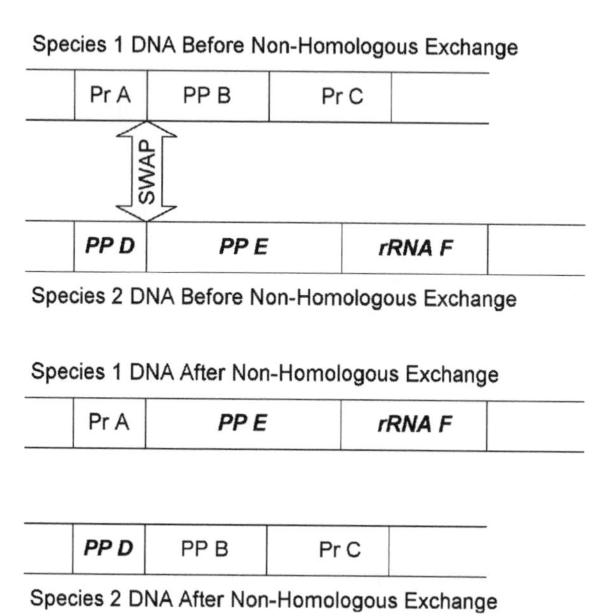

Figure 2.3
Non-homologous exchange

the category of mutation identified above as a sequence rearrangement, and a massive one. Such mutations are normally fatal.

Figure 2.4, on the other hand, illustrates how stability *and* genetic variability may be the result of *homologous* recombination. Note that each DNA strand in the figure is a *little different*. DNA One has Variant 1 of the gene to produce Pr A and Pp B. DNA Two has Variant 2 of the same genes. The Variant 1 genes are *alleles* of the Variant 2 genes in this example. Before the recombination, the two DNA strands have aligned themselves correctly along similar or identical base pair sequences.

When DNA One and DNA Two recombine, the integrity of the gene structure is preserved. More important, each DNA strand gets a working version of a functional gene to make Pr A and Pp B – the versions are just a little different. So the children of this recombination will probably be viable. Finally, the recombination has created a small amount of variability. Before the recombination, Variant 1 of Pr A may never been combined with Variant 2 of Pp B. This testing of different combinations is the essence of genetic variability introduced by crossover.[9]

Even the process of transposon recombination, a non-homologous process in bacteria where entire intact genes are inserted by the trans-

[9]This example is a little simplistic. The swap in Figure 2.4 is intergenic (between genes). Intragenic homologous recombination also occurs but it has the same property of preserving essential structure.

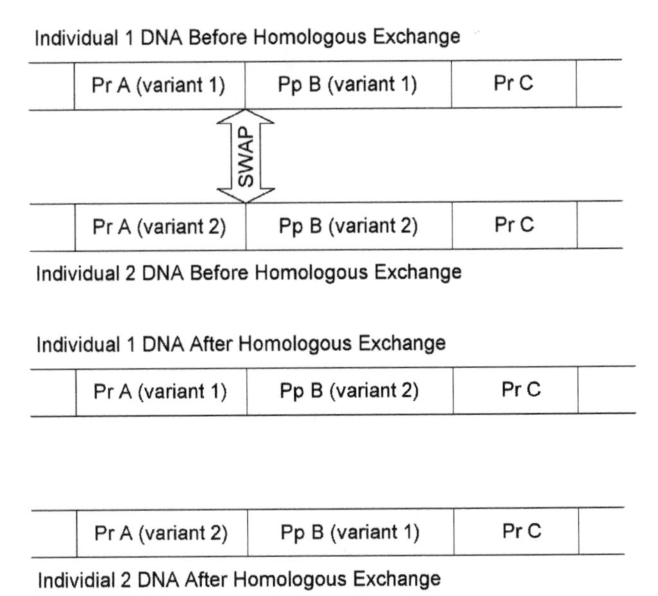

Figure 2.4 *Homologous recombination*

poson into the DNA of another bacterium, is the exception that proves the importance of preserving structure in recombination. An *entire* working gene or set of genes is inserted into the bacterium.[10]

We have addressed homology here because it may be found in almost all forms of genetic exchange, from the lowest bacteria to sexual reproduction in mankind. The ubiquity of homologous exchange and the clear importance of structure preservation raise troubling questions for GP crossover. GP crossover is clearly *not* homologous. We discuss the problems this has led to and solutions in some detail in Chapter 6.

2.6 Species and Sex

Our treatment of biology will end with a brief discussion of species and sex.[11]

Speciation in Asexual Organisms

We classify organisms that reproduce asexually in species. For example, we refer to *E. coli* bacteria and *Shigella* and the like. But in many ways, the concept of species is elusive in asexual organisms. We can tell if two sexual organisms are members of the same species

[10] This is the process by which bacteria are believed to transfer antibiotic immunity among themselves – even among different species. Some transposons carry resistence to as many as five strains of antibiotic.

[11] For example, we shall not delve into the very interesting question of diploidy.

by mating them and seeing if they have viable offspring – that is the definition of a species. On the other hand, absent mating trials, our definition of the *E. coli* "species" is based on appearance and function, not on its genetic structure. In fact, there is often more genetic variation between two different *E. coli* bacteria than there is between either of them and a completely "different" species of bacteria. Population studies of *E. coli* suggest that most *E. coli* bacteria are literal clones of successful *E. coli* variants. In the $Q\beta$ replicase experiments, was the fast RNA the same "species" as the initial RNA template? How would we even begin to answer that question?

Muller's Rachet

In asexual populations, mutation is the primary driving force of evolution. But as we have noted above, most mutations are damaging to the organism, or worse, lethal. In smaller populations, a strong argument may be made that slightly deleterious mutations will tend to accumulate, becoming fixed over time, and that the process will continue to worsen *ad infinitum*. This is referred to as "Muller's rachet" [Muller, 1932] [Haigh, 1978]. In enormous bacterial populations this may be of little consequence and this may be why sexuality has never evolved for many such species. But in higher animals with much smaller populations, Muller's rachet may be an important force.

Speciation in Sexual Populations

By way of contrast, speciation could not be more specific in populations of sexually reproducing organisms. Can two organisms mate and produce viable offspring? That is the question. Why is a well-defined species important for sexual reproduction? Sexual recombination may be of great value to a species for two reasons:

1. Sexual recombination allows the species to combine numerous favorable mutations into one individual much more rapidly than asexual reproduction.

2. Sexual recombination probably ameliorates the effect of Muller's rachet.

But recall our earlier discussion of homology. If the DNA in the two parents does not match closely, most matings will result in disaster, as in Figure 2.3. Thus, to get the benefit of sexual reproduction, a species must maintain a group of mating individuals that have identical or close to identical chromosomal and DNA structures. It will not work any other way.

Genetic programming, from the start, has relied principally on the crossover between two parent programs (sexual reproduction) to cause evolution (see Chapter 5 and [Koza, 1992d]). The difficulty of finding or even defining a homologous species in GP is discussed at some length in Chapter 6.

Exercises

1. Which molecules carry and store the genetic information in nature?
2. Which are the two main sources of genetic variation (the equivalent of genetic operators) in nature?
3. How do bacteria transfer genetic information?
4. Explain the concept of homology. What is homologous recombination?
5. Describe the genetic code in nature. How many "letters" does it use? What do they code for?
6. What is a gene? What is an intron?
7. Explain protein synthesis. Outline the path from gene in DNA to protein.
8. What is a genotype and what is a phenotype?
9. Explain natural selection. Are there other kinds of selection?
10. Explain the concept of fitness.

Further Reading

C. Colby,
INTRODUCTION TO EVOLUTIONARY BIOLOGY.
http://wcl-l.bham.ac.uk/origins/faqs/faq-intro-to-biology.html

P. Coveney and R. Highfield,
THE ARROW OF TIME.
W.H. Allen, London, 1990.

R. Dawkins,
THE BLIND WATCHMAKER.
Norton, New York, NY, 1987.

M. Eigen,
STEPS TOWARD LIFE: A PERSPECTIVE ON EVOLUTION.
Oxford University Press, Oxford, 1992.

B. Lewin,
GENES V.
Oxford University Press, Oxford, 1994.

J. Maynard Smith and E. Szathmáry,
THE MAJOR TRANSITIONS IN EVOLUTION.
W.H. Freeman, Oxford, 1995.

J. Medina,
THE CLOCK OF AGES.
Cambridge University Press, New York, NY, 1996.

L. Orgel,
EVOLUTION OF THE GENETIC APPARATUS.
in JOURNAL OF MOLECULAR BIOLOGY, vol. 38, pp. 381-393, 1968.

J. Watson, N. H. Hopkins, J. W. Roberts, J. Argetsinger-Steitz, A. M. Weiner,
MOLECULAR BIOLOGY OF THE GENE.
Benjamin/Cummings, Menlo Park, CA, 1987.

3 Computer Science and Mathematical Basics

Contents

Anyone who considers arithmetical methods of producing random numbers is, of course, already in a state of sin.

J. VON NEUMANN, 1951

Overview

In this chapter we shall introduce some fundamental notions of computer science and mathematics necessary for understanding the GP approach. The leading question therefore is: What are the mathematical and information-processing contexts of GP, and what are the tools from these contexts that GP has to work with?

3.1 The Importance of Randomness in Evolutionary Learning

As we have seen in the last chapter, organic evolution is one of the effective means of "automatic" learning that we observe in nature. GP is based on a crude model of what we understand to be the mechanisms of organic evolutionary learning. The principle dynamic elements of that model are:

- Innovation caused by mutation, combined with
- Natural selection.

Together and by themselves, these two dynamics appear to be sufficient to cause organic evolution to occur in self-replicating entities. Asexual single cell reproduction, mutation, and natural selection were the sole mechanisms of evolutionary learning for many millions of years [Watson et al., 1987].

In addition to mutation and natural selection, the model on which GP is based also includes sexual reproduction. Sexual reproduction and the related mechanism of gene crossover obviously confer some evolutionary advantage, at least for organisms with diploid gene structures – sexual reproduction has been a very successful strategy for hundreds of millions of years.

Evolution in Nature vs. Evolution in Computers

These different mechanisms of evolutionary learning (mutation and sexual recombination) operate in nature and in the computer. In nature, mutation is basically free. It is a byproduct of entropy. For example, DNA is not always replicated accurately and UV light randomly flips nucleotide "bits" and so forth. In short, the tendency in nature toward disorder will always tend to cause changes in ordered entities and in their offspring. So, random change comes for free in organic evolutionary learning. In contrast, a large amount of the energy expended in evolutionary learning is used to conserve the phenotype despite the entropic tendency toward disorder. This learned stability of the phenotype is an extremely important achievement of evolution.

The Costs of Variation

Sexual reproduction is another matter altogether. The physical mechanisms of biological sexual reproduction are complex and must be self-maintained against the pull of entropy. The mechanism of sexual selection also imposes great cost on the individual and the species. Species evolve "tags" to enable members of the species to identify other members of the species. Some of those tags come at the expense of certain fitness-related aspects like speed. Bird calls can identify the location of a bird to predators. A male peacock's tail slows it down. So, sexual reproduction is decidedly not "free."

In computer programs, on the other hand, stability is an ingrained feature, at least for the time frames relevant to us with respect to computerized evolutionary learning. In such a time frame, we cannot count on entropy causing bits to flip. Of course, if we were willing to wait for the actual physical mechanism of the RAM to deteriorate, we might achieve this effect. But this would be far too slow with respect to our life expectancy. So as a practical matter, computer programs will not change in the time scale required unless we explicitly add a driving force for evolutionary learning.[1] Of course, that separate driving force is simulated mutation and perhaps simulated sexual reproduction.

How can we simulate entropic change in the context of a computer program? After all, entropy measures the effects of random processes. GP simulates entropic change by using a pseudo-random number generator. Increasing order and decreasing entropy (that is, causing learning) by randomly changing computer programs may look counter-intuitive. But driving a search process by a randomness engine is, in a way, the most general procedure that can be designed. When random changes are combined with fitness-based selection, a computerized system is usually able to evolve solutions faster than random search [Eigen, 1992][Koza, 1992d]; see Chapter 2.

The above argument is independent of whether the search space is small enough to allow exhaustive search or so large that only sampling can resonably cover it: on average, it is better to visit locations non-deterministically. By "non-deterministic," we mean that the algorithm, after visiting a location, always has a choice where to go

[1]Note, however, the interesting hypothesis put forward by Belady and Lehman [Belady and Lehman, 1985] that, in fact, a real but unnoticed "natural selection" is acting on computers which results in co-adaptation of components (e.g., periphery and system) or of hardware and software to the effect that most changes in the configuration cause system failure.

next. Non-deterministic algorithms are presently being developed in different areas of computer science with considerable success.

GP as a General Search Process

In conclusion, GP as a general search process in the space of all possible programs/algorithms will depend heavily on randomness in different flavors. For this reason, we shall see in the next section how random numbers can be generated within a computer.

3.2 Mathematical Basics

Randomness and Probability

The mathematical basis for randomness is probability theory. Its origins go back well into the 17th century. Stimulated by problems of gambling, Blaise Pascal and Pierre Fermat were the first to conceptualize probability. The field only later developed into an important branch of mathematics where the laws of random events are studied.

Because random events play such a prominent role in GP, we shall take a closer look at probability. Before going into details of probability, however, we should consider discrete elements and their combinations.

3.2.1 Combinatorics and the Search Space

In GP, it is often necessary to estimate the size of the search space and the difficulty of the problem being solved. The search space is, simply put, the set of all programs that could be found by the GP algorithm, given a certain programming language.

Combinatorics is a mathematical basis for calculating the size of the search space. Combinatorics answers the question: Given a set of discrete elements – function symbols, for instance – in how many different ways can we order them? We are thus concerned, in the simplest case, with linear orderings of a set of elements.

Let us represent elements by lowercase letters, listed in alphabetic order. For instance, consider $E = \{a, b, c\}$, the set of 3 elements a, b, c.

Permutation

Definition 3.1 *A* **permutation** *is an arbitrary ordering of the elements of a set E that uses each element once.*

N different elements constituting the set E can be ordered in $N!$ different permutations.

If we select a subset of K out of N elements from the set we consider a combination of order K.

Combination

Definition 3.2 *A* **combination** *of order K is an arbitrary selection of K out of N elements from the set E without replacement.*

There are

$$C_N^K = \binom{N}{K} = \frac{N!}{(N-K)!K!} \quad (3.1)$$

different combinations if we do not consider replacement and order of selection. C_N^K is called the binomial coefficient.

If, however, the order of those K elements is additionally considered, there is a factor $K!$ between the former combinations and what we call variations.

Variation

Definition 3.3 *A* **variation** *of order K is an ordered selection of K out of N elements from a set E.*

There are

$$V_N^K = \frac{N!}{(N-K)!} \quad (3.2)$$

variations and

$$\bar{C}_N^K = \binom{N+K-1}{K} \quad (3.3)$$

combinations if repetition of elements is allowed.

Combinatorics is important for GP when we want to compute the size of a search space and, therefore, the difficulty of a search problem. How we can use combinatorics to gain information will be shown after we have introduced some basic concepts of computer science.

3.2.2 Random Numbers

As mentioned before, randomness is of utmost importance in genetic programming. This section is devoted to a discussion of the essentials of random numbers and their generation through algorithms. We are interested here in mechanisms which lead to a distribution of numbers within a given interval that looks as much as possible like the outcome of a random process. Since we shall employ deterministic algorithms to generate those numbers, we are in fact not dealing with random numbers but with quasi-random or pseudo-random numbers. Knuth [Knuth, 1981] has studied in detail what we need to consider if we wish to generate random numbers.

Linear Congruential Method

The first message from Knuth is that a complicated and random-looking algorithm does not guarantee the generation of good sequences of random numbers. Instead, very simple algorithms perform surprisingly well in generating random numbers. The simplest is the linear congruential method with suitably chosen parameters.

The linear congruential method of generating a sequence of equally distributed numbers between 0 and 1 goes back to Lehmer

[Lehmer, 1951] and reads:

$$r_{n+1} = (ar_n + b) \bmod m \tag{3.4}$$

Thus, the $(n+1)$th random number, r_{n+1}, is generated by taking its precedessor, r_n, multiplying it by a, adding b and dividing the result modulo m. The parameters to be chosen before the algorithm can be started are

$$m > 0 \tag{3.5}$$

$$0 \leq a < m \tag{3.6}$$

$$0 \leq b < m \tag{3.7}$$

and

$$0 \leq r_0 < m, \tag{3.8}$$

the seed of the algorithm. In order to arrive at a number within the prescribed interval, r_n has to be mapped using

$$r'_{n+1} = r_{n+1}/m \tag{3.9}$$

Knuth [Knuth, 1981] recommends choosing the parameters as follows: m (the period of the generator) should be very large, presumably a power of 2 or 10; a should be picked so that $a \bmod 8 = 5$ or $a \bmod 200 = 21$, respectively; a should be between $0.01m$ and $0.99m$, without having a regular pattern; and b might be taken as 1 or m. With these choices, a good random number generator can be implemented that passes several tests of randomness.

χ^2 Test

The performance of random-number generating algorithms is itself measurable with different metrics. It is difficult to give a general recipe here, because it depends on the application the pseudo-random numbers are used for, but one of the general-purpose tests most often used is a statistical test called the χ^2 test. It is advisable to use at least the χ^2 test with any kind of random-number generating algorithm. If a distribution of numbers does not pass this test, the corresponding number generator does not generate random numbers of sufficient quality and should be discarded.

The mathematical reasoning behind the χ^2 test is easy to understand: If we want to generate N random numbers in the interval between $0 \leq i \leq k-1$, with $k \in \mathbb{N}$, they should appear with nearly equal, but not exactly equal frequency $\frac{N}{k}$. This can be tested by measuring the actual frequencies f_i in a sequence and comparing them to the baseline:

$$\chi^2 = \sum_{i=0}^{k-1} (f_i - \frac{N}{k})^2 / \frac{N}{k} \tag{3.10}$$

If χ^2 is near to k, then the random number generator is good, otherwise it is bad. "Near" means that $2\sqrt{k} \leq \chi^2 \leq k$.

The χ^2 test is by no means the only performance measure for random-number generators, but for the purpose of genetic programming, it is completely sufficient.

In the next section, we shall relax again from the rigors of generating randomness artificially and shall consider some useful notions of probability theory.

3.2.3 Probability

The two most fundamental notions of probability theory are:

- random experiments
- events

Random experiments are experiments that are undetermined as to their final outcome. This, of course, prevents accurate predictions of the result of a particular random experiment. For example, before we flip a coin, the forthcoming act of flipping the coin and reading the result is a random experiment.

Elementary Events

Events, on the other hand, are the observed results of random experiments. So, once the coin has been flipped and we read "heads" or "tails," the result we have read is an event. These events are more specifically called elementary events if they are at the most detailed level of categories into which we classify the results of a particular type of experiment. The set of all possible outcomes of an experiment is called the "sample space."

A series of repeated random experiments results in different outcomes whose relative frequency can be counted.

Relative Frequency

Definition 3.4 *The* **relative frequency** f_A *of an event* A *being* K *times the outcome of a series of* N *random experiments is* $f_A = \frac{K}{N}$.

This assumes independence of all the experiments in the series. Probability is then usually defined as the limit for an infinitely long series of experiments.

Probability

Definition 3.5 *The* **probability** p_A *of finding* A *as the outcome of a random experiment is the limit* $N \to \infty$ *of the relative frequency* f_A *of* A: $p_A = \lim_{N\to\infty} f_A$.

There are other ways of defining probability, notably Kolmogorov's axiomatic way [Kolmogorov, 1950]. The mathematically inclined reader may consult a textbook [Feller, 1968] [Feller, 1971] on probability theory.

Random Variables and Probability Distributions

Now that we have assigned probability to certain events as outcomes of a specific type of random experiment, we can consider the actual outcome of an experiment, represented by a random variable x. In a given experiment, x will assume a value $X_1, X_2, X_3, \ldots$ out of all the values possible in that experiment. The values of random variable x plus their respective probabilities, $P(x = X_1), P(x = X_2), P(x = X_3), \ldots$ are called the probability distribution $p(x)$ of random variable x. $p(x)$ is a discrete distribution obeying the boundedness and normalization rules

$$0 \leq p(x) \leq 1 \tag{3.11}$$

$$\sum_{y=0}^{\infty} p(y) = 1 \tag{3.12}$$

Expectation Value and Variance

Two of the most important quantities in connection with the probability of a random variable are its expectation value and its variance. Equation 3.12 can be formulated as the zeroth order ($m = 0$) of a moment quantity

$$\sum_{y=0}^{\infty} p(y)y^m \tag{3.13}$$

with general moment m, and the expectation value is the first order moment

$$\sum_{y=0}^{\infty} p(y)y^1 = \sum_{y=0}^{\infty} p(y)y = E(y) \tag{3.14}$$

The expectation value measures a random variable's most important tendency, often called the mean of y.

The variance, on the other hand, is closely related to the second order moment of y:

$$\sum_{y=0}^{\infty} p(y)y^2 \tag{3.15}$$

in that this quantity is corrected by the mean of the variable:

$$\sum_{y=0}^{\infty} p(y)(y - E(y))^2 = \mathrm{var}(y) \tag{3.16}$$

The variance or its square root are useful for characterizing the dispersion of a random variable y.

One other form of probability is worth noting here: conditional probability. Often, one event depends on the occurrence of another. Conditional probabilities are introduced to quantify that dependence.

Definition 3.6 *The* **conditional probability** *of an event A, given the occurrence of an event B, $p(A|B)$ is the probability of both events co-occurring $p(A \wedge B)$, normalized by the probability of event B, $p(B)$.*

Bernoulli Process and the Binomial Distribution

As an example of a discrete probability distribution we consider the binomial distribution. The binomial distribution results from the so-called Bernoulli process specifying certain assumptions about a series of random experiments. A random experiment in a Bernoulli process can have one of two possible outcomes, for instance: yes or no, up or down, success or failure. The probabilities of those events is constant over the series of experiments and independent in successive trials.

Let p be the probability for one of the two events. Let N be the number of experiments in the series and K be the occurrence of the event. The binomial distribution is then

$$B_K^N(p) = \binom{N}{K} p^K (1-p)^{N-K} \qquad K = 0, 1, \ldots, N \tag{3.17}$$

For a given number of experiments and a given occurrence probability p of an elementary event, this distribution is a function $B(K)$ of the discrete values K (see Figure 3.1, top).

Probability Density Functions

It often happens that a classification of outcomes of a random experiment into discrete events is quite artificial. For those cases, a continuous random variable should be available that is allowed to assume arbitrary values. The resulting probability distribution function is called a probability density function $p(x)$.

If we wished then to know $P(a \leq x \leq b)$, the probability of finding variable x between values a and b, we could simply integrate the probability density function:

$$P(a \leq x \leq b) = \int_a^b p(x)dx \tag{3.18}$$

Normal Distribution

The Gaussian or normal distribution is the best known of the continuous probability density functions. It can be computed from

$$g(x) = \frac{1}{\sqrt{2\pi}\sigma} \exp\left[-\frac{(x-\mu)^2}{2\sigma^2}\right] \equiv N(\mu, \sigma), \qquad -\infty < x < \infty \tag{3.19}$$

The normal distribution $N(\mu, \sigma)$ with parameters μ (mean) and σ (standard deviation) can be used as an approximation to very many discrete probability distributions as well as in most continuous cases. It has some special mathematical properties that make it particularly easy to handle mathematically (e.g., third and higher order moments disappear). Therefore, it is the tool of choice for many analyses involving random variables.

In the context of evolutionary algorithms, the normal distribution $N(0, \sigma)$ is an important means to provide additive variation. The neutral value of additive variation is 0. Therefore, an expectation value of 0 is usually applied.

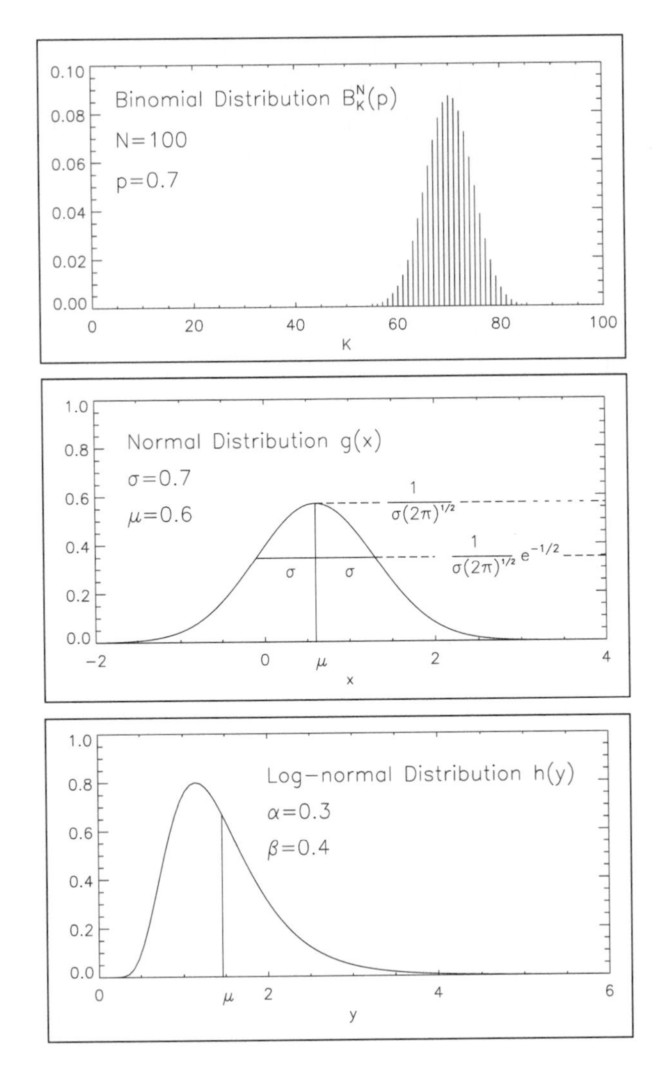

Figure 3.1
Three distributions: binomial, normal, and log-normal. The constants α and β of the log-normal distribution $\exp(-\frac{(\ln y-\alpha)^2}{2\beta^2})$ can be transformed into $\mu = \exp(\alpha + 0.5\beta^2)$ and $\sigma^2 = \exp(2\alpha + \beta^2)$ $(\exp(\beta^2) - 1)$.

Multiplicative Variation and the Log-Normal Distribution

Sometimes, however, there is a need for multiplicative variation. Multiplicative variation plays a role, for example, in adaptive step size control in evolutionary algorithms. It can be achieved with a distribution having 1 as its neutral value. A very natural way to realize this is the so-called log-normal distribution

$$y = \exp(x) \tag{3.20}$$

where x is a normally distributed random variable derived from $N(0, \tau)$. Taking the logarithm of 3.20, we can see that $\ln y$ is distributed normally. The probability distribution $h(y)$ for y is therefore:

$$h(y) = \frac{1}{\sqrt{2\pi}\tau} \exp\left[-\frac{(\ln y)^2}{2\tau^2}\right] \qquad 0 < y < \infty \tag{3.21}$$

3.3 Computer Science Background and Terminology

All computers are created equal.

UNKNOWN

This section provides basic computer science concepts and terminology relevant to genetic programming.

3.3.1 The Turing Machine, Turing Completeness, and Universal Computation

Although the *Turing machine* (TM) is one of the best known types of computing machines, no one has ever built one.[2] The Turing machine is an extremely simple computer, and the surprising fact is that a TM can simulate the behavior of *any* computer. So, any computer that can simulate a TM can also simulate any other computer. However, the TM is a very inefficient computer, and it is generally not useful for calculations. Rather, the concept of a Turing machine has been an invaluable tool for mathematicians analyzing computers and what they can and cannot do.

The TM was introduced by the British mathematician Turing in his milestone 1936 paper "On computable numbers, with an application to the Entscheidungsproblem" [Turing, 1936]. His simplified computer consists of only four parts (see Figure 3.2):

- A long tape of paper where symbols can be read, written and changed (overwritten). The tape is divided into squares that contain exactly one symbol or that are empty.
- A read/write device for reading one symbol from the tape or writing/changing one symbol on the tape. The device can move, one step at a time, over the tape to read the contents of other squares, illustrated with the box in Figure 3.2.
- A finite set of states.
- A set of *state transition rules*. This set can be interpreted as the "program" of the Turing machine that defines a certain *action* the TM must perform depending on the symbol it is currently reading and on what state it is in. An action is a state change along with either a change of the device position or a write on the tape.

[2]However, a well-meaning clerk at the Library of Congress has set aside a whole category number for books on "marketing of Turing machines."

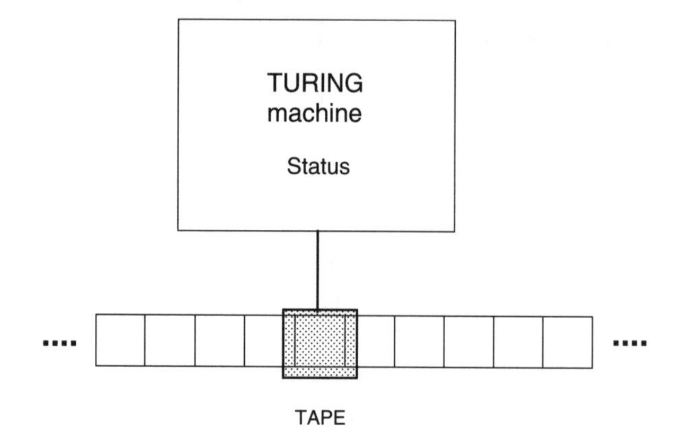

Figure 3.2
Schematic view of a Turing machine

Each rule specifies a combination of state and tape square content and an action to perform if this combination matches the current state and tape square content. Some authors assume the tape to be of infinite length, but all that is needed is that the tape can grow to an arbitrary length by addition of new squares to either end.

Turing Completeness

A programming language is said to be *Turing complete* if it allows to write a program that emulates the behavior of a certain arbitrary TM. A Turing complete language is the most powerful form of a programming language. All commercial programming languages, such as FORTRAN or C, are Turing complete languages. In fact, only a few simple program constructs are needed to make a language Turing complete.

On the one hand, a too-simple language (such as one containing addition and no other operator) is *not* Turing complete. It cannot simulate a Turing machine and is, therefore, unable to represent an arbitrary algorithm. On the other hand, a program language with addition and a conditional jump operator is Turing complete. The expressiveness of a language is an important property to consider when choosing functions for the function set of a GP experiment. In general, it seems desirable to work with a Turing complete language when doing genetic programming.

Teller proved in 1994 [Teller, 1994b] [Teller, 1994c] that a GP system using indexed memory and arithmetic operators is Turing complete Note that the equivalence of different Turing complete languages only holds for functionality, not for efficiency or time complexity of the words from the languages. Languages may also have different properties with respect to evolution of algorithms in a GP system.

Structure and Function of a Turing Machine

Formally, a Turing machine T can be defined as

$$T = (I, O, Q, \delta, q_0, F) \tag{3.22}$$

where I is an input alphabet, O is an output alphabet with $I \subset O$, Q is a finite set of all states of T, $q_0 \in Q$ is the initial state, $F \subset Q$ is the set of finite states, and

$$\delta : (Q - F) \times O \rightarrow Q \times (O \cup \{l, r\}) \tag{3.23}$$

is a state transition function. δ may be a partial function, in which case there is at least one $(q, a) \in (Q - F) \times O$ not mapped on any $(r, c) \in Q \times (O \cup \{l, r\})$. $l, r \notin O$ denote a move of the read/write device – also called the *head* – by one tape square to the left or right. There is a blank $b \in O - I$. Each tape square is considered to be initialized with b.

Given the structure of T, we now consider its *function*. At the outset, the initial word $w \in I^*, w = s_0..s_n$, is on the tape. All tape squares not used by w contain b. T has state q_0, and the head is on s_0. Each processing step of T works as follows. T has state $q_1 \notin F$. The current square, the square the head is on, contains symbol a_1. If $\delta(q_1, a_1)$ is defined and gives (q_2, a_2), T makes a transition into q_2 and processes a_2 as follows. If $a_2 \in O$, then T replaces a_1 by a_2. If $a_2 \in \{l, r\}$, then the head moves one square in the corresponding direction. If $\delta(q_1, a_1)$ is not defined, or if $q_1 \in F$, then T terminates.

Universal Turing Machine and Universal Computation

A *universal Turing machine* U can emulate any Turing machine T. To that end, U is given an initial $w \in I^*$ that defines T and x, which is T's initial word. In this sense, U is said to be able to perform *universal computation*. Obviously, in the case of U, there is no need to change U in order to perform a different computation. Thus, a universal Turing machine is a mathematical model for a modern computer. The working memory is analogous to a TM's tape, a program is analogous to the emulated Turing machine T, and the program's input is analogous to x. Note, however, that a real computer may not have enough memory for some programs, while a Turing machine always has enough tape squares.

3.3.2 The Halting Problem

The motivation for Turing's work with respect to the TM was not so much to find a universal computing machine that could be shown to be as powerful as any other machine. Instead, he wanted to clarify whether such machines had limitations and whether there were problems computers could not solve. Turing's objective in his famous paper mentioned above was to show that there are functions that cannot be computed by a Turing machine.

An important example of a problem that cannot be solved by a TM is the halting problem. It asks if there is a program that can determine whether another program halts or not. One result of

Turing's work is that there is no program that can determine the termination properties of *all* programs. There are many programs whose halting property can be decided with a mechanized procedure, but the halting problem is unsolvable in the general case.

Halting Theorem

The halting theorem also has important implications for the evolution of programs with a GP system. If we are evolving programs with a terminal set and function set that make the underlying programming language Turing complete, then we cannot know beforehand which programs will terminate and which will not. For a GP run, we must therefore use one of the methods available for ensuring time bounded execution of an evolved program. Time bounded execution is further described in Section 10.2.8.

3.3.3 Complexity

The size of a computer program in GP is often referred to as the program's complexity.[3] There are various measures of complexity or size in the GP literature. One of the most natural and commonly used is simply the number of nodes in a tree-based genetic programming system. Other definitions of complexity that have been used are the number of bits needed to express a program in linear form, or the number of instructions, for example, in machine code; see Figure 3.3.

The complexity of a computer program is arguably related to the capability of the program to generalize from a set of given data. A shorter program is more likely to show a feasible behavior with data it has not been explicitly trained on than a longer program would. Chapter 8 will discuss the generalization issue for programs in more depth.

Kolmogorov Complexity and Generalization

It has been argued that a short program with low complexity has a higher probability of generalizing well. In general, the mathematical property "complexity of a computable object" can be said to be the shortest program that produces the object upon execution. Note that this complexity concept differs from the complexity definition discussed above for programs. The complexity of a chunk of information is a property of this information regardless of the type of information, and it is applicable to many other kinds of objects than computer programs. We call this complexity property of an object the *Kolmogorov complexity* of the object. Synonyms for Kolmogorov complexity are *algorithmic information*, *algorithmic complexity*, or *Kolmogorov-Chaitin complexity*.

[3]Very often, a logarithmic function of the size is used.

Complexity

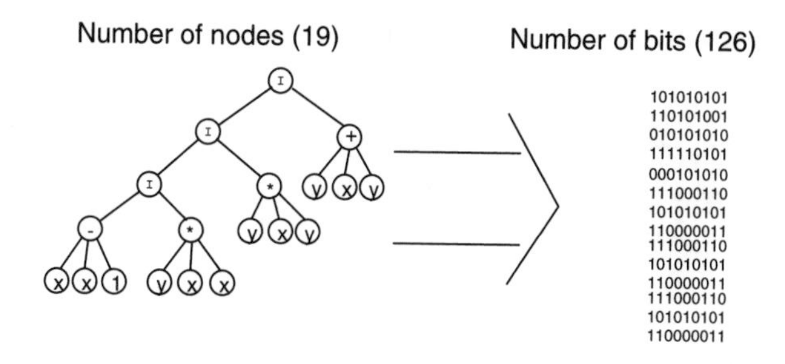

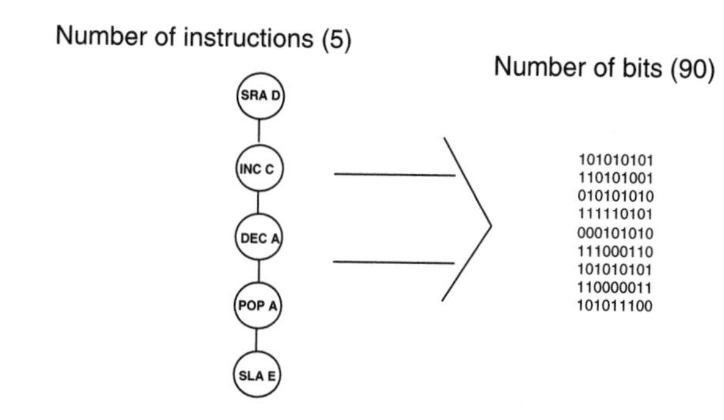

Figure 3.3
Different complexity measures: number of nodes, instructions, or bits

In GP, we are interested in selecting a program – which can be viewed as a model or hypothesis – that fits the data we have observed. For instance, we might want to predict the next number in a data series given the presently known data points. Or, in symbolic regression, we might like to fit a function to a set of fitness cases such that the function – with high probability – accurately models the underlying problem outside of the domain of the given fitness cases.

It can be shown that the probability of guessing a program that correctly models some fitness cases or other observed data is dominated by the probabilities of the shortest programs consistent with these data. In other words, if two programs model the same data, the shorter one can be argued to have a higher probability of being general with respect to the underlying problem. However, this subject is still in debate in the scientific community [Li and Vitanyi, 1997].

An interesting question is the relation between the complexity of a program and that of its environment. In nature, the complexity of phenotypes seems to draw heavily on the complexity of the environment. The genome, in other words, seems to "harness" the complexity of the environment to generate an organism. Can a similar process help to produce complex programs in GP?

3.4 Computer Hardware

3.4.1 Von Neumann Machines

A von Neumann machine is a computer where the program resides in the same storage as the data used by that program. This machine is named after the famous Hungarian-American mathematician von Neumann, and almost all computers today are of the von Neumann type.

The group of which von Neumann was a member at the time of this invention[4] was pondering ways to facilitate programming of computers. They considered it too tedious to reconnect banks of cables and to reset myriads of switches to reprogram a computing device. By contrast, input of data via punched cards or paper tapes was much easier. So the idea was born of simply inputting and storing programs in the same way as data.

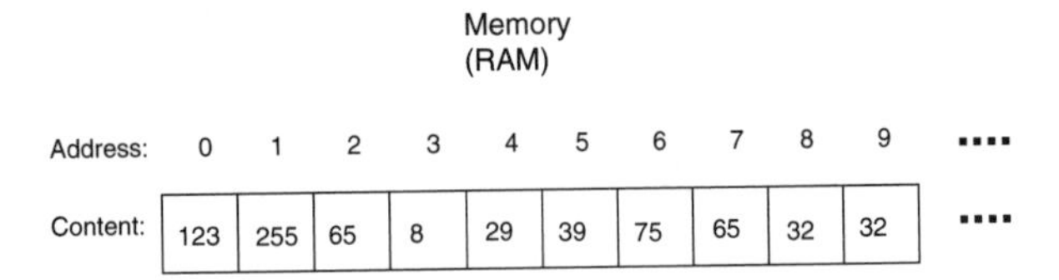

Figure 3.4
Schematic view of a computer's RAM memory

The fact that a program can be regarded as just another kind of data makes it possible to build programs that manipulate programs and – in particular – programs that manipulate themselves. The memory in a machine of this type can be viewed as an indexed array of integer numbers, and thus a program is also an array of integers as depicted in Figure 3.4.

This approach has been considered dangerous in the history of computing. A contemporary book on computer hardware explains:

> The decision of von Neumann and his collaborators to represent programs and data the same way, i.e. interchangeable, was of captivating originality. Now a program was allowed to modify itself, which caused speculations about learning, self-reproducing and, therefore, "living" systems. Although these ideas were enticing and still are, their consequences are dangerous. Not only are programs that modify themselves unfathomable (incomprehensible), but they also lead to an unpredictable behavior of the computer. Fortunately, at least when the output is being printed, this regularly ends in a chaos of confusing columns of numbers and erroneous data which can be thrown away immediately. This

[4]The ENIAC project at Moore School, with J.P. Eckert, J.W. Mauchly, and H.H. Goldstine.

> chaos is, of course, not acceptable in large and sensitive technical systems. Already when programming these systems, manifold security measures have to be taken such that the system does not become autonomous.
>
> H. LIEBIG AND T. FLIK, 1993, translated from German

It is not unreasonable to hope that genetic programming will change this attitude completely.

Different machines use integers of different maximal sizes. Currently, 32-bit processors are the most common type of processor available commercially. The memory of such a machine can be viewed as an array of integers, each with a maximal size of $2^{32} - 1$, which is equal to 4 294 967 295, and a program in such a machine is nothing more than an array of numbers between zero and 4 294 967 295. A program that manipulates another program's binary instructions is just a program that manipulates an array of integers. This is an important fact for manipulating binary machine code in GP and will be taken up again in Chapter 9 and in more detail in Chapter 11. Figure 3.5 illustrates the principle of a von Neumann machine and also how it can be used for meta-manipulation.

The Processor

The processor is the black box doing all the "intelligent" work in a computer. The principles of different present-day processors are surprisingly similar. To simplify somewhat we can say that the processor consists of three parts. First, it has a device for storing and retrieving integers from the memory array. Then it has a number of registers for internal storage of integers and a unit for arithmetic and logic operation between the registers, the ALU, as shown in Figures 3.5 and 3.6.

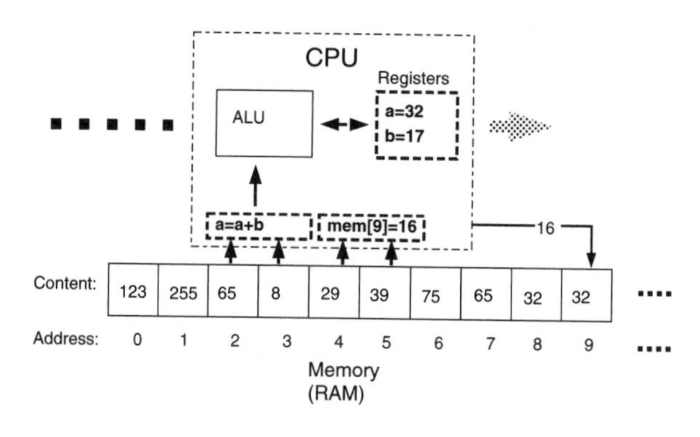

Figure 3.5
The CPU and what it does

A register is a place inside the processor where an integer can be stored. Normally, a register can store an integer with the same size as the word size of the processor. For instance, a 32-bit processor has registers that can store integers between 0 and 4 294 967 295. The most important register is the program counter (PC) which gives the index to the next instruction to be executed by the processor. The processor looks at the contents of the memory array at the position of the program counter and interprets this integer as an instruction, which might be to add the contents of two registers or to place a value from memory into a register. An addition of a number to the program counter itself causes transfer of control to another part of the memory, in other words, a jump to another part of the program. After each instruction the program counter is incremented by one and another instruction is read from memory and executed.

The ALU in the processor performs arithmetic and logic instructions between registers. All processors can do addition, subtraction, logical *and*, logical *or*, etc. More advanced processors do multiplication and division of integers and some have a floating point unit with corresponding floating point registers.

In Figure 3.5 we can see the overall activity of the processor working on memory cells. Integers in memory become instructions to the processor when the CPU reads and interprets them as commands. Figure 3.6 shows more details. The memory retrieval device uses a register or some arithmetic combination of registers to get an index number. The contents of the memory array element with this index number are then placed in one of the registers of the processor.

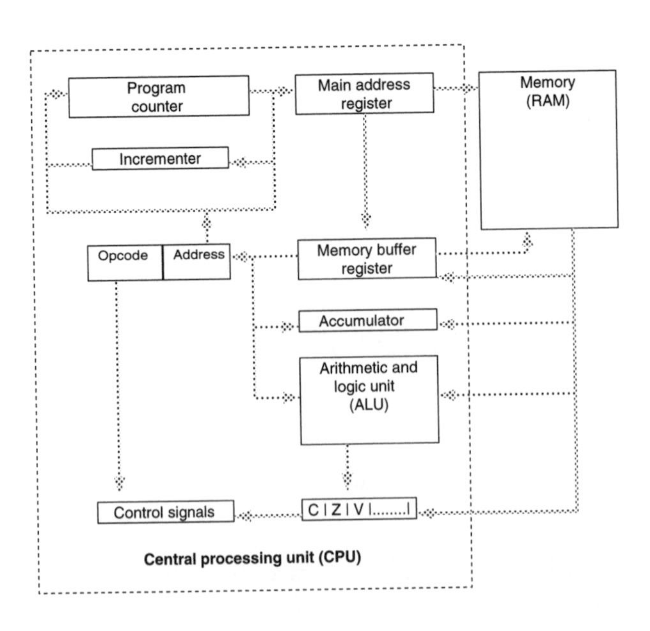

Figure 3.6
Schematic view of the central processing unit (CPU)

All the behavior we see in today's computers is based on these simple principles. Computers doing graphics or animations, controlling a washing machine, or monitoring a car's ignition system all do the same memory manipulations and register operations.

RISC/CISC

At present, both CISC processors (Pentium) and RISC processors (SPARC or PowerPC) are used in commercial computers. CISC is an acronym for Complex Instruction Set while RISC is an acronym for Reduced Instruction Set. As indicated by the acronym, RISC processors have fewer and less complex instructions than CISC processors. This means that a RISC processor can be implemented with less complex hardware and that it will execute its instructions faster. The RISC approach follows some advice on instruction set design given by von Neumann:

> The really decisive consideration in selecting an instruction set is simplicity of the equipment demanded by the [instruction set] and the clarity of its application to the actual important problems, together with [its] speed in handling those problems.
>
> J. von Neumann, 1944

One of the central ideas of RISC architectures is the extensive use of registers in the code. The late Seymour Cray, one of the pioneers of high performance computing, made the following remark:

> [Registers] made the instructions very simple.... That is somewhat unique. Most machines have rather elaborate instruction sets involving many more memory references than the machines I have designed. Simplicity, I guess, is a way of saying it. I am all for simplicity. If it's very complicated, I can't understand it.
>
> S. Cray, 1975

3.4.2 Evolvable Hardware

Recently, interest has grown rapidly in a research field that has evolution of electronic hardware as its topic. With the advent of increasingly cheaper freely programmable gate array chips (FPGAs), a whole new area for evolutionary algorithms is opening up.

For the first time it seems feasible to actually do trials in hardware, always with the possibility in mind that these trials have to be discarded due to their failure. Evolvable hardware (EHW) is one important step, because with it, there is no need to discard the entire hardware; instead one simply reprograms the chip. The EHW field is developing quickly, and the interested reader should consult original literature [DeGaris, 1993] [Hirst, 1996].

3.5 Computer Software

> Of all the elements that make up the technology of computing, none has been more problematic or unpredictable than software. ... The fundamental difficulty in writing software was that, until computers arrived, human beings had never before had to prepare detailed instructions for an automaton – a machine that obeyed unerringly the commands given to it, and for which every possible outcome had to be anticipated by the programmer.
>
> M. Campbell-Kelly and W. Aspray, 1996

With the exception of EHW, hardware is too rigid to represent changing needs for computer functionality. Use of software is the only way to harvest the universality of computers. In this section we discuss the most basic aspects of software.

First, we take a look at the most elementary representation of software: machine language, the "native language" of a processor, which it speaks fast. Also, we shall introduce assembly language. Second, we shall look at classes of higher languages. They are easier to use for humans, but they have to be translated into a machine language program. Third, elementary data structures – structures that allow the storing, reading, and writing of data in certain logical ways – will be presented. Fourth, we shall air some thoughts on manual versus genetic programming.

3.5.1 Machine Language and Assembler Language

A program in machine language is a sequence of integers. These integers are often expressed in different bases – decimal, octal, hexadecimal, or binary – in order to simplify programming and reading machine language programs. By binary machine code, we mean the actual numbers stored in binary format in the computer.

However, it is often impractical to use numbers for instructions when programming in or discussing machine language. Remembering, for instance, that the addition instruction is represented by 2 416 058 368 in the SUN SPARC architecture is not natural to the human mind (what was that number again?). If we represent the same instruction in hexadecimal base (90 022 000), it will be more compact and easier to reason about, but it is still not natural to remember. For that purpose, assembly language was developed. Assembly language uses mnemonics to represent machine code instructions. Addition is represented by the three letters ADD. The grammar for assembly language is very simple, and the translation from assembly language to machine code is simple and straightforward. But assembly language is not machine language, and cannot be executed by the processor directly without the translation step.

It is worth noting that different processors implement different word sizes, instructions, and coding of the instructions, but the differences between instructions for processors of different families are surprisingly small.

3.5.2 Higher Languages

Both machine language and assembler are called low-level language because hardware aspects are important when writing a program. Low-level languages are machine-oriented languages. In contrast, high-level languages or problem-oriented languages do not require detailed knowledge about the underlying hardware. A problem-oriented language, like FORTRAN [Morgan and Schonfelder, 1993], allows for modeling by use of abstract representations of operators and operands.

General and Special Purpose

We can further distinguish between general-purpose and special-purpose high-level languages. The former provide problem-independent language elements, while the latter supply elements that are tailored for modeling situations from a certain problem domain. For instance, Pascal [Sedgewick, 1992] is a general-purpose language, while SQL [Celko, 1995] is a special-purpose language used in the database-query domain. All general-purpose languages are Turing complete, of course.

High-level languages can be classified by the differing sets of principles they obey. We mention a few prominent language classes in the chronological order of their development. Table 3.1 summarizes the situation.

Class	Entity	Principle
imperative	variable	von Neumann architecture
functional	function	lambda calculus
predicative	predicate inference rule	logic
object-oriented	instance, class, state, method	object-oriented reality

Table 3.1
Language classes

Imperative Languages

Ada [Barnes, 1982], BASIC, C [Kernighan et al., 1988], FORTRAN, Pascal, or SIMULA [Kirkerud, 1989] are examples of imperative languages. Program statements explicitly order (Latin *imperare*) the computer *how* to perform a certain task. Since imperative languages are the oldest high-level languages, they are closest to low-level languages.

Principles underlying the von Neumann computer architecture influenced the design of imperative languages. For instance, an essential entity of any imperative language is a *variable*, and typical statements semantically look like:

```
put value 42 into variable universalAnswer
```

The existence of variables results from the von Neumann principle of dividing memory into equal-sized cells, one or more of which may represent a variable.

Functional, Applicative Languages

LISP, LOGO, ML, and BETA are examples of functional or applicative languages. Program statements define a certain task. In mathematics, a function is a relation between a set A and a set B such that each element in A is associated with exactly one element in B. In other words, a function maps A onto B.

A program represents a function that maps input data and internal data into output data. This is the main point of a functional language. Its essential entity is called a *function*. Using a function on its arguments is called *application*, so a functional language is also called *applicative*. Functions can be combined into composed functions. A value can be a function, too. The lambda calculus is the mathematical foundation of functional languages.

Predicative Languages

Like a functional language, a predicative language is based on mathematical principles. PROLOG is such a language. Programming means describing to the computer *what* is wanted as result. It does not mean saying *how* to get to the result. An example from [Stansifer, 1995] should illustrate this.

```
mortal (X) :- man (X) /* if X is a man then X is mortal */
man (socrates). /* Socrates is a man. */

mortal (socrates)? /* question to system */
yes /* system answer */
```

The basic idea of predicative languages is to represent a program by valid statements (predicates) and inference rules. The rules describe how to deduce predicates from predicates. Getting a certain result r from a predicative program implies asking a question like: "for which value(s) of r is a certain predicate true?"

Object-Oriented Languages

SMALLTALK-80, C++, and JAVA are examples of object-oriented programming languages. The principle behind these languages is modeling a system by objects. An object represents a certain phenomenon. It may have states. Objects may communicate by sending messages to each other. An object receiving a message may react to this event by executing a method. As a result of a method execution, the receiving object may, for example, send an object to the other object.

Each object is an instance of a class that defines states and methods of each of its instances. A subclass of a class defines its instances as special cases of the corresponding superclass. Instances of a subclass inherit all methods and states of all corresponding superclasses.

Table 3.2 shows some instances of high-level languages and how they have been used in connection with genetic programming.

Language	Individual	Implementation	Source
C	•	•	[Keller and Banzhaf, 1996]
C++		•	[Keith and Martin, 1994]
FORTRAN	•	•	[Banzhaf, 1994]
LISP	•	•	[Koza, 1992d]
PROLOG	•		[Osborn et al., 1995]

Table 3.2
GP systems that allow for evolving high-level language individuals and/or are implemented in a high-level language

3.5.3 Data Structures

> Any stupid boy can crush a bug. However, all the professors in the world cannot make one.
>
> UNKNOWN

Maybe artifical evolution will be able to come up with a structure as complex as a hemipterous insect some day. Currently, GP is already challenged when trying to manipulate comparatively primitive *data structures* in a meaningful way. This section discusses some prominent data structures offered by many common programming languages.

A data structure is either a value range or a structure of data structures and a set of constructors. In the former case, the data structure is called *simple*; in the latter case, *composed.* A constructor constructs a composed data structure from other data structures.

From the point of view of the von Neumann architecture, program code and data are binary sequences. The primitive organization of eight bits giving a byte and of a machine-dependent byte number giving a memory word are the only forms of structure in this sequence. Most real-world situations, however, are phenomena with complex structure. Data structures for use in modeling these situations should be problem oriented. Just as mathematical concepts underlie high-level languages, data structures in particular are drawn from mathematical structures and operators. Figure 3.7 shows a hierarchy of data structures. An arrow running from a structure A to a structure B means that B is a special case of A.

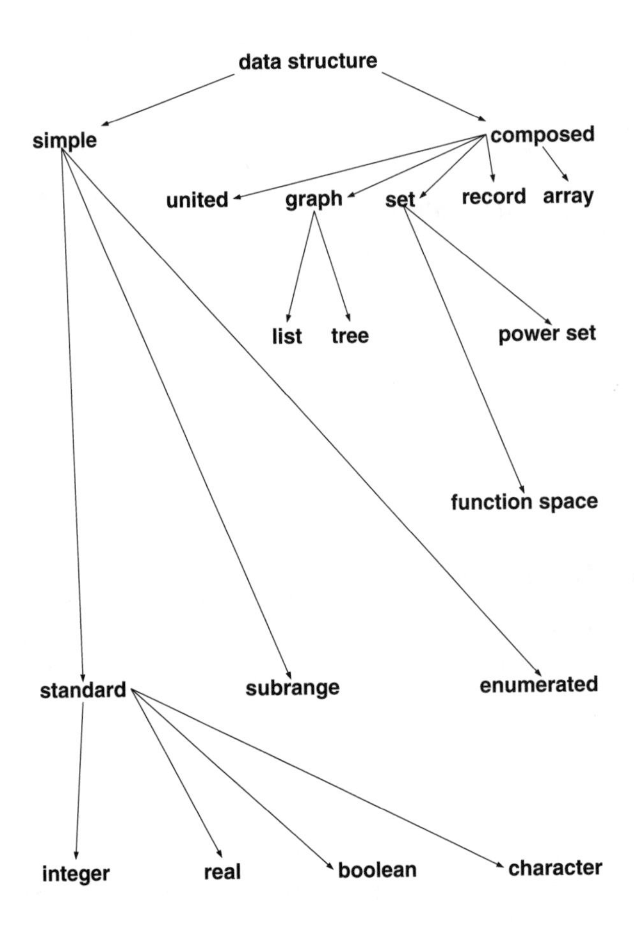

Figure 3.7
Data structure hierarchy. An arrow running from a structure A to a structure B means that B is a special case of A.

Typical simple concrete structures are `integer`, `real`, `Boolean`, `character`, `enumerated`, and `subranges` of these. The kinds of values contained in `integer`, `real`, and `character` are obvious. `Boolean` contains just two values: `true` and `false`. `True` denotes that a certain condition is given, `false` the opposite. `Enumerated` is a simple abstract structure. It contains a finite linear-ordered set of values. For instance, `red, yellow, green` with order `red` > `yellow` > `green` is an enumerated structure `color`. A `subrange` is a contiguous interval of a value range. For instance, `red, yellow` is a subrange of `color`.

There are many potential constructors and composed structures. Five prominent constructor classes and the corresponding structure classes are discussed next. Many high-level languages offer instances of these classes as language elements.

Aggregation

Aggregation builds a Cartesian product of structures, called a record. For instance, $(1, 3.14)$ is an element in the record $\mathbb{N} \times \mathbb{R}$. If all structures of a record are identical, the record is called an *array*. For instance, $(2.718, 3.141, 1.414)$ is an element in the array $\mathbb{R} \times \mathbb{R} \times \mathbb{R}$.

A person profile like

```
first name:         Smith
surname:            John
age (yrs):          32
height (meters):    1.95
weight (kg):        92
```

is an element from a record $N \times N \times A \times H \times W$. N is an array of twenty characters, A is a subrange from 0 to 150 of integer, H is a subrange from 0.3 to 2.72 of real,[5] W is a subrange from 0 to 300.

Generalization, Recursion

Generalization unites structures. $(2.718, 3.141, 1.414)$ and 10, for instance, are elements in the united structure $\mathbb{R} \times \mathbb{R} \times \mathbb{R} \cup \mathbb{N}$. Recursion constructs a structure consisting of an infinite discrete value range. Recursive structures can be constructed in several high-level languages, like LISP or C.

Graph, Tree, List

A structure can be modeled by a structure called a *graph*. A graph is a set of nodes and a set of edges that connect certain nodes with each other. A node represents a part of an instance of the structure.

For example, consider the record $R := \mathbb{N} \times R,\ or\ R := \mathbb{N}$. This is a – potentially infinite – recursive structure $\mathbb{N} \times \mathbb{N} \times \mathbb{N} \times ....$ An instance of this structure is $(1, 836, 947, 37, 7959739793457, ...)$. Such a linear recursive structure can be modeled as a list. A list is a special case of a graph: it is cycle-free and each node is connected, at most, to two different other nodes. In the example, each integer value can be represented by a node.

For another instance, consider the expression $a + b * c$. Let us represent each symbol in this expression by a specific node. Now, connect the PLUS node with the a-node and the MULT node. Then, connect the MULT node with the b-node and the c-node. Finally, explicitly mark exactly one node as a special node. What you get is a tree, which is a very important special case of a graph: it is cycle-free, and one of its nodes is explicitly marked as the root node.

Thus, a tree is a *hierarchical* graph: all nodes connected to the root node are called the *children* of the root node. Of course, these children can be seen as root nodes of their own children, and so on. Obviously, a tree may be composed of subtrees. That is, a tree is a recursive structure. In particular, a list is a special case of a tree.

[5]Memorializing Robert P. Wadlow 1918-1940, whose attested height defines the upper range limit.

In the example of the arithmetic expression, the PLUS node may be the root node. Then, the subtrees a and $b*c$ model subexpressions.

In general, graphs can be used as models of structures and processes from real-world applications. For instance, certain complex industrial processes may be modeled as Petri nets, which are powerful instances of general graphs. A tree can model a decision process: each node represents an option set, and depending on the chosen option, the process branches to a child node. A list may model a stack of engine parts that are successively being popped from the stack by an engine-building robot.

Power Set

A power set of a set S is the set of all subsets of S. The power set constructor gives the power set of a structure as a new structure. For instance, the power set of {`red, yellow, green`} is {∅}, {`red`}, {`yellow`}, {`green`}, {`red, yellow`}, {`red, green`}, {`yellow, green`}, {`red, yellow, green`}. A power set can help in modeling an archetype that implies combinations of entities. For instance, consider a set of chess players. In order to compute all possible player combinations, one simply constructs the power set and isolates all subsets of size 2.

Function Space

A function space is a set of mathematical functions. The function set constructor gives a corresponding structure. For instance, the set of all continuous functions is a function space.

Selector

A selector selects a value of a component of a composed structure. For instance, the selector $[i]$ selects component i in an array and answers i's value.

Data structures are an essential topic in computer science, and it remains to be seen how GP can make use of them. Some studies already indicate that a connection can be made (see Chapter 10).

3.5.4 Manual versus Genetic Programming

Programming by hand is the traditional method of generating useful programs – it is what we call the craftsman approach to programming. After the invention of the von Neumann machine, this programming method was greatly facilitated by allowing programs to be input in the same way as data were handled. By this time, programs consisted of numbers symbolizing machine instructions to be executed directly by the processor.

From Bits to Memo Code

A first step in abstraction took place, when instead of machine instruction bits, assembler code could be written to instruct the processor. A translation step had to be added to the processing of programs that transformed the memo code of assembler into machine-understandable sequences of instructions.

Later, high-level languages were used for writing programs, e.g., ALGOL – a language for defining algorithms – or COBOL and FORTRAN, all of them particularly suited to specific applications of the computers available at that time. Also, LISP appeared at this time, but, for some time, remained a language to be interpreted instead of compiled. Compilation was a technique developed with the arrival of high-level languages.

From Assembler to High-Level Languages

The introduction of high-level languages was a big step in making programmers more efficient, since writing an abstract program that could later be compiled into machine language allowed an average manual programmer to produce more lines of code per time unit in terms of machine language.

From High-Level Languages to Algebraic Specification

An even more abstract language level is provided by tools for algebraic specification. Using algebraic specification, the programmer does not even have to write high-level code. Instead, he or she specifies the desired behavior in an abstract behavioral description system which then generates source code to be compiled by a regular compiler.

One might reasonably ask whether GP could be applied to algebraic specification. In other words: Would it be possible to evolve programs on the level of an algebraic specification that only later on would be translated into code? To our knowledge, nobody has tried GP on this level yet, but it might be a fruitful area to work in.

If we now confront the manual method of writing programs with an automatic one like GP, the similarities are striking.

A Programmer's Heuristics

Consider, for instance, the method programmers use who do not have a good command of a certain programming language. As an example of such a language, let us imagine a macro-assembler language that, without insulting anyone, we can consider to be nearly obsolete in the late 1990s. Nevertheless, suppose that many applications in a certain domain, say databases, are written with this macro-assembler language. In order to adapt such an application – a program – to a new environment, say another query type or other content and user type, the code segments written in macro-assembler have to be rewritten.

How do programmers do this job? The most productive way to do it is by what is known as "cut and paste." They will take useful segments from older applications and put them together for the new one. They will also change those segments slightly in order to make them viable in the new application. Slight changes might mean using different variables or changing a very few lines of subroutines already in existence. Sometimes it might be necessary to write a new subroutine altogether, but those events will be kept to a minimum.

Some older subroutines will get copied over to the new application, but the corresponding calls will never be made since their functionality has ceased to be important in the new environment. Programmers will either see this by analyzing the code, and cut the routines out, or they will comment these routines out, or they will not dare to decide and will simply leave them in for the application to choose whether or not to use them.

After everything has been put together, a programmer will compile the resulting code, debug it, and test whether the new functionality has been reached or not. Typically, this is an iterative process of adding and deleting segments as well as testing different aspects of the application, moving from easy to hard ones.

The last few paragraphs have described in terms of a database programmer [Ludvikson, 1995] a procedure that may be identified with actions taken by a genetic programming system. Cut and paste translates into crossover of applications, generation of new segments or lines of code translates into mutations, and debugging and testing correspond to selection of those programs that function properly. On the other hand, one would hope that a good programmer would take more informed steps through the search space than would random mutation.

Incidentally, the fact that much code is unused or commented out finds its correspondence in so-called introns automatically evolving in GP programs.

The main difference between programming by hand in this way and automatic tools like a GP system is that GP can afford to evolve a population of programs simultaneously, which is something a single programmer could not do.

One major point to understand from this example is that a programmer would only work in this way if

- the environments changed only slightly between applications, or
- the programming language was hard to handle.

We can conclude that it must be very hard for a GP system to generate code without any idea of what a given argument or function could mean to the output.

Exercises

1. Give two elementary components of evolutionary learning.
2. How many permutations of the list of natural numbers smaller than 10 are there?
3. Give the number of programs represented as binary trees with L terminals and K functions.
4. What is the best-known method for generating random numbers? How does it work?
5. What is conditional probability?
6. Explain the term *probability distribution.* Define two different probability distributions.
7. What is a Turing machine, and in what sense is it similar to a modern computer?
8. Is it always possible to decide if an individual in a GP system will end execution? Explain.
9. Give an example of a computer language that is Turing complete and of one that is not.
10. What is a tree data structure, and what are its basic components?
11. What is the difference between assembly language and binary machine code?
12. Give at least two different computer language paradigms.

Further Reading

W. Feller,
AN INTRODUCTION TO PROBABILITY THEORY
AND ITS APPLICATIONS, VOLUMES 1 AND 2.
Wiley, New York, NY., 1968 and 1971.

M. Li and P.M.B. Vitányi,
INDUCTIVE REASONING AND KOLMOGOROV COMPLEXITY.
JCSS 44(2): 343–384, 1992.

R. Motwani and P. Raghavan,
RANDOMIZED ALGORITHMS.
Cambridge University Press, Cambridge, 1995.

I. Niven,
THE MATHEMATICS OF CHOICE.
MMA, Washington, DC., 1965.

E. Sanchez and M. Tomassini (Eds.),
TOWARDS EVOLVABLE HARDWARE.
LNCS 1062, Springer, Berlin, 1996.

R. Stansifer,
THE STUDY OF PROGRAMMING LANGUAGES.
Prentice Hall, London, 1995.

A.M. Turing,
ON COMPUTABLE NUMBERS, WITH AN APPLICATION
TO THE ENTSCHEIDUNGSPROBLEM.
Proceedings of the London Mathematical Society, Series 2, Vol. 42, 1939.

N. Wirth,
ALGORITHMS + DATA STRUCTURES = PROGRAMS.
in PRENTICE-HALL SERIES IN AUTOMATIC COMPUTATION.
Prentice Hall, Englewood Cliffs, NJ, 1976.

4 Genetic Programming as Evolutionary Computation

Contents

> A process which led from amoeba to man appeared to philosophers to be obviously a progress, though whether the amoeba would agree with this opinion is not known.
>
> B. RUSSELL, 1914

The idea of evolving computer programs is almost as old as the computer itself. Pioneering computer scientist Turing envisioned it already in the early 1950s, but the field was not systematically explored for another 40 years. We have already briefly looked at Friedberg's work from the late 1950s [Friedberg, 1958] [Friedberg et al., 1959], and this work can be considered the embryonic stage of program evolution, or genetic programming. Friedberg's work at this time was only one of many efforts toward automatic program induction.

The idea of automatic programming is natural, once you have learned how complex and tedious manual programming is. It has been a part of the artificial intelligence and machine learning fields ever since. The quest has searched down many different roads. One track has lead to the creation of the "art" of compiler writing. In this chapter, however, we will focus on the background of genetic programming in the light of other techniques for simulating evolution. But first a few words on the *name* GP.

The term *genetic programming* was coined independently by Koza and de Garis, who both started using the term in papers in 1990 to label their own, different techniques. When Koza's definition of the term started to dominate after his important 1992 book, de Garis switched to "evolutionary engineering" and after that GP – to many people – represented the evolution of program structures in tree or LISP form. However, as should be clear by now, in this book we use genetic programming as an umbrella term for all forms of evolutionary program induction.

4.1 The Dawn of Genetic Programming — Setting the Stage

Genetic programming is one of many techniques for computer simulation of evolution. Lately the general term *evolutionary algorithms* (*EA*) has emerged for these techniques. EAs mimic aspects of natural evolution, natural selection, and differential reproduction. In Chapter 2 we have seen how Darwin's principle of natural selection is used to explain the evolution of all life forms on Earth. Various aspects of this principle have been simulated in computers, beginning with Friedberg's work.

Until recently, most efforts have been in areas other than program induction, often as methods for optimization. Evolutionary algorithms work by defining a goal in the form of a quality criterion and then use this goal to measure and compare solution candidates in a stepwise refinement of a set of data structures. If successful, an EA will return an optimal or near optimal individual after a number of iterations. In this sense, the algorithms are more similar to breeding of, let us say, dogs than to natural selection, since breeding also works with a well-defined quality criterion.

When dogs are bred to have, for example, long hair and short legs, the breeder selects – from a group of individuals – the best individuals for reproduction according to this quality criterion. In this case, he or she selects the ones with the longest hair and shortest legs for mating. The process is repeated with the offspring over many generations of dogs until a satisfying individual is found – and a Siberian Husky has been turned into an Angora Dachshund. The same method has given our pigs an extra rib in only 30 years.

Fitness and Selection

This approach is very similar to the basic principle of all evolutionary techniques. The process of selecting the best individuals for mating is simply called *selection* or, more accurately, *mating selection*. It will be discussed in detail in the next chapter. The quality criterion is often referred to as *fitness* in EAs and it is with this standard we determine which individuals shall be selected. We also need a technique for mating and reproduction. In the reproduction process it is important to have a mechanism for *variation* – to generate a differential and to make sure that children do not become identical copies of their parents, which would render improvements impossible.

Variation

The two main variation operators in EAs – and in nature – are mutation and exchange of genetic material between individuals. Mutation changes a small part of an individual's genome while crossover (recombination and sexual reproduction) exchanges genetic material usually between two individuals, to create an offspring that is a combination of its parents. Different EA techniques usually emphasize different variation operators – some work mostly with mutation while others work mostly with crossover. An illustration of a basic EA can be seen in Figure 4.1.

Sometimes the boundaries between EAs and other search algorithms are fuzzy. This is also true for the boundaries between EAs that are GP and those that are not. In any case, Friedberg in 1958 can be considered one of the pioneers of EAs and GP, even though his work lacks some of the presently more common EA and GP ingredients and even though he was hampered by the constraints of the computer power available at that time.

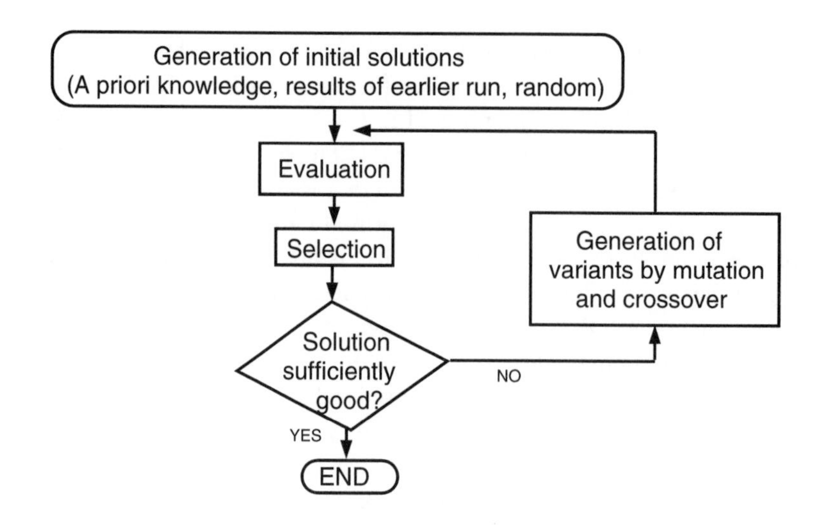

Figure 4.1
Basic evolutionary algorithm

The objective of Friedberg's system was to induce an assembler language on a virtual one-bit register machine. Due to the limited computer resources in his day, the induced structures – his programs – could only tackle modest problems like adding two bits.

The Bremermann Limit

But the situation was not as bad as one might think. At least, this was the argument of another pioneer at the time, Bremermann. Bremermann asked himself whether there is a principal limit to computation that computers will never be able to break through. In his seminal paper [Bremermann, 1962], he noted that no data processing system, whether artificial or living, can process more than 2×10^{47} bits per second per gram of its mass. He arrived at that surprising conclusion by an energy consideration, taking into account Heisenberg's uncertainty principle in combination with Planck's universal constant and the constant speed of light in vacuum. Bremermann's argument continued that if we'd had a computer the size and mass of the Earth at our disposal computing from the Earth's birth, we would anyway not be able to process more than 10^{93} bits of information. Considering combinatorial problems, this so-called Bremermann limit is actually not very large. Concluding his paper, Bremermann wrote:

> The experiences of various groups who work on problem solving, theorem proving, and pattern recognition all seem to point in the same direction: These problems are tough. There does not seem to be a royal road or a simple method which at one stroke will solve all our problems. My discussion of ultimate limitations on the speed and amount of data processing may be summarized like this: Problems involving vast numbers of possibilities will not be solved by sheer data processing quantity. We must look for quality, for refinements, for tricks, for every ingenuity that we can think

of. Computers faster than those of today will be of great help. We will need them. However, when we are concerned with problems in principle, present day computers are about as fast as they will ever be.

H. BREMERMANN, 1962

Returning to Friedberg, his structures were fixed-size virtual assembler programs. His algorithm was started like most EAs with random creation of one or more random structures. As a variation operator he used *mutation* – a random change in a bit of his program structure.

Friedberg's approach has a serious drawback: he employs "binary" fitness, that is, a program is either perfect or it is not. The feedback information from a certain program *cannot* be used for guiding the subsequent search process. Thus, the process has similarities to simple random search.

Though the results from his system were modest, it definitely represents great pioneering work in the field of EAs, GP, and ML. He even considered topics that advanced genetic programming research is concerned with today, like parsimony of programs; see Chapter 10 of this book.

In the next decade – the 1960s – several of today's best known EAs were created. In this chapter we will take a look at three of them: genetic algorithms, evolutionary programming, and evolution strategies. But first we dig deeper into the concept of evolutionary algorithms.

4.2 Evolutionary Algorithms: The General View

Natural evolution has been powerful enough to bring about biological phenomena as complex as mammalian organisms and human consciousness. In addition to generating complexity, biological systems seem so well adapted to their environments and so well equipped with sensory and motor "devices" that the impression of purposeful optimization is evoked. This has caused an ongoing controversy in biology about whether evolution is indeed optimizing structures or not [Duprè, 1987].

Gleaning Recipes from Evolution

For millennia, engineers have been inspired to learn from nature and to apply her recipes in technology. Thus, evolution stimulated two questions:

1. Does copying evolution help in the optimization of technical devices, such as airplane wings, car motors, or receptors and sensors for certain physical signals?

2. Does copying evolution provide us with the creativity to generate new and complex solutions to well-known difficult problems?

Brought into the realm of computer science, these questions could read as:

1. Does simulating evolution (with "evolutionary algorithms") provide us with a tool to optimize problem solutions?

2. Does it provide a tool to build solutions by generating complexity through combination of program constructs?

Basic Ingredients of EAs

Evolutionary algorithms are aimed at answering these questions. Based on very simple models of organic evolution, these algorithms aim to catch the basic success ingredients of natural evolutionary processes. Equipped with those basic ingredients, EAs are applied to various problems in computer science that are not easy to solve by conventional methods, such as combinatorial optimization problems or learning tasks.

Different flavors of EAs have been developed over the years, but their main ingredients can be summarized as:

- Populations of solutions
- Innovation operations
- Conservation operations
- Quality differentials
- Selection

Representation

Consider an optimization problem. The first decision to be made is how to represent a solution. In EAs, solutions are represented by genotypes, genomes, or chromosomes.[1] Once a representation for a solution has been fixed, a judgment of a solution candidate should be possible, based on the problem to be solved. The representation allows us to encode the problem, e.g., by a set of parameters that are to be chosen independently from each other. A particular instantiation of this representation should be judged, giving a quality of the solution under consideration. The quality might be measured by a

[1]Note that genotypes and solutions often are not identical!

physical process, by an evaluation function to be specified in advance, or even by a subjective juror, sitting in front of the computer screen.

1,2, ... n Solutions: A Population

Usually, EAs work with a population of solutions, in order to enable a parallel search process. Indeed, the right choice for the size of a population is sometimes decisive, determining whether a run completes successfully or not. We shall learn more about this in the context of genetic programming. Population size is generally an important parameter of EAs.

Once a representation has been chosen that can be plugged into a decoder, resulting in a rating of individual solutions, corresponding operators have to be defined that can generate variants of the solution. We have mentioned two classes of operators above, innovation operators and conservation operators.

Innovation

Innovation operators ensure that new aspects of a problem are considered. In terms of our optimization problem above, this would mean that new parameter values are tried, in either one or more of the different parameter positions. The innovation operator in EAs is most often called mutation, and it comes with three EA parameters determining:

- its strength within a component of a solution,
- its spread in simultaneous application to components within a solution,
- and its frequency of application within the entire algorithm.

A very strong mutation operator would basically generate a random parameter at a given position within a solution. If applied to all positions within a solution, it would generate a solution completely uncorrelated with its origin, and if applied with maximum frequency, it would erase all information generated in the population during the EA search process so far.

Conservation

Conservation operators are used to consolidate what has already been "learned" by various individuals in the population. Recombination of two or more solutions is the primary tool for achieving this goal. Provided the different parameters in our solution representation are sufficiently independent from each other, combinations of useful pieces of information from different individuals would result in better overall solutions. Thus, in the ideal case, a mixing of the information should take place that will accelerate the search for the globally optimal solution.

There are different ways to achieve a mixing of solutions. For the sake of simplicity, here we shall concentrate on two individuals recombining their information, although multi-recombinant methods also exist in the literature. Well-known recombination methods are one-point, two-point, or n-point crossover for binary (discrete) parameter values between two individuals, as well as discrete and intermediate recombination for n-ary or continuous parameter values. Depending on these features, these operators carry a set of EA parameters governing:

- type of recombination
- its frequency of application within the entire algorithm

Linkage and Epistasis

Given these two means of exploring the search space, a solution should be found quite efficiently. Unfortunately, reality is not usually so simple, and there are many interrelationships between various components of a problem solution. This is called linkage and prohibits efficient search, since the variation of one parameter might have a negative influence on overall fitness due to its linkage with another. In effect, we are dealing with non-linear problems here, with an interaction between components. In the literature, this phenomenon is sometimes called *epistasis.*

Quality Differentials

With the generation of genotypic variants one would expect differences in phenotypic behavior to appear in the population. As Friedberg's work has demonstrated, however, this is not necessarily the case. If there was only a binary fitness function (stating a solution by "1" and no solution by "0"), then there would not be enough difference between individuals in the population to drive an evolutionary search process. An algorithm of this kind would degenerate into a multi-membered blind search. Thus, a very important aspect of EAs is a graded fitness function that distinguishes a better solution from a good one.

Selection

It is on these differentials that selection can work on. Due to the finiteness of a population, not all the variants generated by the means mentioned above can be stored. This forces us to select from the variants both the candidates to be included and the individuals from the population that are due for replacement. Following Darwin, this process is called selection.

EA as Dynamical Systems

From a dynamical systems point of view, the operators of an EA work to destabilize a population, and the selection operator works to stabilize it. Thus, if one is looking for good solutions to an optimization problem, good solutions should tend to be stable whereas bad solutions should tend to be unstable. The art of choosing an appropriate representation and an appropriate set of operators is often a matter of experience and intuition, and can only be mastered by working with the algorithms.

4.3 Flavors of Evolutionary Algorithms

Even if the basic ingredients of EAs are quite similar, there are hundreds of variants to EAs. In this section, we look at three early approaches that were most influential and illustrate the climate in which GP was born.

4.3.1 Genetic Algorithms

One of the best known EAs is the *genetic algorithm* (*GA*) developed by Holland, his students, and his colleagues at the University of Michigan [Holland, 1992]. The GA is an important predecessor of genetic programming, from which the latter derived its name. GAs have proved useful in a wide variety of real-world problems.

The original GA has two main characteristics: it uses a fixed length binary representation and makes heavy use of crossover. The simple representation of individuals as fixed length strings of zeros and ones (Figure 4.2) puts the spotlight on an important issue of all EAs mentioned in Chapter 1 – the encoding of the problem. In GAs we must find a suitable way to code a solution to our problem as a binary string. Finding good coding schemes is still an art, and the success of a GA (and EA) run often depends on the coding of the problem.

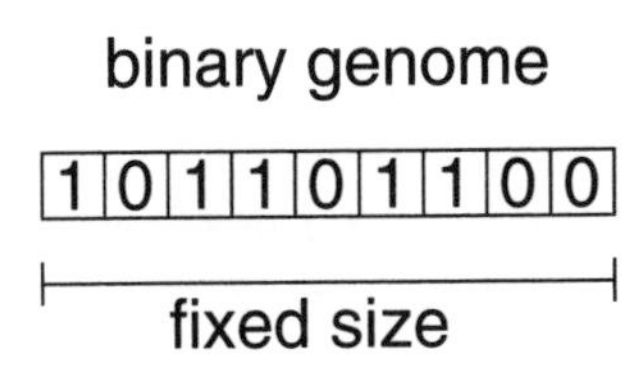

Figure 4.2
Problem representation in the binary string of a GA

The commonest form of crossover is called one-point crossover and is illustrated in Figure 4.3. Two parent individuals of the same length are aligned with each other and a crossover point is chosen at random between any of their component positions. The tails of the two individuals from this point onward are switched, resulting in two new offspring.

Like many GP systems, GAs focus on the crossover operator. In most applications of GAs, 95% of operations are either reproduction, i.e., copying strings, or crossover. Usually, only a small probability is used for mutations.

Fitness-Proportional Selection

Another key ingredient to GAs, at least until the late 1980s, was fitness-proportional selection. Fitness-proportional selection assigns reproduction opportunities to individuals based on their relative fitness in the present population. Thus, it is a stochastic process of

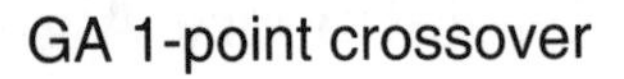

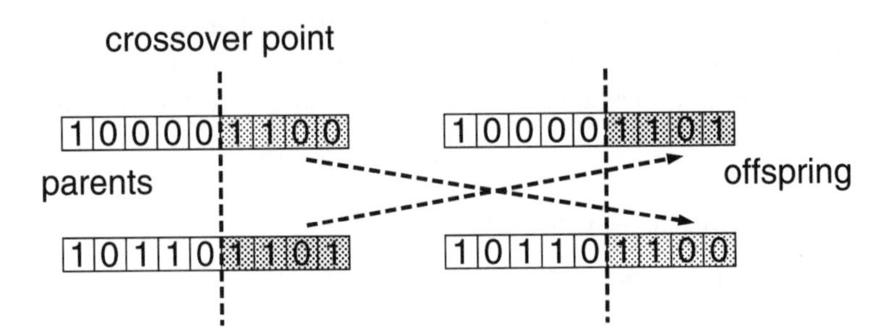

Figure 4.3
The one-point crossover operator in genetic algorithms

selection, which will draw individuals on the basis of their performance as compared to other individuals. It is also called roulette wheel selection. Each individual gets a part of a roulette wheel in proportion to its fitness and the average fitness of all other individuals. The roulette wheel spins, and if an individual has the lucky number it is allowed to reproduce itself into the next generation. This simple selection mechanism contains one of the basic principles of EA selection – more individuals than the best one have a chance to reproduce. This principle has been shown to be essential in guaranteeing genetic diversity and helps keep the search away from local optima.

After reproduction has taken place, a certain percentage of the population is allowed to vary by crossover (mostly) and mutation (rarely); see Section 4.3. Different forms of selection are discussed in more depth in the next chapter.

Theoretical Aspects

Fitness-proportional selection is also a key ingredient in one of the main theoretical achievements of GAs: the notion of building blocks and schemata.

The idea is that, given a certain problem representation, the GA is able through repeated crossover and reproduction to combine those parts of a solution that are necessary to form a globally optimal solution. The argument is that each individual in the population participates in numerous ways in the search process in every given generation by exposing different schemata to the evaluation process.

Schemata

A schema is a string that contains 0, 1, and $*$ (i.e., "don't care") symbols. In fact, * symbols are the characteristic feature of schemata, coming about by projections into different subspaces of the entire search space. Clearly, each individual is always moving in many subspaces at the same time. Holland argued [Holland, 1975] that the GA actually progresses by this sampling of subspaces.

Schema Theorem

Holland formulated a theorem, the schema theorem of GAs, stating that, provided fitness-proportional selection is used, the probability of certain schemata to appear in the population shifts over time in such a way as to approach the overall fitness optimum. The schema theorem is even more specific in giving a lower bound on the speed for selecting better schemata over worse.

The schema theorem has, among other things, motivated the use of binary representation in GAs. It can be argued from this theorem that the alphabet in a GA should be as small as possible. Naturally, the binary alphabet with only two members is the smallest. This principle is disputed, but we can note that nature uses a quite small alphabet of four letters in its DNA code.

The schema theorem has been criticized heavily in recent years for not being able to explain why GAs work. One of the main reasons was its extrapolation of fitness developments from generation n to generation $n + k$. It is generally not possible to apply a difference equation recursively k times and have an accurate measure of the probabilities of schemata occurrence in the long term without very restrictive assumptions about the underlying process.

Representation of Individuals and Genetic Operators

The usefulness of binary strings as representations of optimization problems has been challenged over the years. Within the last decade, more and more variants of genetic algorithms have been put forward that do not use binary representations. For example, constraint optimization problems as they appear in most practical applications have shown a tendency to favor other representations [Michalewicz, 1994].

A representation should always reflect fundamental facts about the problem at hand. This not only makes understanding of the search easier but it is often a precondition of successful GA runs.

Correspondingly, genetic operators have to be chosen that allow unrestricted movement in the problem space spanned by the chosen representation.

Classifier Systems

In his 1975 book [Holland, 1975], Holland mentioned AI as one of the main motivations for the creation of genetic algorithms. He did not experiment with direct use of GAs to evolve programs but contributed to the creation of another research field with the invention of the *classifier systems.* Holland and Reitman proposed this type of system in 1978 [Holland and Reitman, 1978].

Classifier Systems

A classifier system induces a general-purpose Turing complete algorithm comprising three components: a rule-based programming language, a simulated market economy, and a genetic algorithm. The rules are the individuals in the genetic algorithm. Together, all the rules can be seen as a program performing a task. When certain rules fire in response to some input, the system generates some output. Any "reward" for the output gets accredited a fitness proportional to the contributing rules, which are the individuals. The genetic algorithm then operates on the rules. Rules resulting from the genetic algorithm build the next potentially modified classifier system. The metaphor is borrowed form a market economy where many individuals *co-operate and compete* to achieve higher efficiency in solving a goal and each individual is rewarded for its part in the success.

A classifier system is not regarded as evolutionary program induction since the complete program is not evolved in the individuals of the population. The individual rules in a classifier system are not capable of solving the task by themselves.

Variable-Size Individuals

However, in 1980, Smith [Smith, 1980] invented a variant of a classifier systems introducing variable-size strings as individuals. In his approach an individual is a complete rule-based program that can solve the task defined by the fitness function. Since his system uses variable length representation of Turing complete individuals, each aiming alone at solving a problem, his approach can be considered an important step toward genetic programming.[2] Smith applied his technique to the objective of finding good poker playing strategies and rules, with some success.

4.3.2 Evolutionary Strategies

Evolutionary strategies (*ES*), developed in the 1960s, are another paradigm in evolutionary computation. Newer accounts of the work of its pioneers, Rechenberg and Schwefel, can be found in [Rechenberg, 1994] [Schwefel, 1995].

The idea of using evolution as a guiding principle and thus of developing evolutionary strategies arose from problems in experimental optimization. Rechenberg and Schwefel were working with hydrodynamic problems when they hit upon the idea of using random events by throwing dice to decide the direction of an optimization process. Thus, discrete mutations were the first evolutionary variations to be applied within evolutionary strategies. Due to the limitations of the

[2] We shall call those systems "early genetic programming."

basic experimental setup, only one object could be considered at a time, so the population consisted of one individual only. But the selection process was already in place, keeping track of the "fitness" of an experimental configuration and its variation due to the random mutations applied.

Mutation

Soon afterwards, digital computers became valuable tools and evolutionary strategies were devised that were able to operate with continuous variables. Following closely the trains of thought already established, individuals were represented as real-valued vectors, and mutation was performed by adding normally distributed random numbers with expectation value 0. In this approach, small variations are much more frequent than large variations, expressing the state of affairs on the phenotypic level in nature.

Causality

In fact, it was always considered a hallmark of evolutionary strategies to emphasize causality, i.e., the fact that strong causes would generate strong effects. Translated into evolutionary strategies, large mutations should result in large jumps in fitness, and small mutations should result in small changes in fitness.

Recombination

In later years, the benefit of using populations of individuals was recognized in evolutionary strategies by introducing different sorts of recombination operators. Discrete recombination selects the (continuous) features from different parents alternatively, with an additional parameter for the specification of 1-, 2-, or n-point recombination. Intermediate recombination, on the other hand, involves mixing the features stemming from the parents in a different way, shuffling single features component-wise, by taking either the arithmetic mean or other kinds of weighting. Recently, multi-recombinant strategies have been studied as well [Beyer, 1995].

Selection

Although selection will be discussed in more detail later (Section 5.5), a short remark is in order here: In evolutionary strategies, selection is a deterministic operator, which chooses the $\mu < \lambda$ individuals to constitute the population in the next generation. μ denotes the number of (present and future) parents; λ denotes the number of offspring. Thus, selection in ESs is *over-production selection*, not mating selection as in GAs. As such, it is nearer to what Darwin called "natural selection" [Schwefel and Rudolph, 1995].

Self-Adaptation

One other key aspect of advanced ESs is to allow a learning process on the level of "strategy parameters" of the algorithm. Whereas so far evolutionary strategies have concerned themselves with adapting phenotypic variables (object variables), it is possible to assign strategy parameters, like mutation rate(s) or recombination method, to each individual. Doing this results, over time, in a selection of better adapted individuals, in both the domain of object variables and the domain of strategy parameters.

By extending the representation of individuals to include strategy parameters, however, a distinction has been introduced between phenotype and genotype. And although the strategy parameters are subjected to the same variation policy (mutation and recombination) as are the object parameters, information in the domain not expressed in the phenotype does evolve differently than in the other domain. Selection indirectly favors the strategy parameter settings that are beneficial to make progress in the given problem domain, thus developing an internal model of the environment constituted by the problem.

In the realm of evolutionary strategies, structure evolution has been considered by Lohmann [Lohmann, 1992].

4.3.3 Evolutionary Programming

Another important EA and predecessor of GP is *evolutionary programming (EP)* also created in the early 1960s by Fogel, Owens, and Walsh [Fogel et al., 1965] [Fogel et al., 1966]. EP uses the mutation operator to change finite state machines (FSM). A finite state machine or finite automaton is a very simple computer program that consists of a machine moving around in a graph of nodes called states. The state automaton has many similarities with a Turing machine and under the right circumstances an FSM program may be considered to be Turing complete. EP uses a collection of mutations that manipulate specific components of the representation. It operates more directly on the representation than GAs. Evolutionary programming employs random creation, mutation, and *fitness-based* reproduction: on average, a better individual gets reproduced more often.

In its original form, EP was used to solve sequence prediction problems with the help of finite state machines. The FSMs – represented by transition tables and initial states – were allowed to vary through mutation in various aspects, e.g., number of states, initial state, state transition, or output symbol. We can see here that EP has realized a symbolic representation of computer programs, formulated as automata. It has successfully applied the evolutionary principles of variation and selection to a problem from artificial intelligence.

Population

More specifically, EP started with a population of FSMs that were allowed to change state according to input and present state, and were then evaluated according to whether the output symbols produced by the FSMs agreed with the following input symbol or not. In this way, the entire symbol sequence was run and a fitness was assigned to each FSM. Offspring FSMs were generated by copy-

Mutation and Selection

ing machines and applying mutations with uniform probability distribution to these copies. Gaussian mutations were used to modify numeric components similar to how it is done in ES. The Poisson distribution was used to select how many mutations will be applied to create a new FSM. The mutations to FSMs manipulated the available representational components, like, e.g., add state, delete state, change transition, change initial state, change output symbol. In a typical run, the better performing half of the population was kept and the rest was substituted by variants of the better half.

Self-Adaptation

One specific feature of mutation in EP is that, as the optimal value for fitness is approached, the mutation rate is decreased. This is achieved by letting the fitness influence the spread of mutations, for example, by tying it to the variance of the Gaussian distribution. The nearer the optimum, the sharper the distribution becomes around 0.

In recent years, EP has expanded in scope and has taken up other methods, e.g., tournament selection, and has allowed other, different problem domains to be addressed by the algorithm, but still refrains from using recombination as a major operator for generating variants [Fogel, 1995]. Self-adaptation processes are put into place by allowing meta-algorithmic variation of parameters.

4.3.4 Tree-Based GP

With Smith's development of a variant of Holland's classifier systems where each chromosome (solution candidate) was a complete program of variable length we might have had the first real evolutionary system inducing complete programs, even though this approach was still built on the production rule paradigm.

Two researchers, Cramer [Cramer, 1985] and Koza [Koza, 1989], suggested that a tree structure should be used as the program representation in a genome. Cramer was inspired by Smith's crossover operator and published the first method using tree structures and subtree crossover in the evolutionary process. Other innovative implementations followed evolving programs in LISP or PROLOG with similar methods for particular problems [Hicklin, 1986] [Fujiki and Dickinson, 1987] [Dickmanns et al., 1987].

Koza, however, was the first to recognize the importance of the method and demonstrate its feasibility for automatic programming in general. In his 1989 paper, he provided evidence in the form of several problems from five different areas. In his 1992 book [Koza, 1992d], which sparked the rapid growth of genetic programming, he wrote:

> In particular, I describe a single, unified, domain-independent approach to the problem of program induction – namely, genetic

programming. I demonstrate, by example and analogy, that genetic programming is applicable and effective for a wide variety of problems from a surprising variety of fields. It would probably be impossible to solve most of these problems with any one existing paradigm for machine learning, artificial intelligence, self-improving systems, self-organizing systems, neural networks, or induction. Nonetheless, a single approach will be used here – regardless of whether the problem involves optimal control, planning, discovery of game-playing strategies, symbolic regression, automatic programming, or evolving emergent behavior.

J. Koza, 1992

4.4 Summary of Evolutionary Algorithms

Year	Inventor	Technique	Individual
1958	Friedberg	learning machine	virtual assembler
1959	Samuel	mathematics	polynomial
1965	Fogel, Owens and Walsh	evolutionary programming	automaton
1965	Rechenberg, Schwefel	evolutionary strategies	real-numbered vector
1975	Holland	genetic algorithms	fixed-size bit string
1978	Holland and Reitmann	genetic classifier systems	rules
1980	Smith	early genetic programming	var-size bit string
1985	Cramer	early genetic programming	tree
1986	Hicklin	early genetic programming	LISP
1987	Fujiki and Dickinson	early genetic programming	LISP
1987	Dickmanns, Schmidhuber and Winklhofer	early genetic programming	assembler
1992	Koza	genetic programming	tree

Table 4.1
Phylogeny of genetic programming

GP has many predecessors, and above we have looked at the most influential ones. Table 4.1 summarizes the history that led up to the present situation. Today there exists a large set of different genetic programming techniques which can be classified by many criteria, such as abstracting mechanisms, use of memory, genetic operators employed, and more [Langdon and Qureshi, 1995].

In the following chapters we will take a closer look at the commonest GP algorithm, Koza's tree-based system, but will also look into GP's many variants in existence today, each of them with its own specific benefits and drawbacks.

Exercises

1. Which evolutionary algorithms other than GP exist? What was their respective original application area?
2. Which evolutionary algorithm uses only mutation?
3. Give a basic example of a mutation operator in ES.
4. Describe the representation of a GA.
5. What is fitness-proportional selection?
6. What is a schema, and what does the schema theorem state?
7. What is a classifier system? Which are its main components?
8. Explain self-adaptation of parameters in EAs. What does it have to do with the mapping between genotypes and phenotypes?
9. What is epistasis?
10. What would you consider to be the main differences between GP and other EAs?

Further Reading

T. Bäck,
EVOLUTIONARY ALGORITHMS IN THEORY AND PRACTICE.
Oxford University Press, New York, NY, 1996.

D. Fogel,
EVOLUTIONARY COMPUTATION.
IEEE Press, New York, 1995.

L. Fogel, A. Owens, and M. Walsh,
ARTIFICIAL INTELLIGENCE THROUGH SIMULATED EVOLUTION.
John Wiley, New York, NY, 1966.

D. Goldberg,
GENETIC ALGORITHMS IN SEARCH, OPTIMIZATION
& MACHINE LEARNING.
Addison-Wesley Publishing Company, Inc., Reading, MA, 1989.

J. Holland,
ADAPTATION IN NATURAL AND ARTIFICIAL SYSTEMS.
The University of Michigan Press, Ann Arbor, MI, 1975.
New Edition 1992.

M. Mitchell,
AN INTRODUCTION TO GENETIC ALGORITHMS.
MIT Press, Cambridge, MA, 1996.

I. Rechenberg,
EVOLUTIONSSTRATEGIEN.
Holtzmann-Froboog, Stuttgart, Germany, 1975.
New Edition 1994.

H.-P. Schwefel,
NUMERICAL OPTIMIZATION OF COMPUTER MODELS.
John Wiley & Sons, Chichester, UK, 1981.
New Edition 1995.

Part II

Genetic Programming Fundamentals

5 Basic Concepts — The Foundation

Contents

> But Natural Selection, as we shall hereafter see, is a power incessantly ready for action, and is as immeasurably superior to man's feeble efforts as the works of Nature are to those of Art.
>
> C. DARWIN, 1859

In the short time since the publication of Koza's 1992 book, over eight hundred GP papers have been published. Researchers have devised many different systems that may fairly be called genetic programming – systems that use tree, linear, and graph genomes; systems that use high crossover rates; and systems that use high mutation rates. Some even blend genetic programming with linear regression or context free grammars while others use GP to model ontogeny, the development of a single cell into an organism.

The purpose of this chapter is to boil this diversity down to the essential common themes. The important features shared by most GP systems are:

- **Stochastic decision making**. GP uses pseudo-random numbers to mimic the randomness of natural evolution. As a result, GP uses stochastic processes and probabilistic decision making at several stages of program development. The subjects of randomness, probability, and random number generation were discussed in Chapter 3 and we use them here.

- **Program structures**. GP assembles variable length program structures from basic units called *functions* and *terminals*. Functions perform operations on their inputs, which are either terminals or output from other functions. The actual assembly of the programs from functions and terminals occurs at the beginning of a run, when the population is initialized.

- **Genetic operators**. GP transforms the initial programs in the population using genetic operators. Crossover between two individual programs is one principal genetic operator in GP. Other important operators are mutation and reproduction. Specific details and more exotic operators will be discussed in Chapters 6, 9, and 10.

- **Simulated evolution of a population by means of fitness-based selection**. GP evolves a population of programs in parallel. The driving force of this simulated evolution is some form of *fitness-based selection*. Fitness-based selection determines which programs are selected for further improvements.

This chapter will look at these common themes at some length, both theoretically and practically. By the end of the chapter, the reader should have a good idea of how a typical GP run works.

5.1 Terminals and Functions – The Primitives of Genetic Programs

The functions and terminals are the primitives with which a program in genetic programming is built. Functions and terminals play different roles. Loosely speaking, terminals *provide* a value to the system while functions *process* a value already in the system. Together, functions and terminals are referred to as *nodes*. Although this terminology stems from the tree representation of programs, its use has spread to linear and graph structures as well.

5.1.1 The Terminal Set

Terminal Set

Definition 5.1 *The* **terminal set** *is comprised of the inputs to the GP program, the constants supplied to the GP program, and the zero-argument functions with side-effects executed by the GP program.*

It is useful to think for just a moment about the use of the word *terminal* in this context. Input, constant and other zero-argument nodes are called terminals or leafs because they terminate a branch of a tree in tree-based GP. In fact, a terminal lies at the end of every branch in a tree-structured genome. The reason is straightforward. Terminals are inputs to the program, constants or function without argument. In either case, a terminal returns an actual numeric value without, itself, having to take an input. Another way of putting this is that terminal nodes have an *arity* of zero.

Arity

Definition 5.2 *The* **arity** *of a function is the number of inputs to or arguments of that function.*

Inputs as Terminals

The terminal set is comprised, in part, of inputs. Chapter 1 spoke at length about the learning domain and the process of selecting features (inputs) from the learning domain with which to conduct learning. Recall that the selected features (inputs) became the training set – that is, the data upon which the system learns. Viewed this way, GP is no different from any other machine learning system. When we have decided on a set of features (inputs), each of these inputs becomes part of the GP training and test sets as a GP terminal.

Genetic programming is quite different from other machine learning systems in how it represents the features (inputs). Each feature (input) in the training set becomes part of the terminal set in a GP system. Thus, the features of the learning domain are just one of the primitives GP uses to build program structures. The features are not represented in any fixed way or in any particular place. In fact, the GP system can ignore an input altogether.

Constants as Terminals

The terminal set also includes constants. In typical tree-based GP, a set of real-numbered constants is chosen for the entire population at the beginning of the run. These constants do not change their value during the run. They are called *random ephemeral constants*, frequently represented by the symbol $\Re$. Other constants may be constructed within programs by combining random ephemeral constants using arithmetic functions.

By way of contrast, in linear GP systems, the constant portion of the terminal set is comprised of numbers chosen randomly out of a range of integers or reals. In these systems, the constants may be mutated just like any other part of the program. Thus, linear constants can change, unlike typical tree system random ephemeral constants.

5.1.2 The Function Set

Function Set

Definition 5.3 *The* **function set** *is composed of the statements, operators, and functions available to the GP system.*

The Range of Available Functions

The function set may be application-specific and be selected to fit the problem domain. The range of available functions is very broad. This is, after all, genetic *programming*. It may use any programming construct that is available in any programming language. Some examples follow:

- **Boolean Functions**
  For example: `AND, OR, NOT, XOR`.

- **Arithmetic Functions**
  For example: `PLUS, MINUS, MULTIPLY, DIVIDE`.

- **Transcendental Functions**
  For example: `TRIGONOMETRIC` and `LOGARITHMIC FUNCTIONS`.

- **Variable Assignment Functions**
  Let a be a variable available to the GP system. $a := 1$ would be a variable's assignment function in a register machine code approach. The same function would appear in a tree-based system with an S-expression that looked something like this:

  ```
  ( ASSIGN a 1 )
  ```

  where 1 is an input to the `ASSIGN` node. Of course, there would have to be a corresponding `READ` node, which would read whatever value was stored in a and pass it along as the output of the `READ` node.

- **Indexed Memory Functions**
  Some GP systems use indexed memory, i.e., access to memory cells via an index. Chapter 11 will provide details. But note that it is straightforward to manipulate indexed memory in GP.

- **Conditional Statements**
  For example: `IF, THEN, ELSE`; `CASE` or `SWITCH` statements.

- **Control Transfer Statements**
  For example: `GO TO, CALL, JUMP`.

- **Loop Statements**
  For example: `WHILE ...DO, REPEAT ...UNTIL, FOR ...DO`.

- **Subroutines**
  The range of functions is considerably broader than the preceding list. Any function that a programmer can dream up may become a part of the function set in GP. For example, in a robotics application, primitives could be created by the programmer that were specific to the problem, such as **`read sensor`**, **`turn left`**, **`turn right`**, and **`move ahead`**. Each of those primitives would become part of the function set or of the terminal set, if its arity were 0. The freedom to choose the function set in GP often reduces the need for pre- and postprocessing.

5.1.3 Choosing the Function and Terminal Set

Sufficiency and Parsimony

The functions and terminals used for a GP run should be powerful enough to be able to represent a solution to the problem. For example, a function set consisting only of the addition operator will probably not solve many very interesting problems. On the other hand, it is better not to use too large a function set. This enlarges the search space and can sometimes make the search for a solution harder. An approximate starting point for a function set might be the arithmetic and logic operations:

`PLUS, MINUS, TIMES, DIVIDE, OR, AND, XOR.`

The range of problems that can be solved with these functions is astonishing. Good solutions using only this function set have been obtained on several different classification problems, robotics control problems, and symbolic regression problems. This set of primitives does not even include forward or backward conditional jumps! In conclusion: A parsimonious approach to choosing a function set is often wise.

Choosing the Constants

A similar parsimonious approach is also effective in choosing the constants. For example, many implementations use 256 nodes for encoding functions and terminals. If there are 56 node labels used for functions that leaves a maximum of 200 nodes for constants. In many cases, this number of constants has proven to be able to solve difficult problems. GP has a remarkable ability to combine the constants at its disposal into new constants. It is not necessary, therefore, to include all constants that may be needed.

Closure of the Function and Terminal Set

Another important property of the function set is that each function should be able to handle gracefully all values it might receive as input. This is called the *closure* property. The most common example of a function that does not fulfill the closure property is the division operator. The division operator cannot accept zero as an input. Division by zero will normally crash the system, thereby terminating a GP run. This is of course unacceptable. Instead of a standard division operator one may define a new function called protected division. Protected division is just like normal division except for zero denominator inputs. In that case, the function returns something else, i.e., a very big number or zero.[1] All functions (square root and logarithms are other examples) must be able to accept all possible inputs because if there is any way to crash the system, the boiling genetic soup will certainly hit upon it.

Some Practical Advice

One final piece of practical advice about the function and terminal set might be helpful. At the beginning of a project, one should not spend too much time designing specific functions and terminals that seem perfectly attuned to the problem. The experience of the authors is that GP is *very* creative at taking simple functions and creating what it needs by combining them. In fact, GP often ignores the more sophisticated functions in favor of the primitives during evolution. Should it turn out that the simpler set of functions and terminals is not working well enough, then it is time to begin crafting your terminals and functions.

5.2 Executable Program Structures

The primitives of GP – the functions and terminals – are not programs. Functions and terminals must be assembled into a *structure* before they may execute as programs. The evolution of programs is, of course, common to all genetic programming. Programs are structures of functions and terminals together with rules or conventions for when and how each function or terminal is to be executed.

[1] If one works with certain floating point instruction sets, manufacturers have sometimes already built in the protection.

Genome and Phenome

The choice of a program structure in GP affects execution order, use and locality of memory, and the application of genetic operators to the program. There are really two very separate sets of issues here. Execution and memory locality are phenomic issues – that is, issues regarding the *behavior* of the program. On the other hand, mutation and crossover are genomic issues – that is, how the "DNA" of the program is altered. In most tree-based GP systems, there is no separate phenotype. Therefore, it appears that structural issues of execution, memory, and variation are the same. But that similarity exists only because of an implicit choice to blend the genome and the phenome. Chapters 9 and 12 shall treat other approaches in detail.

The three principal program structures used in GP are tree, linear, and graph structures. However, GP program structures are often *virtual structures.* For example, tree and graph structures are executed and altered *as if* they were trees or graphs. But how a program executes or is varied is a completely different question from how it is actually held in computer memory. Many tree-based systems do not actually hold anything that looks like a tree in the computer (see Chapter 11). Here, we will examine the manner in which the *virtual* program behaves. Of the three fundamental structures, tree structures are the commonest in GP. Beginning with trees, we shall describe all three in some detail now.

5.2.1 Tree Structure Execution and Memory

Figure 5.1 is a diagram of a tree-based phenome.[2] It has many different symbols that could be executed in any order. But there is a convention for executing tree structures.

The standard convention for tree execution is that it proceeds by repeatedly evaluating the leftmost node for which all inputs are available. This order of execution is referred to as postfix order because the operators appear after the operands. Another convention for execution is called prefix order. It is the precise opposite of postfix order and executes the nodes close to the root of the tree before it executes the terminal nodes. The advantage of prefix ordering is that a tree containing nodes like `IF/THEN` branches can often save execution time by evaluating first whether the `THEN` tree must be evaluated. Applying postfix order to Figure 5.1, the execution order of the nodes is: `d` → `e` → `OR` → `a` → `b` → `c` → `+` → × → `-`.

This same tree structure also constrains the usage of memory on execution. Figure 5.1 uses only *local* memory during execution. Why?

[2]When using arithmetic operators we shall variously use `mul`, `MUL`, ×, ∗ to mean multiplication.

Figure 5.1
A tree structure phenome

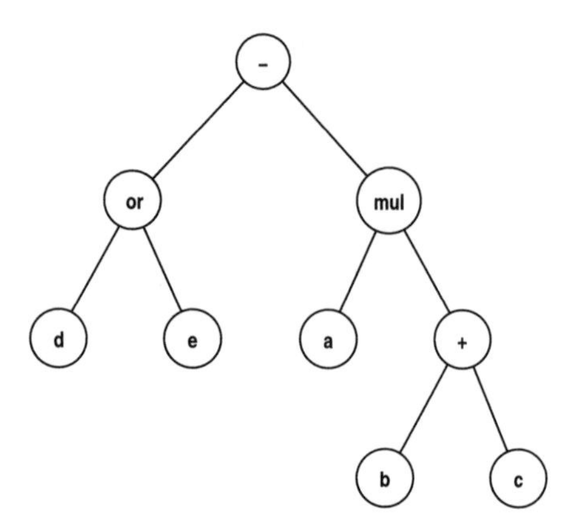

Local memory is built into the tree structure itself. For example, the values of b and c are local to the + node. The values of b and c are not available to any other part of the tree during execution. The same is true for every value in the tree.

5.2.2 Linear Structure Execution and Memory

A linear phenome is simply a chain of instructions that execute from left to right or – depending on how the picture is drawn – from top to bottom. The particular example of a linear genome discussed here is a machine code genome of our AIMGP (for "Automatic Induction of Machine Code with Genetic Programming") system.[3] Figure 5.2 shows such a linear phenome in operation.

The linear program in Figure 5.2 is identical in function to the tree program in Figure 5.1. But unlike a tree structure, the linear phenome has no obvious way for a function to get its inputs. For example, a node in a linear phenome that contained just a + function would be a plus with nothing to add together. What is missing here is memory – a place to hold the inputs to the + and other functions.

There are many ways to give memory to the instructions, but the most prominent in GP is to make the genome a two- or three-address [Nordin, 1994] [Banzhaf and Friedrich, 1994] [Huelsbergen, 1996] register machine. A register machine uses a linear string of instructions operating on a small number of memory registers. The instructions read and write values from and to the registers. The reason a reg-

[3] AIMGP was formerly known as Compiling Genetic Programming System (CGPS).

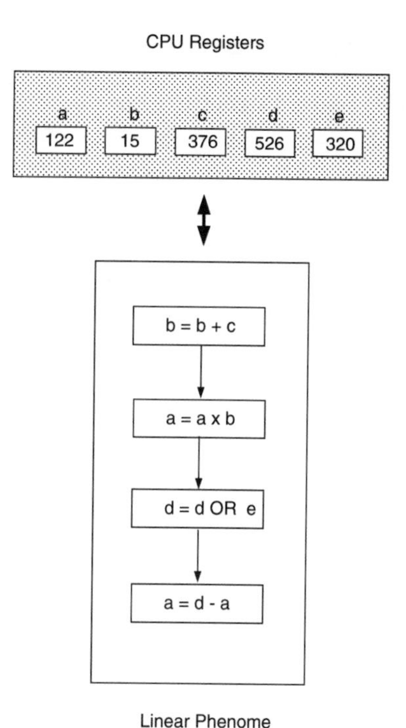

Figure 5.2
AIMGP type linear phenome and five CPU registers. The registers are shown as holding integer values.

ister machine is an excellent way to implement linear phenomes is that every commercial computer in existence contains a CPU that has memory registers operated upon by linear strings of instructions. A register machine represents the most basic workings of a CPU executing machine code instructions. Since we are doing genetic programming, it makes sense to try to use a system that makes direct use of the basic operation of the computer.

In Figure 5.2 the first instruction is `b=b+c`. The effect of this instruction is to add the values in registers `b` and `c` together and to place the sum in register `b`.

The linear program begins execution at the top instruction and proceeds down the instruction list, one at a time. The only exception to this rule is if the program includes jump instructions. Then the execution order becomes very flexible. At the end of the execution, the result is held in register `a`.

There is one other big difference between the linear and the tree approach. The memory in the tree system is, as we said, local. But in a register machine, any of the instructions may access any of the register values. So the values of `b` and `c`, which, as we saw above, are local values in a tree structure, may be accessed by any instruction. Therefore, registers contain *global* memory values.

5.2.3 Graph Structure Execution and Memory

Of the fundamental program structures, graphs are the newest arrival. PADO [Teller and Veloso, 1995b] is the name of the graph-based GP system we shall discuss here. Curiously enough, the name PADO does not have anything to do with the fact that graphs are used for evolution. Graphs are capable of representing very complex program structures compactly. A graph structure is no more than nodes connected by edges. One may think of an edge as a pointer between two nodes indicating the direction of the flow of program control.[4]

PADO does not just permit loops and recursion – it positively embraces them. This is not a trivial point; other GP systems have experimented with loops and recursion only gingerly because of the great difficulties they cause.

Figure 5.3
A small PADO program

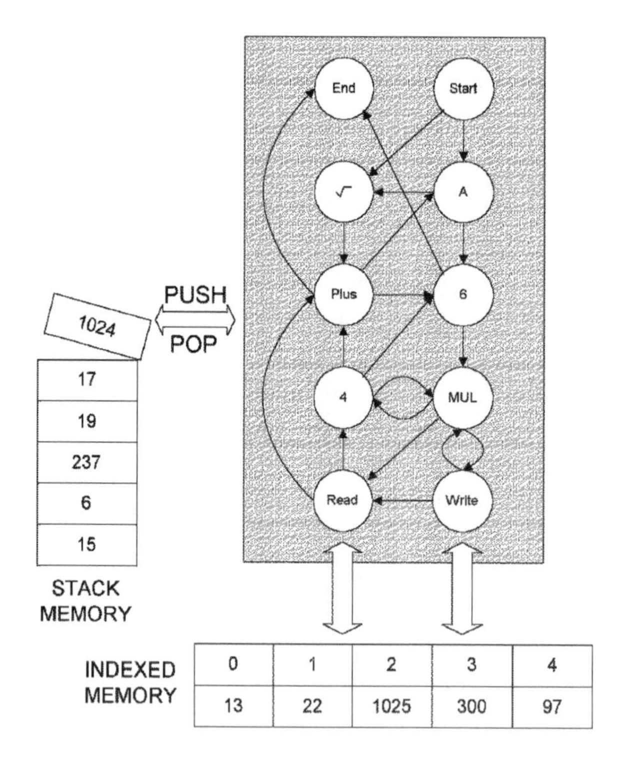

Figure 5.3 is a diagram of a small PADO program. There are two special but self-explanatory nodes in every program. Execution begins at the Start node. When the system hits the End node or another preset condition, execution is over. Thus, the flow of execution

[4]It is well known that tree and linear genomes are also graphs. That is, both have edges and nodes. But trees and linear genomes are graphs with very particular constraints for the edges.

is determined by the edges in the graph. More will be said about that later.

Stack Memory in PADO

Like all GP systems, PADO needs memory to give its nodes the data upon which to operate. Here, data is transferred among nodes by means of a stack. Each of the nodes executes a function that reads from and/or writes to the stack. For example, the node **A** in Figure 5.3 reads the value of the input **A** from RAM and pushes it onto the stack. The node 6 pushes the value 6 onto the stack. The node $\times$ pops two values from the stack, multiplies them, and pushes the result onto the stack. Thus, the system has localized memory. The process may be found in more detail in Chapter 11.

PADO Indexed Memory

Data may also be saved by PADO in indexed memory. The node labeled `Write` pops two arguments from the stack. It writes the value of the first argument into the indexed memory location indicated by the second argument. The `Read` node performs much the same function in reverse. The indexed memory is global memory.

There are two things each node in the graph must do:

1. It must perform some function on the stack and/or the indexed memory; and

2. It must decide which node will be the *next* node to execute.

This latter role is what determines program execution order in a graph. The program itself determines order of execution by choosing between the outgoing edges from the node each time a node is executed. Consider Figure 5.3 again. The $\times$ node may transfer control to the `Write` node, the `Read` node or the 4 node. The system has a decision logic which tests a memory or stack value and, based upon that value, chooses the next node.

5.2.4 Structure as Convention

On the phenomic level, program structure in a virtual tree is just a convention for ordering execution of the nodes and for localizing or globalizing memory. Conventions may be changed as long as they are recognized as conventions. This is a great area of potential flexibility of genetic programming.

This issue is quite different with the register machine system discussed above. That system evolves actual machine code – the program structure is not virtual, nor are the conventions regarding order of execution. The phenomic structure and the execution order are dictated by the CPU. The register machine system is *much* faster than tree or graph systems. But to get that extra speed, it sacrifices the ability to experiment with changes in the conventions regarding order of execution.[5]

The issue is also resolved quite differently by the PADO system. It has discarded traditional GP order of execution conventions. There is no tree or linear structure at all saying "go here next and do this." A PADO program evolves its own execution order. This is another example of the freedom of representation afforded by GP.

5.3 Initializing a GP Population

The first step in actually performing a GP run is to initialize the population. That means creating a variety of program structures for later evolution. The process is somewhat different for the three types of genomes under consideration.

Maximum Program Size

One of the principal parameters of a GP run is the maximum size permitted for a program. For trees in GP, that parameter is expressed as the maximum depth of a tree or the maximum total number of nodes in the tree.

Depth

Definition 5.4 *The* **depth** *of a node is the minimal number of nodes that must be traversed to get from the root node of the tree to the selected node.*

The maximum depth parameter (MDP) is the largest depth that will be permitted between the root node and the outermost terminals in an individual. For the commonest nodes of arity 2, the size of the tree has a maximum number of 2^{MDP} nodes. For linear GP, the parameter is called maximum length and it simply means the maximum number of instructions permitted in a program. For graph GP, the maximum number of nodes is effectively equivalent to the size of the program.[6]

5.3.1 Initializing Tree Structures

The initialization of a tree structure is fairly straightforward. Recall that trees are built from basic units called functions and terminals. We shall assume, now, that the terminals and functions allowable in the program trees have been selected already:

$$T = \{a, b, c, d, e\} \quad (5.1)$$

[5]This system can evolve order of execution but not by changing high-level conventions regarding order of execution. It must do so by including low-level branching or jump instructions [Nordin and Banzhaf, 1995b].

[6]For comparison purposes it might be better to use the maximum number of nodes in a tree as the size parameter.

$$F = \{+, -, \times, \%\} \quad (5.2)$$

There are two different methods for initializing tree structures in common use. They are called *full* and *grow* [Koza, 1992d].

The Grow Method

Figure 5.4 shows a tree that has been initialized using the grow method with a maximum depth of four. Grow produces trees of irregular shape because nodes are selected randomly from the function *and* the terminal set throughout the entire tree (except the root node, which uses only the function set). Once a branch contains a terminal node, that branch has ended, even if the maximum depth has not been reached.

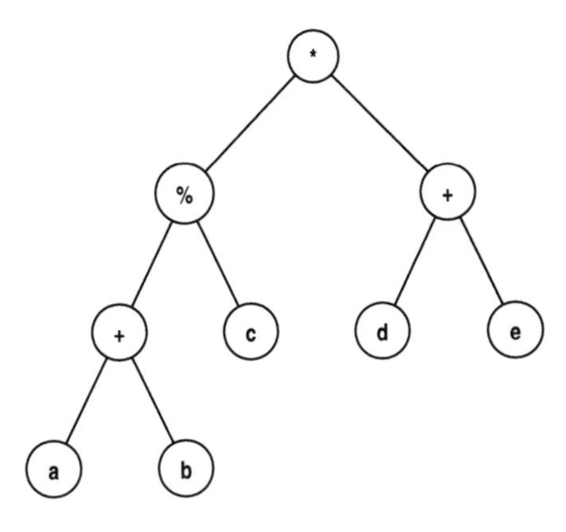

Figure 5.4
Tree of maximum depth four initialized with grow method

In Figure 5.4, the branch that ends with the input d has a depth of only three. Because the incidence of choosing terminals is random throughout initialization, trees initialized using grow are likely to be irregular in shape.

The Full Method

Instead of selecting nodes randomly from the function *and* the terminal set, the full method chooses *only* functions until a node is at the maximum depth. Then it chooses only terminals. The result is that every branch of the tree goes to the full maximum depth.

The tree in Figure 5.5 has been initialized with the full method with a maximum depth of three.

If the number of nodes is used as a size measure, growth stops when the tree has reached the preset size parameter.

5.3.2 The Ramped Half-and-Half Method

Diversity is valuable in GP populations. By itself, the above method could result in a uniform set of structures in the initial population

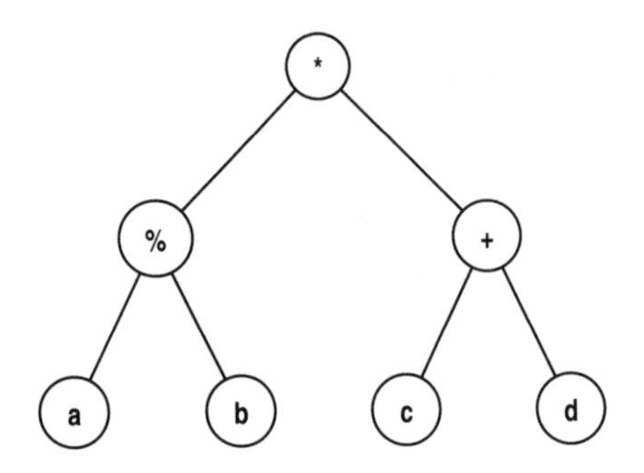

Figure 5.5
Tree of maximum depth three initialized with full method

because the routine is the same for all individuals. To prevent this, the "ramped-half-and-half" technique has been devised. It is intended to enhance population diversity of structure from the outset [Koza, 1992c].

In trees the technique is like this. Suppose the maximum depth parameter is 6. The population is divided equally among individuals to be initialized with trees having depths 2, 3, 4, 5, and 6. For each depth group, half of the trees are initialized with the full technique and half with the grow technique.

5.3.3 Initializing GP Linear Structures

Initializing linear GP structures is somewhat different than the initialization of tree structures. Again, we shall look at the AIMGP system for illustration purposes. AIMGP represents programs as a linear sequence of machine code instructions that operate on CPU registers, as we have seen in Figure 5.2.

Machine code GP individuals have four parts, described as follows:

```
The Header
The Body
The Footer
The Return Instruction
```

For the purpose of this section, the header and the footer may be regarded as housekeeping segments that do not undergo evolution. The return instruction is quite different. Although it, too, may not undergo evolution, it provides the crucial point, for each program, where that program ends. This section will deal primarily with the initialization of the body of the GP individual.

Figure 5.6 shows the basic setup of four registers and one AIMGP individual. The four CPU registers r_0, r_1, r_2, r_3 have been selected for use. These registers are the equivalent of the terminals in trees and might hold either variables (r_0, r_1, r_2) or constants (r_3). The range of constants has to be defined as well. Generally, register r_0 is chosen as the output register. This means that the value that appears in register r_0 at the end of execution is the output of the program.

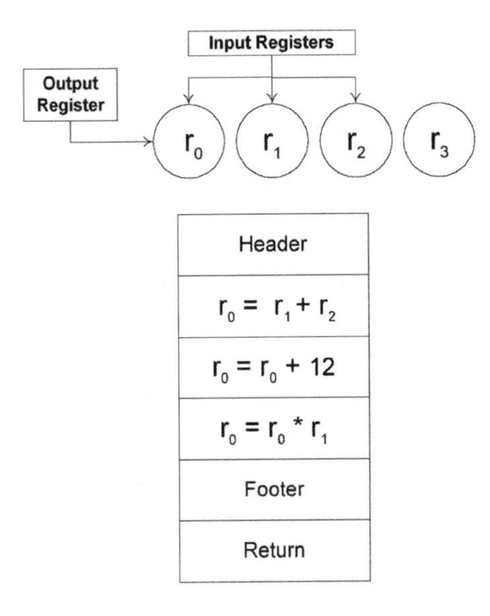

Figure 5.6
Linear AIMGP genome operating on CPU registers. Registers $r_0, r_1, \ldots, r_3$ *are used here.*

Each node in the body of the program is a machine code instruction. Again we shall assume that the number and sort of instructions which are the equivalent of functions in trees have been fixed beforehand. The machine code instructions in the sample program of Figure 5.6 act on three of the four registers. The constant register r_3 is not used in this example program.

The task of initialization in AIMGP now is to choose initial, random instructions that operate on an appropriate subset of the CPU's registers. A AIMGP individual is initialized as follows:

1. Randomly choose a length between two and the maximum length parameter;

2. Copy the predefined header to the beginning of an individual;

3. Initialize and add instructions to the individual until the number of instructions added equals the length chosen in step 1.

The instructions are initialized by randomly choosing an instruction type and then randomly filling out the instruction with references to randomly chosen registers from the register set and/or randomly chosen constants from the constant range;

4. Copy the predefined footer to the end of the individual;
5. Copy the predefined return instruction to the end of the individual.

In this way, the entire population can be initialized. The method described here is used instead of the full and grow methods in trees. An equivalent to those methods might be applied as well.

5.4 Genetic Operators

An initialized population usually has very low fitness. Evolution proceeds by transforming the initial population by the use of genetic operators. In machine learning terms, these are the search operators. While there are many genetic operators, some of which will appear in Chapter 10, the three principal GP genetic operators are:

- Crossover;
- Mutation; and
- Reproduction.

This section will give an introduction to the three basic genetic operators.

5.4.1 Crossover

The crossover operator combines the genetic material of two parents by swapping a part of one parent with a part of the other. Once again, tree linear and graph crossover will be discussed separately.

Tree-Based Crossover

Tree-based crossover is described graphically in Figure 5.7. The parents are shown in the upper half of the figure while the children are shown in the lower half.

More specifically, tree-based crossover proceeds by the following steps:

- Choose two individuals as parents, based on mating selection policy.[7] The two parents are shown at the top of Figure 5.7.

[7]Like, e.g., fitness-proportional selection.

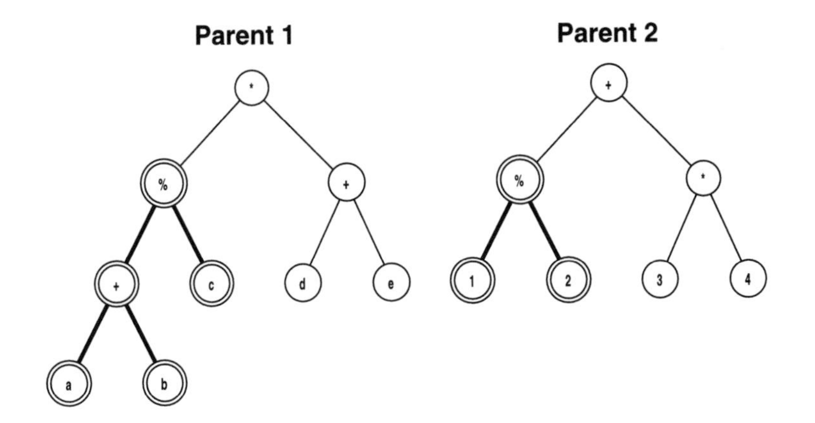

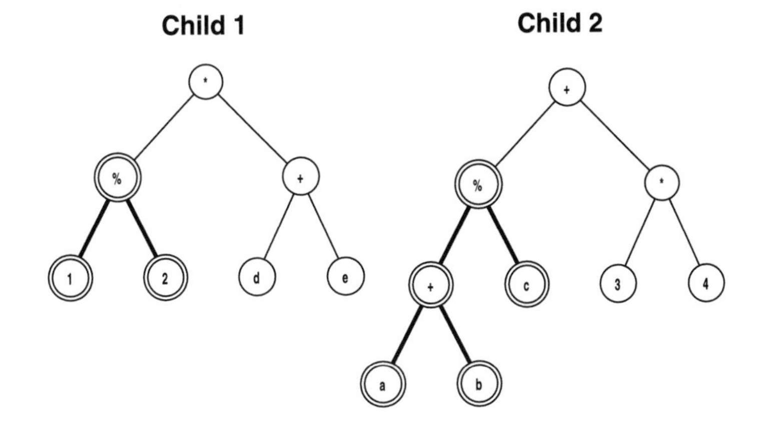

Figure 5.7
Tree-based crossover

- Select a random subtree in each parent. In Figure 5.7, the selected subtrees are shown highlighted with darker lines. The selection of subtrees can be biased so that subtrees constituting terminals are selected with lower probability than other subtrees.

- Swap the selected subtrees between the two parents. The resulting individuals are the children. They are shown at the bottom of Figure 5.7.

Linear Crossover

Linear crossover is also easily demonstrated. Instead of swapping subtrees, linear crossover, not surprisingly, swaps linear segments of code between two parents. Linear crossover is shown graphically in Figure 5.8. The parents are in the left half of the figure while the children are in the right half of the figure.

The steps in linear crossover are as follows:

- Choose two individuals as parents, based on mating selection policy.

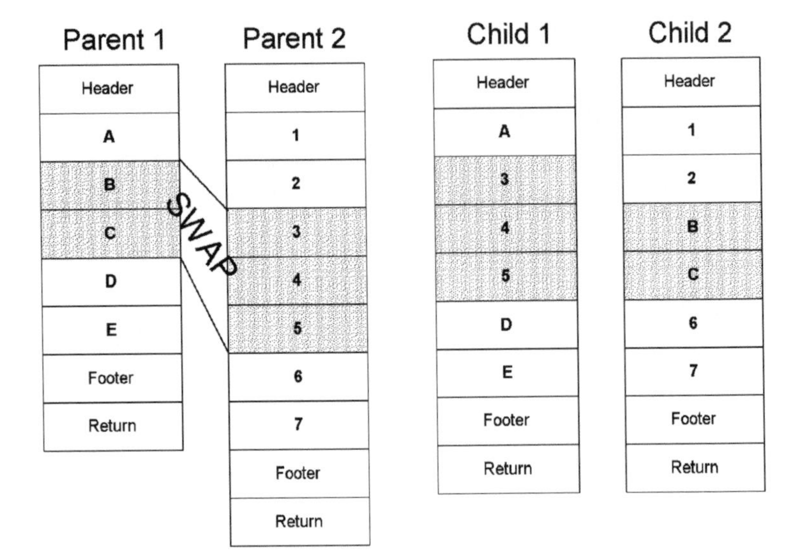

Figure 5.8
Linear crossover

- ❑ Select a random sequence of instructions in each parent. In Figure 5.8, the selected instructions are shown highlighted with light gray.
- ❑ Swap the selected sequences between the two parents. The resulting individuals are the children. They are shown at the right of Figure 5.8.

Graph Crossover

Graph crossover is somewhat more complicated. The following procedure is employed by Teller [Teller, 1996]:

- ❑ Choose two individuals as parents, based on mating selection policy.
- ❑ Divide each graph into two node sets.
  - ❑ Label all edges (pointers, arcs) internal if they connect nodes within a fragment, label them otherwise as external.
  - ❑ Label nodes in each fragment as output if they are the source of an external edge and as input if they are the destination of an external edge.
- ❑ Swap the selected fragments between the two parents.
- ❑ Recombine edges so that all external edges in the fragments now belonging together point to randomly selected input nodes of the other fragments.

With this method, all edges are assured to have connections in the new individual and valid graphs have been generated.

This brief treatment of crossover demonstrates the basics. More advanced crossover topics will be treated in Chapter 6.

5.4.2 Mutation

Mutation operates on only one individual. Normally, after crossover has occurred, each child produced by the crossover undergoes mutation with a low probability. The probability of mutation is a parameter of the run. A separate application of crossover and mutation, however, is also possible and provides another reasonable procedure.

Mutation in Tree Structures

When an individual has been selected for mutation, one type of mutation operator in tree GP selects a point in the tree randomly and replaces the existing subtree at that point with a new randomly generated subtree. The new randomly generated subtree is created in the same way, and subject to the same limitations (on depth or size) as programs in the initial random population. The altered individual is then placed back into the population. There are other types of mutation operators which will be discussed in Chapter 9.

Mutation in Linear Structures

In linear GP, mutation is a bit different. When an individual is chosen for mutation, the mutation operator first selects one instruction from that individual for mutation. It then makes one or more changes in that instruction. The type of change is chosen randomly from the following list:

- Any of the register designations may be changed to another randomly chosen register designation that is in the register set.
- The operator in the instruction may be changed to another operator that is in the function set.
- A constant may be changed to another randomly chosen constant in the designated constant range.

Suppose the instruction

$$r_0 = r_1 + r_2$$

has been selected for mutation. Here are samples of acceptable mutations in this instruction:

$$r_1 = r_1 + r_2$$
$$r_0 = r_2 + r_2$$
$$r_0 = r_1 \texttt{ OR } r_2$$
$$r_0 = r_1 + r_0$$

Tree vs. Linear Mutation

Many of the apparent differences between tree and linear mutation are entirely historical. Tree mutation can alter a single node as linear mutation alters a single instruction. Linear mutation can replace all instructions that occur after a randomly chosen instruction with another randomly chosen sequence of instructions – a procedure similar to replacing a subtree introduced above. Graph mutation is possible as well, but is not treated here.

5.4.3 Reproduction

The reproduction operator is straightforward. An individual is selected. It is copied, and the copy is placed into the population. There are now two versions of the same individual in the population.

5.5 Fitness and Selection

As noted in Chapter 1, genetic programming neither is a hill climbing system (which searches only one path through the search space) nor does it conduct an exhaustive search of the space of all possible computer programs. Rather, GP is a type of beam search. The GP population is the beam – the collection of points in the search space from which further search may be conducted.

Of course, GP must choose which members of the population will be subject to genetic operators such as crossover, reproduction, and mutation. In making that choice, GP implements one of the most important parts of its model of organic evolutionary learning, *fitness-based selection.* Fitness-based selection affects both the ordering of the individuals in the beam and the contents of the beam.

GP's evaluation metric is called a *fitness function* and the manner in which the fitness function affects the selection of individuals for genetic operators may be referred to as the GP selection algorithm. Fitness functions are very problem specific. There are a number of different selection algorithms used in GP.

5.5.1 The Fitness Function

Fitness

Definition 5.5 Fitness *is the measure used by GP during simulated evolution of how well a program has learned to predict the output(s) from the input(s) – that is, the features of the learning domain.*

The goal of having a fitness evaluation is to give feedback to the learning algorithm regarding which individuals should have a higher probability of being allowed to multiply and reproduce and which individuals should have a higher probability of being removed from the population. The fitness function is calculated on what we have earlier referred to as the training set.

The fitness function should be designed to give graded and continuous feedback about how well a program performs on the training set.

Continuous Fitness Function

Definition 5.6 *A* **continuous fitness function** *is any manner of calculating fitness in which smaller improvements in how well a program has learned the learning domain are related to smaller improvements in the measured fitness of the program, and larger improvements in how well a program has learned the learning domain are related to larger improvements in its measured fitness.*

Such continuity is an important property of a fitness function because it allows GP to improve programs iteratively. Two more definitions will be useful before we go into more detail about fitness functions.

Standardized Fitness

Definition 5.7 **Standardized fitness** *is a fitness function or a transformed fitness function in which zero is the value assigned to the fittest individual.*

Standardized fitness has the administrative feature that the best fitness is always the same value (zero), regardless of what problem one is working on.

Normalized Fitness

Definition 5.8 **Normalized fitness** *is a fitness function or a transformed fitness function where fitness is always between zero and one.*

With these definitions in hand, let us look at an example. Suppose we want to find a function satisfying the fitness cases in Table 5.1. Each input/output pair constitutes a training instance or fitness case. Collectively, all of the fitness cases constitute the training set.

	Input	Output
Fitness Case 1	1	2
Fitness Case 2	2	6
Fitness Case 3	4	20
Fitness Case 4	7	56
Fitness Case 5	9	90

Table 5.1
Input and output values in a training set

Suppose that GP was to evolve a program that learned the patterns in the Table 5.1 training set – that is, a program that could predict the output column by knowing only the value in the input column. The reader will probably note that this example is trivially simple and that a program representing the function $f(x) = x^2 + x$ would be a perfect match on this training set.

Error Fitness Function

One simple and continuous fitness function that we could use for this problem would be to calculate the sum of the absolute value of the differences between actual output of the program and the output given by the training set (the error). More formally, let the output of the ith example in the training set be o_i. Let the output from a GP program p on the ith example from the training set be p_i. In that case, for a training set of n examples the fitness f_p of p would be:

$$f_p = \sum_{i=1}^{n} |p_i - o_i| \tag{5.3}$$

This fitness function is continuous. As p_i gets a little closer to o_i, the fitness gets a little better. It is also standardized because any perfect solution, like $f(x) = x^2 + x$, would have zero fitness.

Squared Error Fitness Function

A common alternative fitness function is to calculate the sum of the squared differences between p_i and o_i, called the *squared error*:

$$f_p = \sum_{i=1}^{n} (p_i - o_i)^2 \tag{5.4}$$

Scaled Fitness Functions

In some applications a squared or otherwise scaled fitness measurement can result in better search results. Scaling refers to the fact that one can amplify or damp smaller deviations from the target output. A square function damps small deviations, whereas a square root or inverse function amplifies them.

How do these different fitness functions affect the fitness calculation? Suppose that one individual, Q, in a GP population is equivalent to x^2. Table 5.2 shows the output values of Q for the same training instances used in Table 5.1. The last two columns of Table 5.2 are the fitness for Q calculated by the error and the squared error methods, respectively.

	Input	Output	Q Output	Error fitness	Squared error fitness
Fitness Case 1	1	2	1	1	2
Fitness Case 2	2	6	4	2	4
Fitness Case 3	4	20	16	4	16
Fitness Case 4	7	56	49	7	49
Fitness Case 5	9	90	81	9	81
Total fitness	-	-	-	23	151

Table 5.2
Two different fitness calculations

Symbolic Regression

Where the learning domain is comprised of numeric inputs and outputs, the process of inducing programs that have learned the numeric examples is called symbolic regression. Many GP applications can be reformulated as instances of symbolic regression. The above problem is an example of symbolic regression.

There are many other ways to cast a fitness function. Examples of fitness functions similar to symbolic regression are:

- The number of matching pixels in an image matching application.
- The number of wall hits for a robot controlled by GP and learning obstacle avoidance.
- The number of correctly classified examples in a classification task.
- The deviation between prediction and reality in a prediction application.
- The money won by a GP-controlled agent in a betting game.
- The amount of food found and eaten by an artificial agent in an artificial life application.

There are also other methods for calculating fitness. In co-evolution methods for fitness evaluation [Angeline and Pollack, 1993] [Hillis, 1992], individuals compete against each other without an explicit fitness value. In a game-playing application, the winner in a game may be given a higher probability of reproduction than the loser. In some cases, two different populations may be evolved simultaneously with conflicting goals. For example, one population might try to evolve programs that sort lists of numbers while the other population tries to evolve lists of numbers that are hard to sort. This method is inspired by arms races in nature where, for example, predators and prey evolve together with conflicting goals.

Multiobjective Fitness

In some cases, it might be advantageous to combine very different concepts in the fitness criteria. We could add terms for the length of the evolved programs or their execution speed, etc. Such a fitness function is referred to as a *multiobjective fitness function.*

5.5.2 The Selection Algorithm

After the quality of an individual has been determined by applying a fitness function, we have to decide whether to apply genetic operators to that individual and whether to keep it in the population or allow it to be replaced. This task is called selection and assigned to a special operator, the selection operator.

There are various different selection operators, and a decision about the method of selection to be applied under particular circumstances is one of the most important decisions to be made in a GP run. Selection is responsible for the speed of evolution and is often cited as the culprit in cases where premature convergence stalls the success of an evolutionary algorithm.

We shall discuss selection in a very general context here, including some details of what has been developed in the ES community. Selection in general is a consequence of competition between individuals in a population. This competition results from an overproduction of individuals which can withstand the competition to varying degrees. The degree to which they can withstand the competition is regulated by the selection pressure, which depends on the ratio of offspring to individuals in the population.

Two main scenarios for generational selection have been established since evolutionary algorithms were first studied in the 1960s: (i) the GA scenario, and (ii) the ES scenario.

The GA Scenario

The GA scenario starts with a population of individuals with known fitness and performs a selection of individuals based on their fitness. These are then subjected to variation operations like crossover and mutation or passed on untouched via reproduction into the next generation. In this way, the pool of the following generation is filled with individuals. The next generation usually consists of the same number of individuals as the former one, and fitness computation follows in preparation for another round of selection and breeding. Figure 5.9a outlines the procedure, also known as mating selection.

The ES Scenario

The ES scenario is different. Starting from a given population, a usually larger set of offspring is generated by randomly selecting parents. After fitness evaluation, this population is then reduced by selection to the size of the original population. Thus, a smaller population can be used, as the selection is applied to the pool of offspring (possibly including even the parents). Figure 5.9b outlines the procedure, also known as overproduction selection.

The difference between the two generational scenarios may be seen in the ability in the ES type scenario to tune selection pressure by adjusting the ratio of the number of offspring to the number of parents. The larger this ratio, the higher the selection pressure. A corresponding pressure can be introduced into GAs if the requirement is relaxed that an equal number of offspring be produced after selection. If the size of the offspring pool is larger than the size of the parent pool, then again a larger selection pressure is exerted.

Fitness-Proportional Selection

Fitness-proportional selection is employed in a GA scenario for generational selection and specifies probabilities for individuals to be given

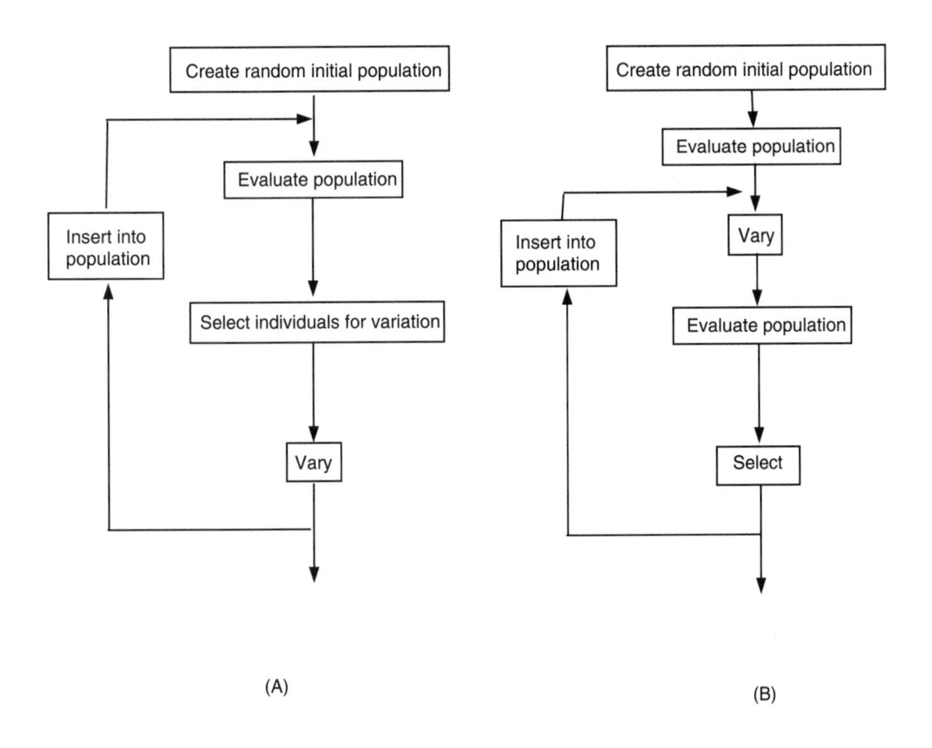

Figure 5.9
Different selection schemes in EAs of type **A** *GA and* **B** *ES*

a chance to pass offspring into the next generation. An individual i is given a probability of

$$p_i = f_i / \sum_j f_j \qquad (5.5)$$

for being able to pass on traits. Depending on the variation operator used, this might result (i) in a copy of that individual, or (ii) in a mutated copy, or (iii) in case two individuals have been selected in the way mentioned, two offspring with mixed traits being passed into the next generation.

Following Holland [Holland, 1975], fitness-proportional selection has been the tool of choice for a long time in the GA community. It has been heavily criticized in recent times for attaching differential probabilities to the absolute values of fitness [Blickle and Thiele, 1995]. Early remedies for this situation were introduced through fitness scaling, a method by which absolute fitness values were made to adapt to the population average [Grefenstette and Baker, 1989], and other methods [Koza, 1992d].

Truncation or (μ, λ) Selection

The second most popular method for selection comes from ES-type algorithms [Schwefel, 1995] where it is known as (μ, λ) selection. A number μ of parents are allowed to breed λ offspring, out of which the μ best are used as parents for the next generation. The same method has been used for a long time in population genetics and by breeders [Crow and Kimura, 1970] [Bulmer, 1980] under the name truncation selection [Mühlenbein and Schlierkamp-Voosen, 1994].

A variant of ES selection is $(\mu + \lambda)$ selection [Rechenberg, 1994] where, in addition to offspring, the parents participate in the selection process.

Neither (μ, λ) / truncation selection nor the following selection procedures are dependent on the absolute fitness values of individuals in the population. The μ best will always be the best, regardless of the absolute fitness differences between individuals.

Ranking Selection

Ranking selection [Grefenstette and Baker, 1989] [Whitley, 1989] is based on the fitness order, into which the individuals can be sorted. The selection probability is assigned to individuals as a function of their rank in the population. Mainly, linear and exponential ranking are used. For linear ranking, the probability is a linear function of the rank:

$$p_i = \frac{1}{N}\left[p^- + (p^+ - p^-)\frac{i-1}{N-1}\right] \tag{5.6}$$

where p^-/N is the probability of the worst individual being selected, and p^+/N the probability of the best individual being selected, and

$$p^- + p^+ = 2 \tag{5.7}$$

should hold in order for the population size to stay constant.

For exponential ranking, the probability can be computed using a selection bias constant c,

$$p_i = \frac{c-1}{c^{N-1}}c^N - i \tag{5.8}$$

with $0 < c < 1$.

Tournament Selection

Tournament selection is not based on competition within the full generation but in a subset of the population. A number of individuals, called the tournament size, is selected randomly, and a selective competition takes place. The traits of the better individuals in the tournament are then allowed to replace those of the worse individuals. In the smallest possible tournament, two individuals compete. The better of the two is allowed to reproduce with mutation. The result of that reproduction is returned to the population, replacing the loser of the tournament.

The tournament size allows researchers to adjust selection pressure. A small tournament size causes a low selection pressure, and a large tournament size causes high pressure.

Tournament selection has recently become a mainstream method for selection, mainly because it does not require a centralized fitness comparison between all individuals. This not only accelerates evolution considerably, but also provides an easy way to parallelize the algorithm. With fitness-proportional selection, the communication overhead between evaluations would be rather large.

5.6 The Basic GP Algorithm

It is now possible to assemble all of the individual elements (functions, terminals, fitness-based selection, genetic operators, variable length programs, and population initialization) into an overall algorithm for a basic GP run. There are two ways to conduct a GP run, a generational approach and a steady-state approach. In generational GP, an entire new generation is created from the old generation in one cycle. The new generation replaces the old generation and the cycle continues. In steady-state GP, there are no generations. We will present an algorithm for each approach.

First, however, we will review the preparatory steps for making a GP run. Then we will discuss the two basic ways to approach the GP run algorithm itself.

Summary of Preparatory Steps

Here are the preliminary steps in a GP run, which we have already described in detail in this chapter.

1. Define the terminal set.
2. Define the function set.
3. Define the fitness function.
4. Define parameters such as population size, maximum individual size, crossover probability, selection method, and termination criterion (e.g., maximum number of generations).

Once these steps are completed, the GP run can commence. How it proceeds depends on whether it is generational or steady state.

Generational GP Algorithm

Traditionally, genetic programming uses a *generational* evolutionary algorithm. In generational GP, there exist well-defined and distinct generations. Each generation is represented by a complete population of individuals. The newer population is created from and then replaces the older population. The execution cycle of the generational GP algorithm includes the following steps:

1. Initialize the population.
2. Evaluate the individual programs in the existing population. Assign a numerical rating or fitness to each individual.
3. Until the new population is fully populated, repeat the following steps:
   - Select an individual or individuals in the population using the selection algorithm.
   - Perform genetic operations on the selected individual or individuals.
   - Insert the result of the genetic operations into the new population.
4. If the termination criterion is fulfilled, then continue. Otherwise, replace the existing population with the new population and repeat steps 2–4.
5. Present the best individual in the population as the output from the algorithm.

Steady-State GP Algorithm

The steady-state or tournament selection model is the principal alternative to generational GP. In this approach there are no fixed generation intervals. Instead, there is a continuous flow of individuals meeting, mating, and producing offspring. The offspring replace existing individuals in the same population. The method is simple to implement and has some efficiency benefits together with benefits from parallelization. Good general convergence results have been reported with the method, and it is currently gaining ground in the research community. Here is an example of a basic GP algorithm using the steady-state method and a small tournament size for selection.

1. Initialize the population.
2. Randomly choose a subset of the population to take part in the tournament (the competitors).
3. Evaluate the fitness value of each competitor in the tournament.
4. Select the winner or winners from the competitors in the tournament using the selection algorithm.
5. Apply genetic operators to the winner or winners of the tournament.
6. Replace the losers in the tournament with the results of the application of the genetic operators to the winners of the tournament.
7. Repeat steps 2–7 until the termination criterion is met.
8. Choose the best individual in the population as the output from the algorithm.

Generation Equivalents

The approach is called steady state because the genetic operators are applied asynchronously and there is no centralized mechanism for explicit generations. Nevertheless, it is customary in presenting results with steady-state GP to talk about generations. In fact, steady-state generations are the intervals during training which can be said to correspond to generations in a generational GP algorithm. These intervals are often when fitness is evaluated for the same number of individuals as the population size. For experiments and detailed references on generational versus steady-state GP see [Kinnear, Jr., 1993b].

5.7 An Example Run

This section demonstrates some examples of individuals and measurements from a typical GP run. The task was a function regression with the simple function:

$$y = f(x) = \frac{x^2}{2} \tag{5.9}$$

Ten fitness cases were used for this function regression task, taken from the x-interval $[0, 1]$ and shown in Table 5.3.

Following the steps of Section 5.6 we prepare the run by first deciding on the following issues:

1. Terminal set: Variable x, integer constants between –5 and +5.

Table 5.3
Fitness cases (input and output values) in the training set

	Input	Output
Fitness Case 1	0.000	0.000
Fitness Case 2	0.100	0.005
Fitness Case 3	0.200	0.020
Fitness Case 4	0.300	0.045
Fitness Case 5	0.400	0.080
Fitness Case 6	0.500	0.125
Fitness Case 7	0.600	0.180
Fitness Case 8	0.700	0.245
Fitness Case 9	0.800	0.320
Fitness Case 10	0.900	0.405

2. Function set: Arithmetic functions `+, -, *, %`.
3. Fitness function: Standardized fitness, based on root mean square error over 10 fitness cases.
4. Parameters of individual and population, the initialization and selection method, operator probabilities.

Koza has introduced a very lucid form of listing parameters in the tableau of Table 5.4 named after him. From there, we can read off that $P = 600$ individuals were used for this GP run in a tree-based system. Crossover probability was $p_c = 0.9$. Fitness was based on the error that an individual produced when fed with the input of these fitness cases. More details are listed in the table.

Let us inspect some selected individuals from a GP run. In initial generation 0, the distribution of fitness values was broad. Figure 5.10 shows the best individual in generation 0. We shall call the function resulting from the best individual in generation i $f_i(x)$. So f_0 reads:

$$f_0(x) = \frac{x}{3} \tag{5.10}$$

Figures 5.11–5.15 show best individuals from subsequent generations 1, 2, 3, and from generation 5. So f_1 reads:

$$f_1(x) = \frac{x}{6-3x} \tag{5.11}$$

As we can see, the tree size first expands on its way to an optimal solution and then shrinks again. f_2 reads:

$$f_2(x) = \frac{x}{x(x-4) - 1 + \frac{4}{x} - \frac{\frac{9(x+1)}{5x} + x}{6-3x}} \tag{5.12}$$

Parameters	Values
objective:	evolve function fitting the values of the fitness case table
terminal set:	x, integers from -5 to +5
function set:	ADD, SUB, MUL, DIV
population size:	600
crossover probability:	90 percent
mutation probability:	5 percent
selection:	tournament selection, size 4
termination criterion:	none
maximum number of generations:	100
maximum depth of tree after crossover:	200
maximum mutant depth:	4
initialization method:	grow

Table 5.4
Koza Tableau

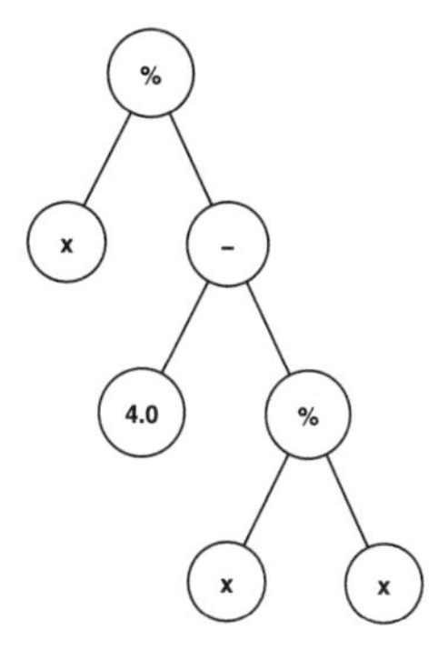

Figure 5.10
Best individual in generation 0. % is protected division.
$f_0(x) = \frac{x}{3}$

In generation 3, the best individual has found the correct solution in its simplest form.

$$f_3(x) = \frac{x^2}{2} \tag{5.13}$$

Subsequent generations start to combine this correct solution with others, with the consequence that the size of the best individual increases again. Because we did not always conserve the best individual found so far (a strategy that is called *elitist* and could be used in a GP run), quality fell again in later generations.

Table 5.5 shows how the function of the best individual program approaches desired outputs over the course of generations 0–3. Fig-

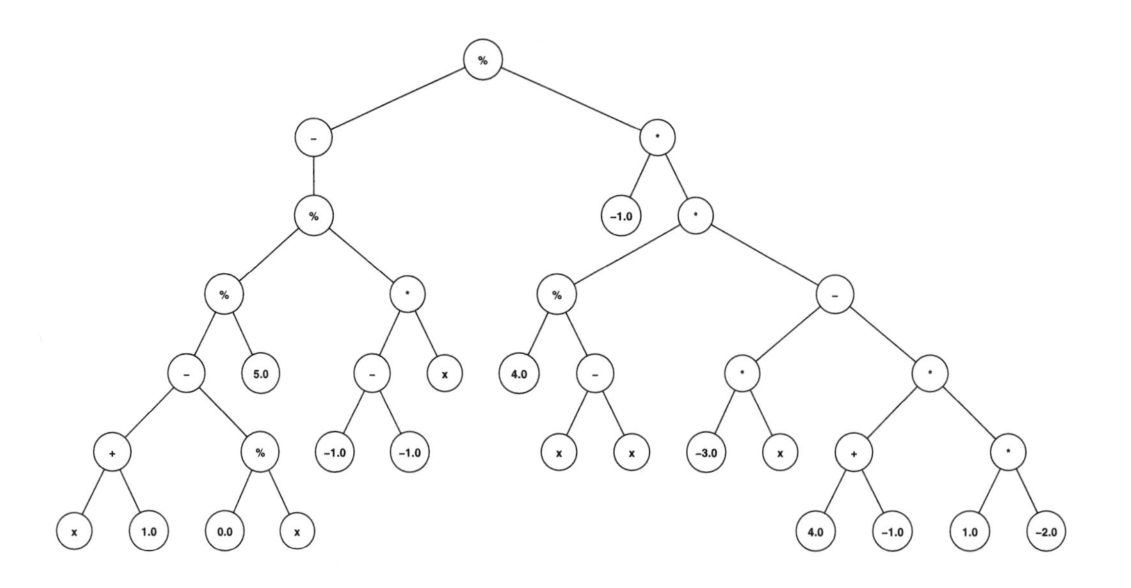

Figure 5.11
Best individual in generation 1. Tree has expanded considerably.
$f(x) = \frac{x}{6-3x}$

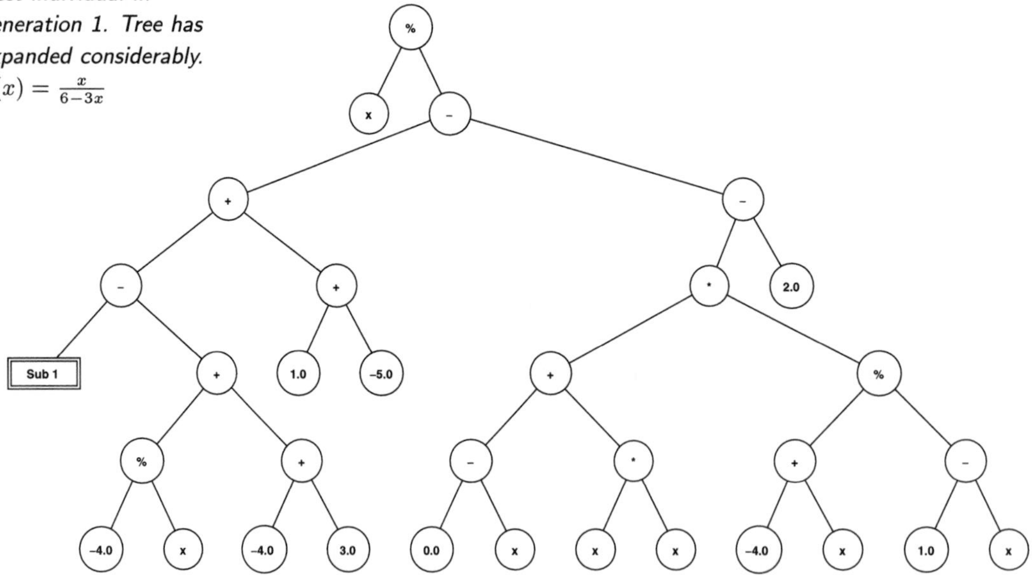

Figure 5.12
Best individual in generation 2. Tree has expanded again. (See text)

ure 5.16 shows the corresponding behavior of the functions for the best individuals.

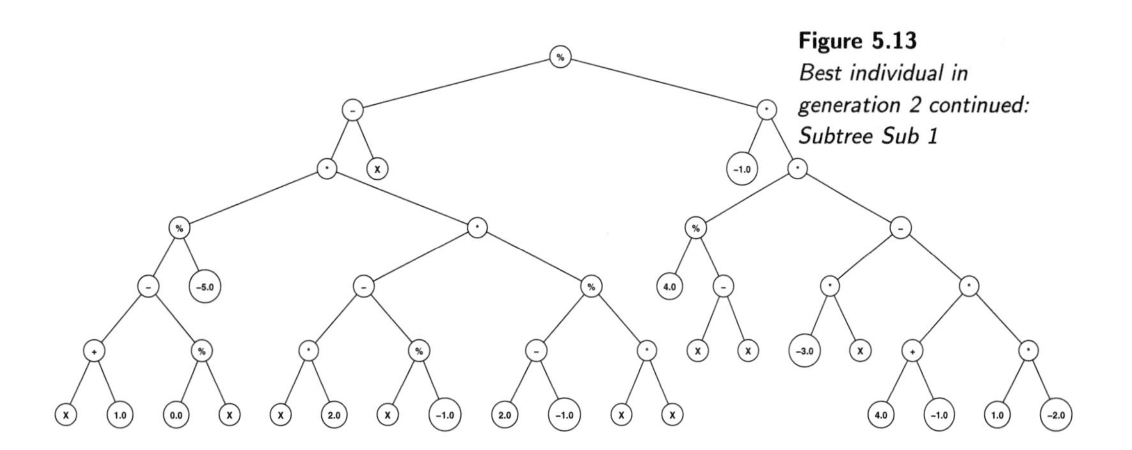

Figure 5.13
Best individual in generation 2 continued: Subtree Sub 1

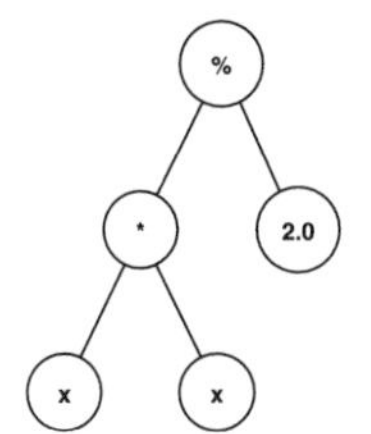

Figure 5.14
Best individual in generation 3. Perfect individual of simplest form found. $f(x) = \frac{x^2}{2}$

	Target output	Gen. 0	Gen. 1	Gen. 2	Gen. 3
Fitness Case 1	0.000000	0.000000	0.000000	0.000000	0.000000
Fitness Case 2	0.005000	0.033333	0.017544	0.002375	0.005000
Fitness Case 3	0.020000	0.066667	0.037037	0.009863	0.020000
Fitness Case 4	0.045000	0.100000	0.058824	0.023416	0.045000
Fitness Case 5	0.080000	0.133333	0.083333	0.044664	0.080000
Fitness Case 6	0.125000	0.166667	0.111111	0.076207	0.125000
Fitness Case 7	0.180000	0.200000	0.142857	0.122140	0.180000
Fitness Case 8	0.245000	0.233333	0.179487	0.188952	0.245000
Fitness Case 9	0.320000	0.266667	0.222222	0.287024	0.320000
Fitness Case 10	0.405000	0.300000	0.272727	0.432966	0.405000

Table 5.5
Target output and best individual output for generations 0 to 3

The above individuals are from a single GP run – a dynamic process that changes profoundly during its execution. Figure 5.17

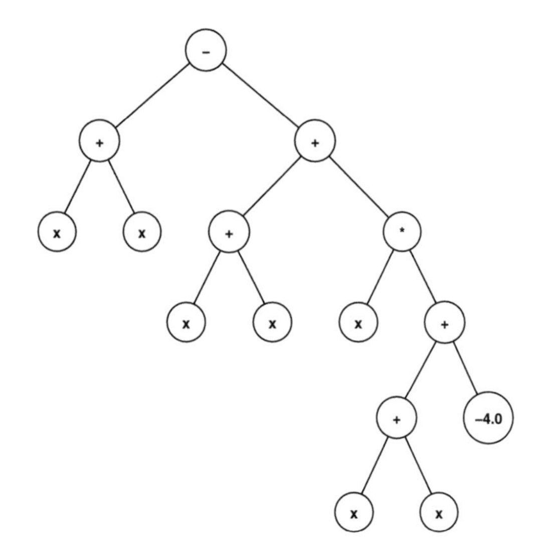

Figure 5.15
Best individual in generation 5. Length increases again. Fitness score is still perfect.
$f(x) = \frac{x^2}{2}$

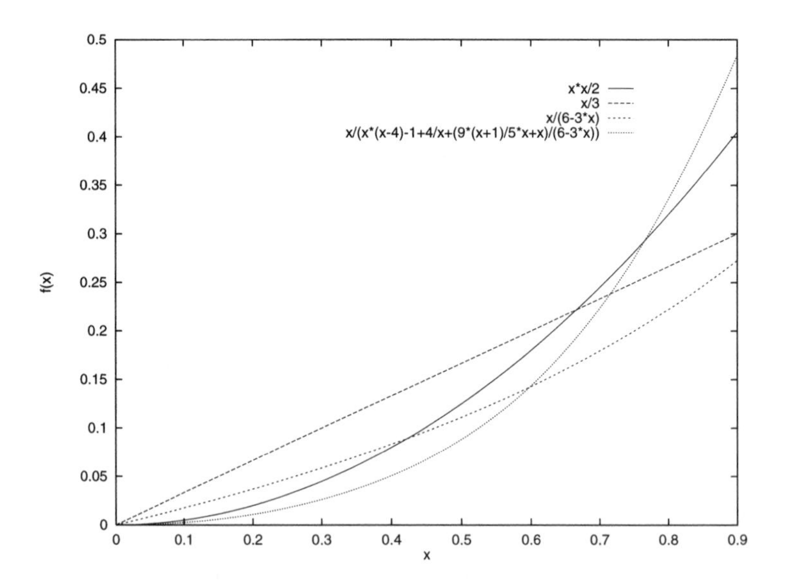

Figure 5.16
Behavior of best individuals of generations 0, 1, 2, 3. Generation 3 individual is identical to function itself.

shows how the average fitness of the entire population and the fitness of the best individual change as the run progresses.

As we shall discuss in Chapter 7, it is instructive to observe the development of program complexity during a run. We have therefore included a complexity measure in Figure 5.17 for illustrative purposes. We can see that average length of programs begins to increase quickly after the best fitness has arrived at its optimal value.

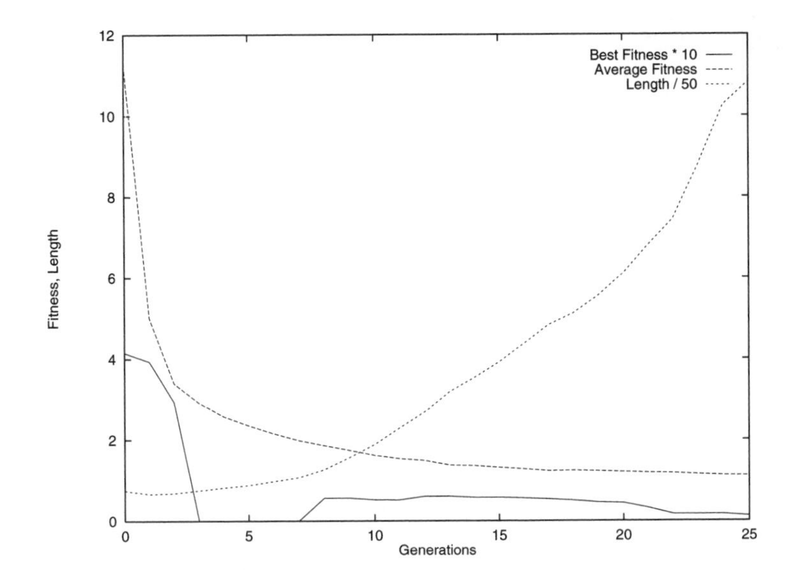

Figure 5.17
Development of fitness over generations. Best individual, average fitness, average length of individuals, scaled

What is interesting is how the simple process of a GP run – some terminals, some functions, crossover, mutation, and iteration – could have such a profound effect. This dynamic nature of the GP process and the many ways in which it manifests itself will prove to be a recurring theme in this book.

Exercises

1. Mention three important features shared by most GP systems.
2. What are the two different basic components of a GP program structure?
3. Give two different types of terminals.
4. What is the arity of a function?
5. Describe three different types of GP genomes.
6. Describe two methods for initialization of a tree structure individual.
7. Name the principal operations of a GP system.
8. How does a basic crossover operator work in a tree structure?
9. Describe three different selection algorithms.
10. Which are the preparatory steps before a GP experiment?
11. What is the difference between generational GP and steady-state GP?
12. Why do we not use only the best individuals as a source for the next generation?
13. Design a terminal and function set in order to classify dogs as terriers, standard poodles, toy poodles, or German shepherds. Which terminals and functions would you not include and why?
14. Design a terminal and function set to derive one of Kepler's laws. Did you consider and reject the inclusion of any terminals or functions?
15. Formulate the grow method of initialization in linear GP.

6 Crossover — The Center of the Storm

Contents

Search operators are among the most important parts of any machine learning system – they define the manner in which the system moves through the space of candidate solutions (see Chapter 1). In most GP systems, crossover is the predominant search operator. For example, in [Koza, 1992d] the crossover operator is applied 90% of the time. Most GP researchers have followed suit.

Crossover and Building Blocks

GP's heavy use of crossover is more than just a preference. Chapter 2 analyzed the mechanism of biological sexual reproduction, including crossover, at some length. The remarkable amount of biological energy that goes into maintaining species and homologous crossover suggests that crossover may well be an effective search operator in population-based machine learning systems like GP. The analogy with biological crossover is, of course, the original inspiration for the use of crossover in machine learning systems. Simply put, GP crossover attempts to mimic the process of sexual reproduction.

The crossover operator has been used as basis for the claim that GP search is more effective than systems based on random transformations (mutations) of the candidate solutions, like simulated annealing. Essentially, Koza has argued that a GP population contains *building blocks.* Simply put, a building block could be any GP tree or subtree that shows up in a fraction of the population. The building block hypothesis of GP follows the same line of argument as does the building block hypothesis from genetic algorithms [Holland, 1975]. Good building blocks improve the fitness of individuals that include them. Therefore, individuals with good building blocks are more likely to be selected for reproduction or breeding. Hence, good building blocks are likely to multiply and spread as they are duplicated and exchanged among individuals.

GP works faster than systems just based on mutations, according to this hypothesis, because good building blocks get combined into ever larger and better building blocks to form better individuals. This argument is based on the schema theorem, which is one of the theoretical underpinnings of genetic algorithms [Goldberg, 1989].

Crossover – The Controversy

The argument about the effectiveness of crossover has generated a good deal of controversy in other parts of the machine learning community [Lang, 1995] and has also provoked some thoughtful analysis in the GP community about just what a building block in GP is and whether we can realistically expect good GP building blocks to be selected by the crossover operator and to be combined into larger and better solutions [Altenberg, 1994b] [O'Reilly and Oppacher, 1992].

This chapter will focus on the central dispute regarding the crossover operator by posing the following two questions:

- Does the GP crossover operator outperform mutation-based systems by locating and combining good building blocks or is GP crossover, itself, a form of macromutation?
- What sorts of improvements may be made to the crossover operator to improve its performance?

Our discussion of other important aspects of the crossover operator, such as its role in creating so-called *introns* will be deferred to later chapters.

A Caveat

This chapter will focus at length on the undoubted shortcomings of the GP crossover operator. It is important, nevertheless, to remember that *something* is going on with GP crossover. GP has amassed an impressive record of achievements in only a few years. Whether crossover acts as a macromutation operator or whether it can, in addition, locate good schemata and combine them into better schemata, GP crossover already has a substantial record of accomplishment.

Chapter Overview

The next several sections of this chapter will be devoted to a critical analysis of GP crossover. First, we will look at the theoretical bases for both the building block hypothesis and the notion that GP crossover is really a macromutation operator. Second, we will survey the empirical evidence about the effect of crossover. Third, we will compare and contrast GP crossover with biological crossover. Finally, we will look at several promising directions for improving GP crossover.

6.1 The Theoretical Basis for the Building Block Hypothesis in GP

The *schema theorem* of Holland [Holland, 1975] is one of the most influential and debated theorems in evolutionary algorithms in general and genetic algorithms in particular. The schema theorem addresses the central question of why these algorithms work robustly in such a broad range of domains. Essentially, the schema theorem for fixed length genetic algorithms states that good schemata (partial building blocks that tend to assist in solving the problem) will tend to multiply exponentially in the population as the genetic search progresses and will thereby be combined into good overall solutions with other such schemata. Thus, it is argued, fixed length genetic algorithms will devote most of their search to areas of the search space that contain promising partial solutions to the problem at hand.

Recently, questions have been raised about the validity of the schema theorem for fixed length genetic algorithms. Nevertheless, the schema theorem remains the best starting point for a mathematically based analysis of the mechanisms at work in genetic algorithms using crossover. There have been several attempts to transfer the schema theorem from genetic algorithms to GP. However, the GP case is much more complex because GP uses representations of varying length and allows genetic material to move from one place to another in the genome.

The crucial issue in the schema theorem is the extent to which crossover tends to disrupt or to preserve good schemata. All of the theoretical and empirical analyses of the crossover operator depend, in one way or another, on this balance between disruption and preservation of schemata.

Koza's Schema Theorem Analysis

Koza was the first to address the schema theorem in GP. In his first book [Koza, 1992d, pages 116–119], Koza provides a line of reasoning explaining why the schema theorem applies to variable length GP. In his argument, a schema is a set of subtrees that contains (somewhere) one or many subtrees from a special schema defining set. For example, if the schema defining set is the set of S-expressions $H_1 = \{(-\ 2\ y), (+\ 2\ 3)\}$ then all subtrees that contain $(-\ 2\ y)$ or $(+\ 2\ 3)$ are instances of H_1. Koza's argument is informal and he does not suggest an ordering or length definition for his schemata.

Koza's statement that GP crossover tends to preserve, rather than disrupt, good schemata depends crucially on the GP reproduction operator, which creates additional copies of an individual in the population. Individuals that contain good schemata are more likely to be highly fit than other individuals. Therefore, they are more likely to be reproduced. Thus, good schemata will be tested and combined by the crossover operator more often than poorer schemata. This process results in the combination of smaller but good schemata into bigger schemata and, ultimately, good overall solutions.

O'Reilly's Schema Theorem Analysis

These ideas are formalized and extended considerably in [O'Reilly and Oppacher, 1995b] [O'Reilly and Oppacher, 1994b] by defining a schema as a multiset of subtrees and tree fragments under fitness-proportional selection and tree-based crossover. Fragments of trees are defined with a method similar to Holland's original schema theorem using a don't care symbol (#). O'Reilly defines her schemata similarly to Koza but with the presence of a don't care symbol in one or more subtrees, $H_1 = \{(-\ \#\ y), (+\ 2\ \#)\}$. Thus, if the defining set for H_1 contains several identical instances of a tree fragment, then the tree must contain the same number of matching subtrees in order to belong to the schema H_1.

O'Reilly's use of the don't care symbols is a major contribution to GP schema theory. It makes it possible to define an *order* and a *length* of the schemata. The order of a schema is the number of nodes which are *not* # symbols and the length is the number of *links* in the tree fragments plus the number of links connecting them. The sum of all links in a tree is variable and the probability of disruption depends on the size and shape of the tree matching a schema. O'Reilly therefore estimates the probability of disruption by the *maximum* probability of disruption, $P_d(H,t)$, producing the following schema theorem:

$$E[m(H,t+1)] \geq m(H,t)\frac{f(H,t)}{\overline{f}(t)}(1 - p_c P_d(H,t)) \qquad (6.1)$$

$f(H,t)$ is mean fitness of all instances of a certain schema H and $\overline{f}(t)$ is average fitness in generation t, while $E[m(H,t+1)]$ is the expected value of the number of instances of H and p_c is crossover probability.

The disadvantage of using the maximum probability is that it may produce a very conservative measure of the number of schemata in the next generation. Even the maximum probability of disruption varies with size, according to O'Reilly's analysis. While this is not a surprising result in variable length GP, it makes it very difficult to predict whether good schemata will tend to multiply in the population or will, instead, be disrupted by crossover.

Whigham's Schema Theorem Analysis

Whigham has formulated a definition of schemata in his grammar-based GP system (see Chapter 9) [Whigham, 1995c]. In Whigham's approach, a schema is defined as a partial derivation tree. This approach leads to a simpler equation for the probability of disruption than does O'Reilly's approach. However, like O'Reilly's derivation, Whigham's predicted probability of disruption also depends on the size of the tree.

Newer Schema Theorems

Recently, Poli and Langdon [Langdon and Poli, 1997] have formulated a new schema theorem that asymptotically converges to the GA schema theorem. They employed 1-point crossover and point mutation as GP operators. The result of their study suggests that there might be two different phases in a GP run: a first phase completely depending on fitness, and a second phase depending on fitness and structure of the individual (e.g., schema defining length). Whereas this work has introduced new operators to make schema dynamics more transparent, Rosca has concentrated on the structural aspect [Rosca, 1997]. He recently derived a version of the schema theorem for *rooted-tree* schemata. A rooted-tree schema is a subset of the set of program trees that matches an identical tree fragment which includes the tree root.

Inconclusive Schema Theorem Results for GP

None of the existing formulations of a GP schema theorem predicts with any certainty that good schemata will propagate during a GP run. The principal problem is the variable length of the GP representation. In the absence of a strong theoretical basis for the claim that GP crossover is more than a macromutation operator, it is necessary to turn to other approaches. In the next two sections, we will first look at the probability of disruption issue with a *gedanken experiment* and then analyze the empirical studies of the crossover operator.

6.2 Preservation and Disruption of Building Blocks: A Gedanken Experiment

We begin with an intuitive discussion of the difficulties of the crossover operator. First, we observe that the crossover operator is a destructive force as well as a constructive one. We have discussed the inconclusive theoretical basis for crossover's possible role in assembling good building blocks into complete solutions. This section describes a way to think about the balance of the constructive and destructive aspects of crossover.

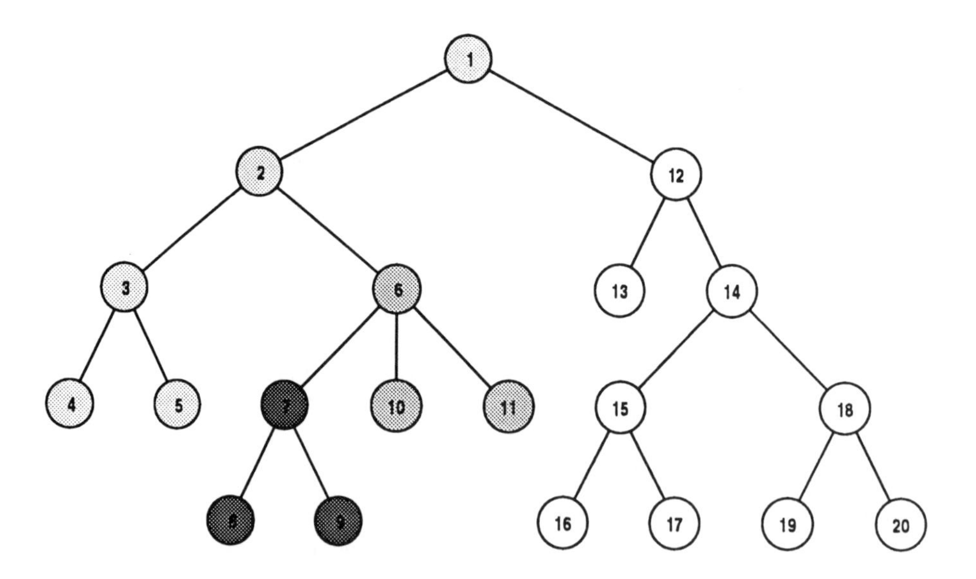

Figure 6.1
A tree assembling building blocks

6.2.1 Crossover as a Disruptive Force

Consider Figure 6.1. It may serve to illustrate what happens when

GP assembles good building blocks into a good program. We will look at the assembly of a good program block by block.

Let us first assume that the dark nodes (7–9) constitute a good building block. At this point there are 19 crossover points in the individual in Figure 6.1. If crossover is distributed randomly among these nodes, the probability of our good building block being disrupted by crossover is 2/19 or about 10.5%.

Now let us assume that crossover finds a new, good building block containing our original good block. That larger block is designated in Figure 6.1 by the black and dark gray nodes (nodes 6–11). The probability that this new block will be disrupted by a crossover event is 5/19 or 26.3%.

Now assume that the light gray nodes are part of a newly found good building block that has been assembled by GP crossover (nodes 1–11). What happens when this larger but better building block faces the crossover operator again? The answer is straightforward. The probability that this new block will be disrupted by a crossover event is 10/19 or 52.6%.

As the reader can see, as GP becomes more and more successful in assembling small building blocks into larger and larger blocks, the whole structure becomes more and more fragile because it is more prone to being broken up by subsequent crossover. In fact, assume that our building block (now, all of the colored nodes) is almost a perfect program. All that needs to be done now is to get rid of the code represented by the white nodes in Figure 6.1. The individual in Figure 6.1 is crossed over and the resulting individual is shown in Figure 6.2. This solution is almost there. There is only one node to get rid of before the solution is perfect. But in this case, just before success, the probability that the perfect solution will be disrupted by crossover is 10/11 or 90.9%.

Crossover can damage fitness in ways other than breaking up good building blocks. Assume that our good building block in Figure 6.2 survives crossover. It may, nevertheless, be moved into a new individual that does not use the good building block in any useful way – in other words, it may be moved to an inhospitable context.

Consider the numbers. If constructive crossover occurs between 10% and 15% of the time, this means that, on average, a good building block that is large must be involved in many crossover events before it is involved in a crossover event that does not break it up.

But that is not the end of the story. In the one event where the good block is not disrupted by crossover, what are the chances that

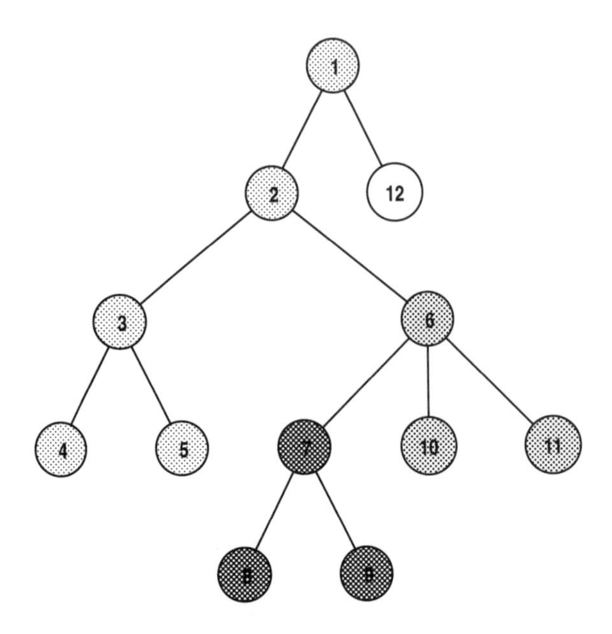

Figure 6.2
A tree that has almost assembled a perfect solution to a problem

it will be inserted into another individual where it is used to good effect and where other code does not cancel out the effect of this good block? The insertion is *context dependent.*

The conclusion is inevitable; crossover is a disruptive force as well as a constructive force – putting building blocks together and then tearing them apart. The balance is impossible to measure with today's techniques. It is undoubtedly a dynamic equilibrium that changes during the course of evolution. We note, however, that for most runs, measured destructive crossover rates stay high until the very end.

6.2.2 Reproduction and Crossover

In standard GP the reproduction operator takes the fittest individuals in the population and duplicates them. Thus, the argument goes, the good building blocks in those duplicated individuals will have many chances to try to find crossovers that are not disruptive.

This argument depends, of course, on the assumption that the high quality of the building block shown by the darkened nodes in Figure 6.1 will somehow be reflected in the quality of the individual in which it appears. But if the nodes around the good code ignore it or transform it in an unacceptable way, that will not be the case.

It also depends on the balance between the reproduction operator and the destructive effects of crossover at any given time in a run. Sometimes good blocks will improve the fitness of the individual they are in. Other times not. So the balance between the constructive and the destructive aspects of crossover is still the dominant theme here. Knowing the correct reproduction parameter and how to adjust it during a run to deal with the dynamic aspects of this problem is something on which no research exists.

Schema Theorem Analysis Is Still Inconclusive

After theoretical considerations and this gedanken experiment, our analysis is still inconclusive. It is impossible to predict with any certainty yet whether GP crossover is only a macromutation operator or something more. Therefore, it is time to consider empirical measurements of the crossover operator.

6.3 Empirical Evidence of Crossover Effects

Two sets of empirical studies bear on the effect of crossover. The first suggests that crossover normally results in severe damage to the programs to which it is applied. The second suggests that well-designed hill climbing or simulated annealing systems, which do not use population-based crossover, are very competitive with GP systems. We shall have a look at them both in turn.

6.3.1 The Effect of Crossover on the Fitness of Offspring

We began measuring the effect of crossover on the relative fitness of parents and their offspring in 1995 [Nordin and Banzhaf, 1995a] [Nordin et al., 1995]. There are two important issues in this regard:

- How can we measure the effect of crossover?
  Measuring the effect of crossover is not as straightforward as it might seem. The problem is that there are always at least two parents and one or more children. So GP systems are never measuring a simple one-to-one relationship.

- Likewise, it is not entirely clear what should be measured.

  Figure 6.3 shows a graph of the effect of crossover on fitness of offspring during the course of a run in symbolic regression in linear GP. Fitness change $\Delta f_{percent}$ is defined as

$$\Delta f_{percent} = \frac{f_{before} - f_{after}}{f_{before}} \cdot 100 \qquad (6.2)$$

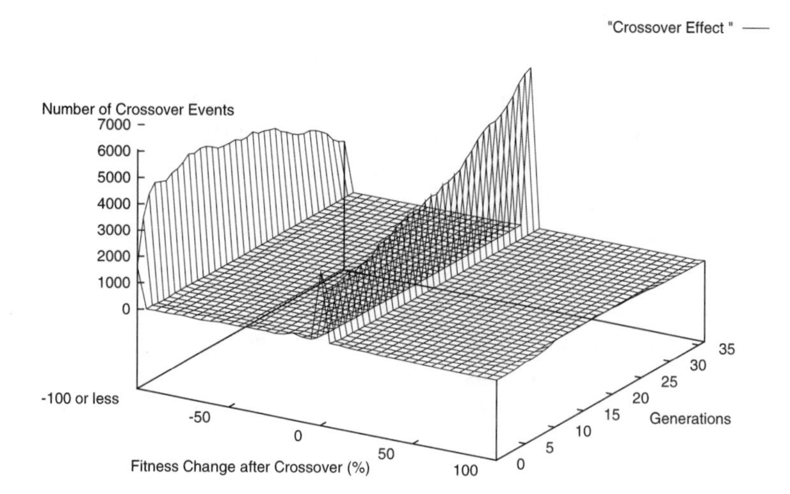

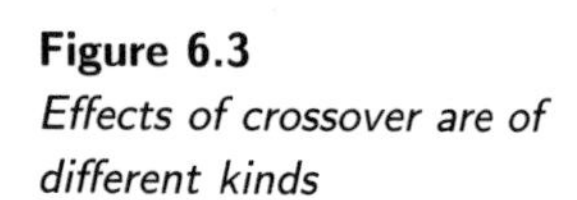
Figure 6.3
Effects of crossover are of different kinds

with f_{before} fitness before crossover and f_{after} after crossover under the assumption of a standardized fitness function, $f_{best} = 0$, $f_{worst} = \infty$.

Individuals with a fitness decrease of more than 100% are accumulated at the left side. Note that throughout training, there are two dominant forms of crossover – very destructive crossover (left) and neutral crossover (middle). There is also a low level of constructive crossover. Although it is possible to measure positive and negative crossover effects exactly, we have found it beneficial to do categorizing measurements. A substantial amount of information may be gleaned already from measuring neutral crossover as a separate category.[1]

Two basic approaches to measuring the effect of crossover have been used in the literature:

- The average fitness of all parents has been compared with the average fitness of all offspring [Nordin and Banzhaf, 1995a]. Note that the effect of this is that both parents and offspring are counted as *one* crossover event. Thus no special cases, like 2 parents and 3 children versus 3 parents and 2 children, need to be treated separately.

[1] Neutral crossover in existing studies has been defined as any crossover event where the fitness of the children is within $\pm$ 2.5% of the fitness of the parents.

- The fitness of children and parents is compared on an individual basis [Teller, 1996] [Francone et al., 1996]. In this approach, one child is assigned to one parent.

  Therefore, each such pairing is counted as one crossover event. In this case, a further specialization is necessary. Do we compare offspring to the best or worst parent? Do we compare them one by one (only possible if numbers are equal)?

Regardless how the measurement is conducted, it is important to separate destructive, neutral, and constructive crossover and to provide a separate measurement for each. The resulting measurements are very informative no matter what technique is used for measurement.

The Results of Measuring the Effect of Crossover

The effect of crossover has been measured for tree-based GP, linear (machine code) GP [Nordin and Banzhaf, 1995a], and graph GP [Teller, 1996]. In all three cases, crossover has an overwhelmingly *negative* effect on the fitness of the offspring of the crossover. For example, in linear genomes, the fitness of the children is less than half the fitness of the parents in about 75% of all crossover events. Similar measurements apply to tree-based crossover [Nordin et al., 1996]. Recently, Teller has measured similar figures for crossover in the graph-based system PADO. Although his measurements were not precisely equivalent to the tree and linear measurements of crossover cited above, his findings are quite consistent – less than 10% of crossover events in graph-based GP result in an improvement in the fitness of offspring relative to their parents. Note that these are global numbers over entire runs. There is certainly a change over the course of a run that cannot be reflected in these numbers at all.

What is remarkable is that three very different ways of measuring crossover on three completely different representations have yielded such similar results. The conclusion is compelling: crossover routinely reduces the fitness of offspring substantially relative to their parents in almost every GP system. This stands in stark contrast to biological crossover.

6.3.2 The Relative Merits of Program Induction via Crossover versus Hill Climbing or Annealing

Lang [Lang, 1995] launched a controversial attack on GP crossover in 1995. Lang's study argued that crossover in a population did not perform nearly as well as a macromutation operator that has been whimsically dubbed *headless chicken crossover.*

Headless Chicken Crossover

In headless chicken crossover, *only one* parent is selected from preexisting learned solutions. An entirely new individual is created randomly. The selected parent is then crossed over with the new and randomly created individual. The offspring is kept if it is better than or equal to the parent in fitness. Otherwise, it is discarded. Thus, headless chicken crossover is a form of hill climbing. In one genetic algorithm study, headless chicken crossover slightly outperformed both traditional genetic algorithms and hill climbing using bit-flipping mutations [Jones, 1995].

Lang's study went considerably further in its conclusion. Lang argued that headless chicken crossover was *much better* than GP crossover. However, his study was based on one small problem (3-argument Boolean functions) and the broad interpretation of his limited results is dubious [O'Reilly, 1995].

Lang's results are of questionable usefulness because of the single problem he chose to show that hill climbing outperformed genetic programming. Every machine learning technique has a bias – a tendency to perform better on certain types of problems than on others. Lang picked only one test problem, the Boolean 3-multiplexer problem. Boolean multiplexer problems have the property that there are no strict local minima. That is [Juels and Wattenberg, 1995]:

> ... from any point in the search space, the graph defining the neighborhood structure contains a path to some optimal solution such that every transition in the path leads to a state with an equal or greater fitness. A. JUELS AND M. WATTENBERG, 1995

In other words, Boolean multiplexer problems like the simple one Lang used are well suited to be solved by hill climbing algorithms like headless chicken crossover. That is, the bias of a hill climbing algorithm is particularly well suited to solving Boolean multiplexer problems.

However, other more thorough studies have raised significant questions about whether the crossover operator may be said to be better than mutation-oriented techniques. One set of those studies is discussed above – crossover is highly destructive to offspring. Other studies suggest that mutation techniques may perform as well as and sometimes slightly better than traditional GP crossover.

Crossover vs. Non-Population-Based Operators

For example, O'Reilly measured GP crossover against several other program induction algorithms that did not rely on population-based crossover. Two were of particular interest to her, mutate–simulated annealing and crossover–hill climbing. Each algorithm starts with a current solution. This is then changed to generate a new solution. For example, the crossover–hill climbing algorithm changes the current candidate solution by crossing it over with a randomly generated program, a form of headless chicken crossover. If the new solution has higher fitness, it replaces the original solution.[2] If the new solution has lower fitness, it is discarded in crossover–hill climbing but kept probabilistically in mutate–simulated annealing.

O'Reilly found [O'Reilly and Oppacher, 1994a] that the mutate–simulated annealing and crossover–hill climbing algorithms performed as well as or slightly better than standard GP on a test suite of six different problems; see also [O'Reilly and Oppacher, 1995a].

In another recent study, O'Reilly extended her results by comparing GP with other operators, most notably a hierarchical variable length mutation, which is an operator explicitly constructed for keeping distances between parent and offspring low.[3] She concluded that crossover seems to create children with large syntactic differences between parents and offspring, at least relative to offspring generated by hierarchical variable length mutation. This adds weight to the crossover-is-macromutation theory.

More recently, Angeline tested population-based and selection-driven headless chicken crossover against standard GP subtree crossover (see Chapters 5 and 9). He compared the two types of operators over three different problem sets. His conclusion: macromutation of subtrees (headless chicken crossover in GP) produces results that are about the same or possibly a little better than standard GP subtree crossover [Angeline, 1997].

6.3.3 Conclusions about Crossover as Macromutation

The empirical evidence lends little credence to the notion that traditional GP crossover is, somehow, a more efficient or better search operator than mutation-based techniques. On the other hand, there is no serious support for Lang's conclusion that hill climbing outperforms GP. On the state of the evidence as it exists today, one must conclude that traditional GP crossover acts primarily as a macromutation operator.

That said, several caveats should be mentioned before the case on crossover may be closed. To begin with, one could easily conclude that traditional GP crossover is an excellent search operator as is. What is remarkable about GP crossover is that, although it is lethal to the offspring over 75% of the time, standard GP nevertheless performs as well as or almost as well as techniques based on long established and successful algorithms such as simulated annealing.

---

[2] In evolutionary strategies, this is known as a $(1 + 1)$ strategy.

[3] This, of course, requires a measure of distance in trees, which O'Reilly took from [Sankoff and Kruskal, 1983]. The same measure was used earlier in connection with the measurement of diversity in GP [Keller and Banzhaf, 1994].

Furthermore, the failure of the standard GP crossover operator to improve on the various mutation operators discussed above may be due to the stagnation of GP runs that occurs as a result of what is referred to as "bloat" – in other words, the exponential growth of GP "introns." In Chapter 7, we will discuss the exponential growth of GP introns at some length. One of our conclusions there will be that the destructive effect of standard GP crossover is the principal suspect as the cause of the GP bloating effect. By way of contrast, the studies comparing crossover and macromutation operators use macromutation operators that are much less likely to cause bloat than standard subtree crossover. So it may be that by avoiding or postponing bloat, macromutation permits a GP run to last longer and, therefore, to engage in a more extensive exploration of the search space [Banzhaf et al., 1996].

Finally, the evidence suggests there may be room for substantial improvement of the crossover operator. Crossover performs quite well even given its highly disruptive effect on offspring. If it were possible to mitigate that disruptive effect to some degree, crossover would likely perform a faster and more effective search. A reexamination of the analogy between biological crossover and GP crossover and several recent studies on crossover suggest various directions for such improvements in crossover.

6.4 Improving Crossover – The Argument from Biology

Although there are many differences between the GP crossover operator and biological crossover in sexual reproduction, one difference stands out from all others. To wit, biological crossover works in a highly constrained and highly controlled context that has evolved over billions of years. Put another way, crossover in nature is itself an evolved operator. In Altenberg's terms [Altenberg, 1994b], crossover may be seen as the result of the evolution of evolvability.

There are three principal constraints on biological crossover:

1. Biological crossover takes place only between members of the *same* species. In fact, living creatures put much energy into identifying other members of their species – often putting their own survival at risk to do so. Bird songs, for example, attract mates of the same species . . . and predators. Restricting mating to intraspecies mating and having a high percentage of viable offspring must be very important in nature.

2. Biological crossover occurs with remarkable attention to preservation of "semantics." Thus, crossover usually results in the same gene from the father being matched with the same gene from the mother. In other words, the hair color gene does not get swapped for the tallness gene.

3. Biological crossover is *homologous*. The two DNA strands are able to line up identical or very similar base pair sequences so that their crossover is accurate (usually) almost down to the molecular level. But this does not exclude crossover at duplicate gene sites or other variations, as long as very similar sequences are available.

In nature, most crossover events are successful – that is, they result in viable offspring. This is a sharp contrast to GP crossover, where over 75% of the crossover events are what would be termed in biology "lethal."

What causes this difference? In a sense, GP takes on an enormous chore. It must evolve genes (building blocks) so that crossover makes sense *and* it must evolve a solution to the problem all in a few hundred generations. It took nature billions of years to come up with the preconditions so that crossover itself could evolve.

GP crossover is very different from biological crossover. Crossover in standard GP is unconstrained and uncontrolled. Crossover points are selected randomly in both parents. There are no predefined building blocks (genes). Crossover is expected to find the good building blocks and not to disrupt them even while the building blocks grow.

Let's look more closely at the details:

- In the basic GP system, any subtree may be crossed over with any other subtree. There is no requirement that the two subtrees fulfill similar functions. In biology, because of homology, the different alleles of the swapped genes make only minor changes in the same basic function.

- There is no requirement that a subtree, after being swapped, is in a context in the new individual that has any relation to the context in the old individual. In biology, the genes swapped are swapped with the corresponding gene in the other parent.

- Were GP to develop a good subtree building block, it would be very likely to be disrupted by crossover rather than preserved and spread. In biology, crossover happens mostly between similar genetic material. It takes place so as to preserve gene function with only minor changes.

- There is no reason to suppose that randomly initialized individuals in a GP population are members of the same species – they are created randomly.

Given these differences, why should we expect crossover among GP individuals to have anything like the effect of biological crossover? Indeed, crossing over two programs is a little like taking two highly fit word processing programs, Word for Windows and WordPerfect, cutting the executables in half and swapping the two cut segments. Would anyone expect this to work? Of course not. Yet the indisputible fact is that crossover has produced some remarkable results. So each difference between biological and GP crossover should be regarded as a possible way to improve GP crossover – some of which, as we will see below, have already been implemented.

6.5 Improving Crossover – New Directions

Our critique of crossover suggests areas in which the crossover operator might be improved. We regard the most basic and promising approach to be modification of GP crossover so that it acts more like homologous crossover in nature. Nature has gone to great lengths to avoid macromutation in crossover (see Chapter 2). There is likely to be a reason for the energy nature devotes to avoiding macromutation. We submit, therefore, that homology should be the central issue in redefining the crossover operator.

However, most of the efforts to improve crossover to date have focussed on the preservation of building blocks, not the preservation of homology. Some of those efforts have been successful, others have intriguing empirical implications regarding the building block hypothesis, and others represent a first look by the GP community at the issue of homology. The remainder of this section will look at these studies.

6.5.1 Brood Recombination

Drawing on work by Altenberg [Altenberg, 1994a], Tackett devised a method for reducing the destructive effect of the crossover operator called brood recombination [Tackett, 1994]. Tackett attempted to model the observed fact that many animal species produce far more offspring than are expected to live. Although there are many different mechanisms, the excess offspring die. This is a hard but effective way to cull out the results of bad crossover.

Tackett created a "brood" each time crossover was performed. One of the key parameters of his system was a parameter called brood

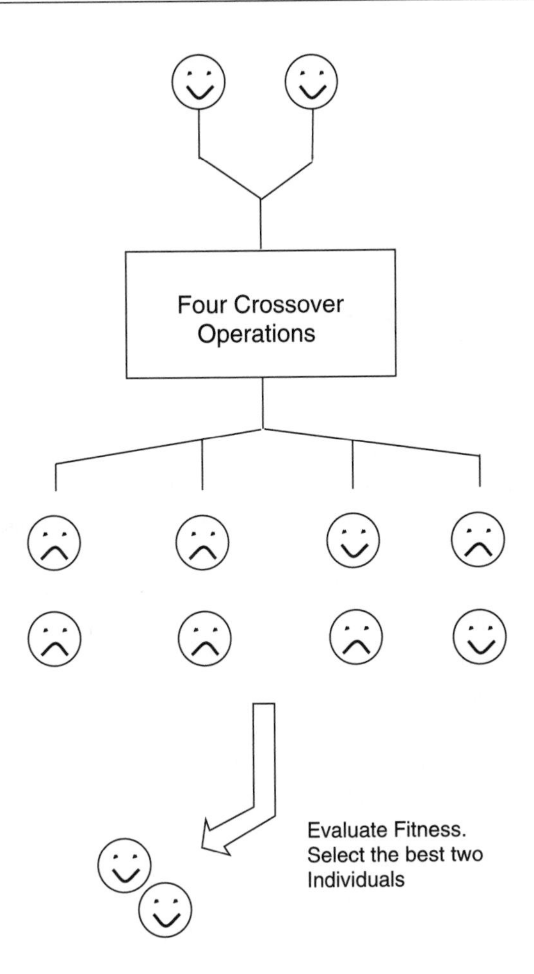

Figure 6.4
Brood recombination illustrated

size N. Figure 6.4 shows the creation and evaluation of a brood where $N = 4$, which took place in the following steps:

1. Pick two parents from the population.
2. Perform random crossover on the parents N times, each time creating a pair of children as a result of crossover. In this case there are eight children resulting from $N = 4$ crossover operations.
3. Evaluate each of the children for fitness. Sort them by fitness. Select the best two. They are considered the children of the parents. The remainder of the children are discarded.

Time-Saving Evaluation Methods

There is one big problem with this approach. GP is usually slow in performing evaluations. Instead of having two children to evaluate, as in standard crossover, brood recombination appears to make it necessary to evaluate $2N$ children. Will this not slow down GP tremendously? The answer is no, because of a clever approach Altenberg and Tackett take to evaluation. Tackett reasons that it is only important that the selection on the brood selects children that are "in the ballpark" – not that they are certainly the best of the brood. So he evaluates them on only a small portion of the training set. Because the entire brood is the offspring of one set of parents, selection among the brood members is selecting for effective crossovers – good recombinations.

Brood recombination is similar to ES-style selection (see Section 4.3.2). There, immediately after applying the genetic operators to μ parents, a selection step chooses the best offspring. However, the number of offspring in (μ, λ) selection, λ, is greater than μ, whereas in brood recombination the number of offspring is an integer multiple of μ.[4]

In Chapter 1 we looked at GP as a type of beam search. The beam is the population. Viewed another way, the beam is the genetic material in the population, not just the individuals. In this light, crossover combined with selection could be viewed as part of the evaluation metric to select genetic material from the beam as the next search point. Brood recombination would then be a discriminating addition to the evaluation metric for the beam.

Is Brood Recombination Effective?

However we view it, we would expect brood recombination to be less disruptive to good building blocks than ordinary crossover because the children of destructive crossover events would tend to be discarded when the brood is evaluated. Therefore, we would predict that GP could build larger building blocks with brood recombination before disruptive forces began to break them up. That being the case, we would expect GP with brood recombination to search for solutions more efficiently than regular GP. That is, for each unit of CPU time, brood recombination should produce a better result.

In fact, that is exactly what happened. Tackett found that brood recombination performed significantly better than standard GP on a suite of problems. Indeed, he found that there was only a small reduction in performance by using 30 out of the 360 training instances to evaluate the brood. In all cases measured, the results achieved by particular amounts of computation and diversity in the population

[4]The trick of not using all fitness cases for program evaluation has been applied successfully completely independent of brood recombination. Gathercole used this technique in 1994 [Gathercole and Ross, 1994]. We refer to the technique as "stochastic sampling" because it uses a stochastically selected subset of all the training cases. In an extreme case, stochastic sampling can be used to train a population on only a single member from the training set at a time[Nordin and Banzhaf, 1995c].

both improved when brood recombination was added. He also found that it was possible to reduce the population size when he used brood recombination.

Brood recombination raises two interesting questions. Does brood recombination work by *not* disrupting building blocks or by adding a different form of search process to the GP algorithm – in machine learning terms, giving the GP algorithm the ability to look ahead when adjusting the beam? Tackett's results may be regarded as consistent with either premise.

All that we would expect the brood recombination operator to do is to change the balance between destructive crossover and constructive crossover. In short, we would predict that, with brood recombination, GP would be able to maintain larger building blocks against destructive crossover. Eventually, as the building blocks got larger, the probability of destructive crossover would increase and we would predict that bloat would set in to protect the larger building blocks. This suggests that a dynamic form of brood recombination, where the size of the brood grows as evolution proceeds, may yield best results.

6.5.2 "Intelligent" Crossover

Recently, researchers have attempted to add intelligence to the crossover operator by letting it select the crossover points in a way that is less destructive to the offspring.

A Crossover Operator That Learns

The PADO system was discussed (Section 5.2.3) as a prototypic graph-based GP system. Simple crossover in GP systems is rather straightforward. But it is far from obvious how to cause intelligent crossover, especially in a graph GP system. Teller has devised a complex but surprisingly effective technique for improving the rate of constructive crossover in PADO by letting an intelligent crossover operator *learn* how to select good crossover points. Teller gives his intelligent crossover operator access to information about the execution path in an evolved program, among other things. This information is then used to guide crossover [Teller and Veloso, 1995b] [Teller, 1996].

Teller's intelligent recombination operator significantly improved the performance of traditional GP crossover. The percentage of recombination events that resulted in offspring better than the parents approximately doubled. Zannoni used a cultural algorithm (a more traditional machine learning algorithm) to select crossover points with similar results [Zannoni and Reynolds, 1996].

A Crossover Operator Guided by Heuristics

Iba [Iba and de Garis, 1996] has proposed a form of intelligent heuristic guidance for the GP crossover operator. Iba computes a so-called performance value for subtrees. The performance value is used to decide which subtrees are potential building blocks to be inserted into other trees, and which subtrees are to be replaced. This heuristic recombinative guidance improves the effect of crossover substantially.

These results are quite consistent with the building block theory. The intelligent operators discussed here obviously found *regularities* in the program structures of very different GP systems and devised rules to exploit those regularities. As a result, the systems were able to choose better crossover sites than systems based on standard GP crossover, which chooses crossover sites randomly. These studies therefore suggest that in GP:

- There are blocks of code that are best left together – perhaps these are building blocks.
- These blocks of code have characteristics that can be identified by heuristics or a learning algorithm.
- GP produces higher constructive crossover rates and better results when these blocks of code are probabilistically kept together.

These points are not a statement of the building block hypothesis in its strong form – they do not prove that the blocks of code are assembled into better combinations. However, the points are certainly strong support for a weak building block hypothesis, namely, that there are blocks of code evolved in GP that are best not disrupted.

Both of the above techniques, brood recombination and smart crossover, attack the crossover problem as a black box. In essence, they take the position that it does not matter how you get better crossover. What is important is to improve the final overall result. The next few techniques we will discuss are based on attempting to model biological crossover so that better results from crossover will emerge from the evolutionary process itself. Although the results of this approach to date are less impressive than brood recombination and smart crossover in the short run, they may hold the most promise in the long run.

6.5.3 Context-Sensitive Crossover

D'haeseleer [D'haeseleer, 1994] took an emergent approach to improving crossover. His work was based on the idea that most crossover does not preserve the context of the code – yet context is crucial to the meaning of computer code. He devised an operator called strong context preserving crossover (SCPC) that only permitted crossover between nodes that occupied exactly the same position in the two parents. D'haeseleer found modest improvements in results by mixing regular crossover and SCPC.

In a way, this approach introduced an element of homology into the crossover operator. It is not strong homology as in the biological example, but requiring crossover to swap between trees at identical locations is somewhat homologous.

6.5.4 Explicit Multiple Gene Systems

Evolution in nature seems to proceed by making small improvements on existing solutions. One of the most ambitious attempts to create an emergent system in which crossover is more effective is Altenberg's constructional selection system [Altenberg, 1995]. He proposes a system in which fitness components are affected by all or some of the genes. Altenberg's system is highly theoretical because the fitness of the individual is just the sum of the fitness components. Figure 6.5 is a stylized diagram of the system.

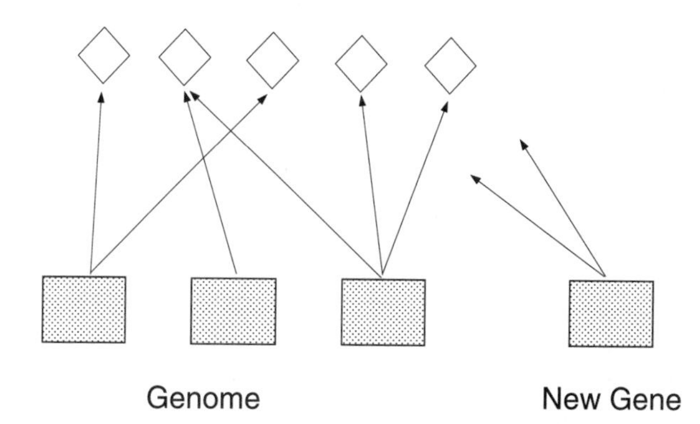

Figure 6.5
Altenberg's constructional selection in operation (adopted from [Altenberg, 1995])

During evolution, a gene is periodically added. If it improves the fitness of the individual, it is kept; otherwise, it is discarded. Between gene additions, the population evolves by *intergene* crossover. Having multiple fitness functions allows the genes to be more independent or, in biological terms, to be less epistatic. Altenberg's results suggest that the population maintains its evolvability because of the selection when adding a new gene. Also, his results clearly suggest that the system is less epistatic with constructional selection.

Altenberg's work was extended and applied to real-world modeling of industrial processes by Hinchliffe [Hinchliffe et al., 1996], who used multiple trees to create a multiple gene model.

Using Regression

Let T represent the output of a GP tree. Hinchliffe's model calculates its output as follows:

$$O_{3D} = aT_1 + bT_2 + cT_3 + ...nT_n + K \tag{6.3}$$

where $T_1, T_2, \ldots, T_n$ are outputs of separate trees which are his genes. Hinchliffe calculates the parameters $a, b, c, \ldots, n$ using pseudo-inverse linear regression. There is, in a sense, a hill climbing step between the GP results and the calculation of the output. The linear regression tends to reduce the impact of negative crossover by assigning higher values to trees that are highly correlated with the actual output.

In this multiple gene system, crossover occurs in two ways:

- As high-level crossover. Only T_is would be exchanged with T_is in the other individual.
- As low-level crossover. Subexpressions may be exchanged from anywhere in either parent.

This algorithm clearly outperformed standard GP. It is unclear, however, whether the multiple gene structure or the addition of hill climbing was the cause of the improved performance. Nevertheless, Hinchliffe's high-level crossover operator bears a strong resemblance to base-pair matching in biological homologous crossover.

6.5.5 Explicitly Defined Introns

Recently, we introduced explicitly defined introns (EDI) into GP.[5] An integer value is stored between every two nodes in the GP individual. This integer value is referred to as the EDI value (EDIV). The crossover operator is changed so that the probability that crossover occurs between any two nodes in the GP program is proportional to the integer value between the nodes. That is, the EDIV integer value strongly influences the crossover sites chosen by the modified GP algorithm during crossover [Nordin et al., 1996].

The Idea

The idea behind EDIVs was to allow the EDIV vector to evolve during the GP run to identify the building blocks in the individual as an *emergent* phenomenon. Nature may have managed to identify genes and to protect them against crossover in a similar manner. Perhaps if we gave the GP algorithm the tools to do the same thing, GP, too, would learn how to identify and protect the building blocks. If so, we would predict that the EDIV values within a good building block should become low and, outside the good block, high. Our results were modestly positive in linear genomes and inconclusive in tree-based genomes.

[5] Why we used the term *intron* will become clear in the next chapter.

Despite our modest results [Nordin et al., 1995], the EDI technique seems to be a promising direction for GP research. Emergent learning is powerful and, if designed properly, permits learning that reflects as few of the biases of the user as possible.

Angeline devised a much better implementation of EDIs in 1996 using real-valued EDIVs and constraining changes in the EDIVs by a Gaussian distribution of permissible mutation to the EDIVs [Angeline, 1996]. The result of applying this new form of EDIs to GP crossover was a substantial improvement in GP performance.

The conclusion that may be drawn about crossover from Angeline's EDI results is that crossover may be improved substantially by allowing the GP algorithm to protect some groups of code from crossover preferentially over other groups of code. This suggests that there are building blocks in GP individuals, that their existence in the solution, on average, does create better fitness in the individual, and that protecting such blocks of code helps crossover to propagate the better building blocks throughout the population.

6.5.6 Modeling Other Forms of Genetic Exchange

So far we have discussed variants of the crossover operator that were inspired by biological sexual recombination. These variants dominate both genetic algorithms and genetic programming.[6] However, there are several ways in which individuals exchange genetic material in nature. These forms include conjugation, transduction, and transformation (see the discussion in Chapter 2). Certainly, in the future it will be important to do research on GP operators inspired by such forms of genetic exchange.

Conjugation

Conjugation is used to describe the transfer of genetic information from one bacterial cell to another. Smith has proposed a conjugation operator for GAs he deems useful for GP as well [Smith, 1996]. Simple conjugation in GAs works like this: two genotypes are selected for conjugation. One of them is the donor, the other one is the recipient. A starting point (with identical position in each genotype) and an end point (again with identical position) are chosen. Then the donor's selected substring is copied from beginning to end to the recipient, replacing the corresponding substring in the recipient.

In GP, we need to distinguish between different representations. In a tree-based GP system, start and end points are nodes in the trees with the same position. In a linear representation, start and end points are the position of an instruction in the programs. In graph-based GP, they are the coordinates of a node in the graphs.

[6] Often, operators of this kind are called recombination operators, although in nature recombination is a much more general term.

To foster the spread of potentially advantageous genetic information, conjugation might be combined with tournament selection. In a 2-tournament scheme, the winner would become the donor, the loser the recipient of genetic material. Multiple conjugation could be done by several different donors copying genetic information to the same recipient. Multiple conjugation involving n donors could be combined preferentially with $n + 1$-tournament selection.

6.6 Improving Crossover – A Proposal

In the course of the discussion of GP crossover, we have seen its main weaknesses. To remedy the situation, we present here a homologous crossover operator for tree-based GP that shares what we identified as important advantages of natural crossover.

When we say we want to achieve "homologous" crossover in GP, we could be speaking about one of two things:

1. The *mechanism* by which biology causes homology, i.e., speciation, almost identical length or structure of DNA between parents, and strict base pairing during crossover; or

2. The *reason* that mechanism of homology has evolved. The reason the mechanism has evolved makes the actual mechanism somewhat irrelevant when changing the medium. In other words, new media like GP may imply new mechanisms.

Our proposal will fall into category 2. That is, we would like to propose a mechanism for crossover that fits the medium of GP and that may achieve the same results as homologous crossover in biology.

So the question is what result does homologous crossover have?

- ❏ Two parents have a child that combines some of the genome of each parent.

- ❏ The exchange is *strongly* biased toward experimenting with exchanging very similar chunks of the genome – specific genes performing specific functions – that have *small* variations among them, e.g., red eyes would be exchanged against green eyes, but not against a poor immune system.

Note that this second issue has two aspects, structure and function. Obviously, in DNA, similarity of structure is closely related to similarity of function. Otherwise homologous crossover would not work. DNA crossover relies entirely on structure (base pair bonding) so far as we know – it does not measure function. It is speciation that assures similarity of genetic function during homologous biological crossover. So if we are proposing a mechanism to achieve a result similar to biological crossover, either we must make sure that structure alone may be used to approximate function or we must separately measure functional homology and use those measurements to guide the crossover.

Here is what we suggest. We measure structural distances by comparing genotypes and functional distances by comparing phenotypic behavior. This way we have two different sources of homology that we can use either separately or combined. In tree-based GP, homologous crossover would work like this:

1. **Mating selection**
   Two trees are selected randomly.

2. **Measurement of structural similarity**
   Structural similarity is defined by using edit distances, already applied by [Sankoff and Kruskal, 1983]. This is a method for comparing variable length structures for similarity. We only need to fix an order in which to traverse the tree: depth first.[7] We number all edges between nodes in both trees according to depth-first traversal. Each edge will subsequently serve as the "coordinate" of the subtree starting from the node it points to.

   Once we have found, for each edge k in the larger tree, a subtree with smallest distance (and therefore an edge $imin(k)$) in the other tree – a distance we call $D_S(k, imin(k))$ – we add up all these minimal distances

   $$\bar{D}_S = \sum_k D_S(k, imin(k)) \qquad (6.4)$$

   and normalize each $D_S(k, imin(k))$ through division by $\bar{D}_S$ to yield a quantity $D_S^N(k, imin(k))$.[8]

3. **Measurement of functional similarity**
   We measure the output of each subtree (again, in the smaller tree) for a (small) sample of the fitness cases. We compare the outputs to those of the other tree and its subtrees and calculate

[7] Efficient search techniques need to be used for the algorithm. It is best to traverse one tree (say, the larger) and to check with all subtrees of the other (say, the smaller) the distance between the two. Since we are always looking for the minimal distance only, distance computation can be stopped as soon as it assumes a larger value than the presently known nearest subtree.

[8] We can do the same for the other tree, in order to act symmetrically.

the functional difference,

$$D_F(k, jmin(k)) = \sum_{\alpha} \mid O_k^{(\alpha)} - O_{jmin(k)}^{(\alpha)} \mid \qquad (6.5)$$

for the sample of α fitness cases. The resulting distances are again normalized by dividing by their sum $\bar{D}_F$

$$D_F^N(k, jmin(k)) = D_F(k, jmin(k)) / \sum_j D_F(j, jmin(j)) \qquad (6.6)$$

4. **Selection of crossover points**
We use these two measures to determine the probability that the trees are crossed over at a specific edge according to a chosen policy. Table 6.1 gives some ideas for possible policies.

Table 6.1
Different policies for crossover point selection with the homologous crossover operator. SMD: structurally most distinct; FMS: functionally most similar. P is the probability of crossover at edge k. n is a normalizing factor assuring that P is a probability.

Type	P(Crossover at edge k)
SMD	$D_S^N(k, imin(k))$
FMS	$1 - D_F^N(k, jmin(k))$
FMS/SMD	$D_S^N(k, imin(k))[1 - D_F^N(k, jmin(k))]/n$

To this end, values in Table 6.1 are interpreted as probabilities, and a roulette wheel is used to select one of the edges (subtrees) for crossover. Suppose we have selected k_s in parent 1 by this method. We then act either deterministically by taking the corresponding edge (subtree) $imin(k_s)$ or $jmin(k_s)$, respectively, from parent 2, or we proceed by selecting a subtree in parent 2 by employing the roulette wheel again. This way we have biased crossover probabilities by structural and functional features of the trees.

An analogous method can be used for linear programs. Here, we could even apply a much simpler method, based on position in the linear sequence only. There is a problem for graph-based systems, though, since we do not have a clear order of execution.

A GP 1-Point Crossover Operator

Recently, Poli and Langdon have formulated a new crossover operator for tree-based GP that has distinctly homologous overtones [Poli and Langdon, 1997b]. It is based on an analogous formulation of the one-point crossover of GAs for genetic programming. One-point crossover for a GA selects one point only in both parents to exchange genetic material at this point (see Figure 4.3). In the work of the authors, this selection process involves checking for structural similarity of the trees in order to find a corresponding point in the second parent, once it has been determined in the first. The experimental evidence so far achieved for this new operator [Poli and Langdon, 1997a] suggests that its behavior is in line with our expectations for a homologous crossover operator: destructive crossover diminishes over the course of generations. Note that Poli and Langdon have based their operator on finding structural homology alone.

6.7 Improving Crossover – The Tradeoffs

Tradeoffs

We have concluded that, in its present state, standard GP crossover acts mainly as a macromutation operator. Indeed, much of our discussion in this chapter has focused on how to improve crossover – how to make it more than a simple macromutation operator. All this seems to assume that simple macromutation is not enough. But it is important not to underestimate the power of a simple mutation operator. Orgel, Tuerk/Gold, and Bartel/Szostak's biological test tube evolution experiments demonstrate the power of simple population-based evolutionary search using *only* mutation, selection, and replication (see Chapter 2). GP crossover may do no more than replicate Orgel, Tuerk/Gold, and Bartel/Szostak's search techniques in digital form. By itself, this would suggest that GP is a powerful algorithm for program induction.

Digital Overhead and Homology

However, because of the claims of the schema theorem, we have expected more of crossover. All indications are that crossover can be improved substantially in both the quality and efficiency of the search it conducts. But there is a cost associated with improving crossover in GP. Each of the techniques we have discussed for enhancing crossover – brood recombination, explicitly defined introns, and so forth – may carry additional digital overhead such as less efficient use of memory and CPU time. This digital overhead may be likened to the large amount of biological energy expended to maintain homologous crossover in nature. Crossover that acts as something other than a macromutation operator does not come free – in biology or in GP. This suggests two important issues:

- Duplicating homologous crossover is probably well worth trying. Nature would not waste so much energy on homologous crossover unless it played an important role in evolutionary learning.
- We probably should not expect the benefits of homologous crossover at any lesser cost than is paid by nature.

Locating the Threshold

It remains to be seen whether we are best off with a minimal cost approach similar to Orgel's fast RNA, or Tuerk and Gold's SELEX algorithm from biology, or whether incurring the significant overhead of implementing more homologous GP crossover will put GP over a threshold that yields disproportionately large results. This threshold appears to exist in nature – natural evolution has incurred the overhead of inventing and maintaining species, peacock tails, and huge dysfunctional antlers on male deer as part of the process of maintaining homology in crossover. Whether we can find a way to cross that threshold in GP is one of the great unanswered questions in evolutionary algorithms.

6.8 Conclusion

The same arguments we have raised for the crossover operator might apply to the mutation operator as well. The mutation operator is stochastic. It certainly stands to benefit from improvements, for example, through smart mutation or other types of added mechanisms.

For the sake of the argument, we have concentrated here on the crossover operator because it is at the heart of standard genetic programming. It is clearly an imperfect operator in the current state of the art. However, recent developments suggest that it will become a much more powerful and robust operator over the next few years as researchers incorporate the approaches discussed in this chapter into their systems, combine the approaches, and devise new ways to improve the operator.

What is the future of the building block hypothesis? Ironically, one of the strongest arguments for the building block hypothesis is the manner in which a GP population adapts to the destructive effects of crossover. GP individuals tend to accumulate code that does nothing during a run – we refer to such code as introns. Recent experimental results strongly suggest that the buildup of introns is primarily an emergent response by a GP population to the destructive effects of crossover [Nordin et al., 1996] [Soule and Foster, 1997a].[9]

We will discuss this phenomenon at greater length in the following chapter, but here the important point is that the presence of introns underlines how important prevention of destructive crossover is in the GP system. Indications are strong that there is something valuable to protect from crossover – probably good building blocks.

[9]There has been some controversy about whether bloat was caused by a GP system defending itself against the destructive effect of crossover. Some researchers have reasoned that bloat could not be due to defense against crossover [Altenberg, 1994a]. Others have argued to the contrary [Angeline, 1996]. The empirical study referred to in the text suggests that Angeline is correct. There is a very strong correlation between bloat and the reduction of destructive crossover in GP systems.

So the challenge in GP for the next few years is to tame the crossover operator and to find the building blocks.

Exercises

1. What is the commonest effect of a crossover event in GP?
2. Give two methods for improving crossover in GP.
3. What is brood recombination?
4. Describe the crossover operator of a graph-based system like PADO.
5. In the text, we mention that the effect of crossover may be measured by pairing one child with one parent. How would you pair the parents of Table 6.2 with the children? How would you score the crossover event where the parents and the children had the following fitnesses (assume that higher is more fit)? You may wish to consult Figure 6.3 in justifying your decision.

Table 6.2
Fitness distribution for exercise 5

Parent 1	Parent 2	Child 1	Child 2
100	1000	500	900
900	1000	1400	900
900	1000	900	1000
900	1000	1000	900
900	1000	450	1100
900	1000	800	900

6. Prove to yourself that it would be better to have a system that uses natural selection *and* combines good building blocks than a system that uses only natural selection.
7. If you were trying to improve crossover in tree-based GP by using a higher-level learner as in [Zannoni and Reynolds, 1996] or [Teller, 1996], what kind of information would you give to the high-level learning system?
8. Devise an algorithm to let tree-based GP grow "one gene at a time." How would you keep track of the genes as the population evolved?
9. How would you introduce species to GP?
10. Design a conjugation operator for GP with linear genomes.
11. What effects could result from using GP conjugation instead of crossover?
12. Suggest a GP operator for transduction and transformation.

Further Reading

P.J. Angeline,
EVOLUTIONARY ALGORITHMS AND
EMERGENT INTELLIGENCE.
Ohio State University, Columbus, OH, 1995.

H. Iba, H. de Garis, and T. Sato,
RECOMBINATION GUIDANCE FOR
NUMERICAL GENETIC PROGRAMMING.
In 1995 IEEE CONFERENCE ON EVOLUTIONARY COMPUTATION.
IEEE Press, Piscataway, NJ, 1995.

U.-M. O'Reilly,
AN ANALYSIS OF GENETIC PROGRAMMING.
Ottawa-Carleton Institute for Computer Science, Ottawa, Ontario, Canada, 1995.

W.A. Tackett,
RECOMBINATION, SELECTION, AND THE GENETIC CONSTRUCTION OF COMPUTER PROGRAMS.
University of Southern California, Department of Electrical Engineering Systems, 1995.

7 Genetic Programming and Emergent Order

Contents

7.1 Introduction

The dynamics of a GP run are similar to the changes that take place during the $Q\beta$ replicase RNA experiments discussed in Chapter 2.[1] Both feature substantial variety in large populations at the outset of the evolutionary process. Both effect profound structural and functional changes in the population during the run, and both eventually stagnate. That is, both GP and $Q\beta$ replicase runs eventually reach a point where further evolution is impossible (see Figure 5.17).

Similarity between GP and DNA / RNA

Changes in a GP run reflect the fact that GP, like natural evolution, is a complex process. It has emergent properties that may not be obvious until observed in action. This chapter focuses on emergent properties arising from GP's freedom of representation of the problem space. GP programs share many features with DNA and RNA (see Chapters 2 and 5). All have variable length structures, all have elements that code for particular actions (functions and codons), and over time, these elements may be moved around or combined with other elements during evolution.

Representation of the Problem

The reader may recall from Chapter 1 that the GP problem representation is, theoretically, a superset of the representations of all other machine learning systems. This stems from both its variable length structure and its freedom of choice of functions and terminals – if a computer can do it, GP can use it in its representation.

GP's enormous freedom of representation is a mixed blessing. With such a huge search space, an algorithm might have to search for a long time. There are benefits to narrowing the search space as long as the researcher has grounds for believing that the answer to the problem lies somewhere in that narrow space. But the price of narrowing the search space is that the problem representation cannot evolve outside this range if the solution does not lie there.

In this chapter, we will look closely at two important emergent properties of GP:

- GP's ability to search the space of the problem representation.
- The problem (or promise) of *introns* or *bloat*.

We group these issues together because both present important and unresolved questions for the future of GP and both appear to emerge in GP runs as a result of one of the most distinctive features of GP – variable length genotypes.[2]

[1] With one important exception: The selection pressure is much more constant in conventional GP systems than in the $Q\beta$ experiments.

[2] Angeline has argued correctly that the distinction between GP and fixed-length GAs is not a real distinction [Angeline, 1994]. That is, every

7.2 Evolution of Structure and Variable Length Genomes

The capability to evolve a representation of the problem depends on the ability of a learning algorithm to modify the structure of its own solutions. We shall illustrate this point with a simple gedanken experiment about how *evolution of representation* is possible with variable length genotypes. We contrast that result with fixed length genotypes.

Fixed Length Genotypes

Typical fixed length evolutionary algorithms (EA) come in many flavors. They may optimize a *fixed length vector* of real numbers subject to constraints (evolutionary programming or evolution strategies). Or they may evolve fixed length bit strings (as in a standard genetic algorithm). As we will see, such fixed length structures have little capacity for evolution of their genotype (their structure) because the length and meaning of each element have been determined in advance.

Variable Length Genotypes

By evolving structure, a variable length genotype may be able to learn not only the parameters of the solution, but also how many parameters there should be, what they mean, and how they interrelate. This introduces many degrees of freedom into the evolutionary search that are missing in fixed length structures.

A Gedanken Experiment

Let us look now at how two very simple EAs – one fixed and one variable in length – would evolve a rod of a certain target length, say, $L_t = 9$ cm. In this very simple form, the problem could be solved using very simple fixed length parameter optimization. The genome would consist of just one gene (a fixed length of 1), which coded for the "length" of the rod. An evolutionary search minimizing the deviation of the actual length $d = (L - L_t)^n$ with $n > 0$ would solve the problem easily.

Let us make the task more difficult. Now, the target rod must be assembled from other rods of three different lengths, say,

$$l_1 = 4 \text{ cm} \tag{7.1}$$

$$l_2 = 2 \text{ cm} \tag{7.2}$$

$$l_3 = 1 \text{ cm} \tag{7.3}$$

GP system sets some maximum size on its representation. So in that sense, it has a fixed length. Likewise, any GA system could set its fixed length to a very high value and then allow the GA to be effectively shorter, as is exemplified in [Wineberg and Oppacher, 1996]. Nevertheless, as a practical matter, GAs rarely assume this form and GP almost always does. So when we refer to GP's "unusual" property of being variable in length, it is a practical argument.

A rod length L of varying size could be generated by putting together a number N of these elements in arbitrary sequence:

$$L = \sum_{i=1}^{N} l^{(i)} \tag{7.4}$$

where each element $l^{(i)}$ is chosen from one of the available lengths

$$l^{(i)} \in \{l_1, l_2, l_3\}, i = 1, \ldots, N \tag{7.5}$$

Now the problem is strongly constrained because the EA may only use "quantized" pieces of rod. A fixed length genome may assemble only a fixed number of rods into the target rod. The quality of such a fixed length system is entirely dependent on what value of N the researcher picks. Table 7.1 shows how the choice of N affects the probable success of the search.

Table 7.1 gives the number of correct solutions possible for different genome sizes, together with examples for each size. There is no

Table 7.1
Solutions to rod assembly problem, $L_t = 9$, given copies of three elementary pieces $l_1 = 1, l_2 = 2, l_3 = 4$

Genome size	# perfect solutions	Sample solution
2	0	-
3	3	4 4 1
4	12	4 2 2 1
5	20	4 2 1 1 1
	5	2 2 2 2 1
6	6	4 1 1 1 1 1
	120	2 2 2 1 1 1
7	42	2 2 1 1 1 1 1
8	8	2 1 1 1 1 1 1 1
9	1	1 1 1 1 1 1 1 1 1
10	0	-

solution where $N < 3$ or $N > 9$. Further, if one chooses $N = 5$, the system can find only a subset of all possible solutions. This subset might even be disconnected in the sense that there is no smooth path through the search space on the way to a perfect solution.[3]

[3]One possible fixed length genome coding that would work in this particular case would be to have three genes, each a natural number that defines how many of one of the rod lengths are to be used in a solution. This cleverly allows the genotype to stay fixed but the phenotype to change. This approach depends on the researcher knowing the problem domain well and using a clever device to solve the problem. It would work only as long as the number of pieces to be assembled remains constant. But what if it

Of course, if a variable length solution were used to represent the problem in Table 7.1, the EA would have access to every possible solution in the table and would not be dependent on the researcher's up-front choice of N. Thus, our gedanken experiment with the rods and the variable length genome may be seen in one aspect as very similar to the $Q\beta$ replicase RNA experiments discussed in Chapter 2. The original RNA template was over 4000 base pairs long. If the RNA had not been able to change its very structure, it would never have found the fast RNA solution, which was only 218 base pairs long. The variable length genotype is perhaps genetic programming's most radical practical innovation compared to its EA roots.

7.3 Iteration, Selection, and Variable Length Program Structures

Until now, our argument may have appeared to assume that variable length solutions, by themselves, have some magical property of changing their own structure. In reality, this property emerges only when the variable length structure is part of a larger dynamic of *iteration* and *selection*. That is, the unique properties of variable length structures appear in an evolutionary context. This result is perhaps not surprising. DNA itself varies greatly in length and structure from species to species. Even the $Q\beta$ replicase RNA experiments involved the evolution of variable length structures.

The Essence of Evolution

The essence of evolution might be subsumed as being (1) iterative insofar as generation after generation of populations are assigned reproduction opportunities; and (2) selective, insofar as the better performing variants get a better chance to use these opportunities.

It is thus the interplay between iteration and selection that makes evolution possible. Dawkins has called this aspect *cumulative selection* [Dawkins, 1987]. It is cumulative because the effects of selection acting on one generation are inherited by the next. The process does not have to start again and again under random conditions. Whatever has been achieved up to the present generation will be a starting point for the variants of the next.

So far we have dealt with abstract notions about variable length structures, problem representation, iteration, and emergent properties. Now it is time to look at how these notions have important prac-

---

turns out that for some reason it would be good to form another piece of, say, length $l_4 = 5$ cm by combining two original pieces together? There would be no gene for such a piece in the genome just introduced; therefore, no modules could be formed that would by themselves constitute further elements for construction of rods.

tical implications in GP. Simply put, the emergent GP phenomena of (1) evolvable problem representations and (2) introns are among the most interesting and troubling results of combining variable length structures with iterative selection. The remainder of this chapter will be devoted to these two issues.

7.4 Evolvable Representations

The reader will recall that the *problem representation* is one of the crucial defining facts of a machine learning system (see Chapter 1). Most ML paradigms are based upon a fairly constrained problem representation – Boolean, threshold, decision trees, case-based representations and so forth. Constraints sometimes have the advantage of making the traversal of the solution space more tractable – as long as the solution space is well tailored to the problem domain. By way of contrast, GP may evolve any solution that can be calculated using a Turing complete language. Therefore, GP search space includes not only the problem space but also the space of the representation of the problem. Thus, it may be that GP can evolve its own problem representations.[4]

7.4.1 Ignoring Operators or Terminals

A simple example is illustrative. Suppose a GP system had the following functions to work with: `Plus`, `Minus`, `Times`, `Divide`. GP can change this representation by ignoring any of these functions, thereby reducing the function set. Why should this happen? When iteration and selection are applied to variable length structures, the system should magnify the exploration of sections of the representation space that produce better results [Eigen, 1992]. So if solutions that use the divide operator were, in general, producing worse results than others, we could expect the system to reduce and, eventually, to eliminate the divide operator from the population.

7.4.2 Finding Solutions of the Correct Length

Elimination of operators is not the only way by which GP may evolve representations. Let us go back to the example above about evolving a rod of a particular length. There is an area of the search space

[4]This is not to say that GP always discovers good representations. There is some evidence that overloading a GP system with functions can cause the system to stick in a local minimum [Koza, 1992d]. Nevertheless, unlike other machine learning paradigms, it is *able* to evolve representations.

between a genome size of three and eight that is a better area to check than, say, a genome size of twenty. The multiplicative effect of iteration, selection, and variable length solutions should draw the search away from genomes that have a size of twenty very quickly and should focus it on the promising area.

Where we do not know the size of the best solution to a problem in advance, the ability of GP to examine different size solutions is very important. If limited to a fixed length, a solution may be too big or too small. Both situations cause problems. If the largest permitted solution is less than the optimal solution size, a fixed length system simply lacks the firepower to find the best answer. On the other hand, if a fixed length solution is too long, it may not generalize well (see Chapter 8). Genetic programming can find a short or a long solution where a fixed length representation cannot.

7.4.3 Modularization and Meta-Learning

The capability of GP to search the space of the problem representation is an area where research has only begun. Modularization research (see Chapter 10) and Koza's automatically defined functions are two promising directions in this regard.

Another very exciting approach to this issue in GP is a *meta-learning* approach. Meta-learning occurs when information about the problem representation from one GP run is used to bias the search in later GP runs. In 1995, Whigham designed an *evolvable context-free grammar*. The grammar defined what types of programs could be created by the GP system on initialization. His system was designed to capture aspects of the function and terminal set that were associated with successful individuals during a run and to use that information to bias the grammar used in future runs. His results were very encouraging [Whigham, 1995a]. This work is probably the best example to date of how GP's flexibility may be used explicitly to evolve problem representations. The authors regard this direction as one of the most promising areas for future GP research.

While variable length solutions have potentially great advantages, they also appear to be the cause of what may be troubling emergent properties in GP runs called, variously, *introns* or *bloat*.

7.5 The Emergence of Introns, Junk DNA, and Bloat

In 1994, Angeline noted that many of the evolved solutions in Koza's book contained code segments that were extraneous. By extraneous, he meant that if those code segments were removed from the solution, this would not alter the result produced by the solution. Examples of such code would be

```
a = a + 0
b = b * 1
```

Angeline noted that this extra code seemed to emerge spontaneously from the process of evolution as a result of the variable length of GP structures and that this emergent property may be important to successful evolution [Angeline, 1994].

GP Intron Terminology

Angeline was the first GP researcher to associate this emergent "extra code" in GP with the concept of biological introns. In Chapter 2, we briefly discussed biological introns. Questions may be raised as to the appropriateness of the analogy between this extra code and biological introns, as we will note below. Nevertheless, the use of the term *intron* to refer to this extra code has become widespread in the GP community and we will, therefore, use the term *introns* or *GP introns* to refer to the extra code in GP that was first identified by Angeline in 1994.

GP Bloat

Also in 1994, Tackett observed that GP runs tend to "bloat." That is, the evolved individuals apparently grow uncontrollably until they reach the maximum tree depth allowed. Tackett hypothesized that GP bloat was caused by blocks of code in GP individuals that, while they had little merit on their own, happened to be in close proximity to high fitness blocks of code. Tackett referred to this phenomenon as hitchhiking [Tackett, 1994].

The Persistence of Introns

A body of research since 1994 has meanwhile established that GP bloat is, in reality, caused by GP introns [McPhee and Miller, 1995] [Nordin and Banzhaf, 1995a] [Nordin et al., 1995] [Soule et al., 1996] [Nordin et al., 1996] [Soule and Foster, 1997b] [Rosca, 1997]. Though Angeline saw the extraneous code as an "occasional occurrence," subsequent research has revealed that introns are a persistent and problematic part of the GP process. For example, studies in 1995 and 1996 suggest that in the early to middle sections of GP runs, introns comprise between 40% and 60% of *all* code. Later in the run, emergent GP introns tend to grow exponentially and to comprise almost all of the code in an entire population [Nordin et al., 1995] [Nordin et al., 1996] [Soule et al., 1996] [Soule and Foster, 1997b] [Rosca, 1997]. The evidence is strong that evolution selects for the existence of GP introns.

Bloat and Run Stagnation

Bloat is a serious problem in GP. Once the exponential growth of introns associated with bloat occurs, a GP run almost always stagnates – it is unable to undergo any further evolution [Tackett, 1994] [Nordin et al., 1995] [Nordin et al., 1996]. On the other hand, Angeline suggested in his seminal article that emergent introns may have a beneficial effect on evolution in GP. Was Angeline wrong or are there two different but closely related phenomena at work here?

It may well be that during the early and middle part of a run, introns have beneficial effects. For example, theory and some experimental evidence suggests introns may make it more likely that good building blocks will be able to protect themselves against the damaging effects of crossover [Angeline, 1996] [Nordin et al., 1995] [Nordin et al., 1996]. On the other hand, the exponential growth of introns (bloat) at the end of the run is probably deleterious.

Overview of Remainder of Chapter

The remainder of this chapter will be devoted to the major issues in GP regarding introns. In brief summary, they are as follows:

- How valid is the analogy between GP introns and biological introns? Can we learn anything from biology about GP introns?
- How should the term *intron* be defined in GP?
- What causes the emergence of GP introns?
- What are the relative costs and benefits of introns in GP runs?
- How may GP researchers deal with bloat?

With this introduction in hand, it is now possible to move on to a detailed analysis of the phenomenon of introns.

7.5.1 What Can We Learn from Biological Introns?

In Chapter 2, we briefly discussed introns, exons, and junk DNA. In a greatly oversimplified overview, DNA is comprised of alternating sequences of:

- Base pairs that have no apparent effect on the organism called junk DNA;
- Base pairs that exert control over the timing and conditions of protein production, often referred to as control segments; and
- Base pairs that comprise genes. Genes are the active DNA sequences responsible for manufacturing proteins and polypeptides (protein fragments).

The gene sequences themselves may be further divided into:

- Exons, base sequences that actively manufacture proteins or polypeptides; and

- Introns, base sequences that are not involved in the manufacture of proteins.

It is tempting to conclude that, because biological introns are not involved in protein manufacture, they must be the biological equivalent of the GP introns. They are there but they do nothing. While there are undoubted similarities between GP introns and biological introns, there are also real differences that must be taken into account when making analogies between the two. The following are some of the more important similarities and differences.

No Direct Effect on the Genotype

A gene produces a protein by first transcribing an mRNA (messenger RNA) molecule that is effectively a copy of the base sequence of the gene. Then the mRNA molecule is translated into a protein or protein fragment. What happens to the introns during this process is fascinating. Each intron in a gene has a base pair sequence at the beginning and end of the intron that identifies it as an intron. Before the mRNA is translated to a protein, all of the intron sequences are neatly snipped out of the translating sequence. In other words, biological introns do not directly translate into proteins. Therefore, they do not directly translate into the phenotype of the organism. In this sense they are quite similar to GP introns. The "extraneous code" in GP introns does not affect the behavior of a GP individual at all. Recall that best analogy in GP to the biological phenotype is the *behavior* of the GP individual. Thus, neither biological nor GP introns directly translate to their respective phenotypes.

Protection Role of GP and Biological Introns

We will describe in detail, later, the theoretical and experimental results that suggest that GP introns may play some role in protecting good building blocks against destructive crossover. Watson describes a similar role for biological introns:

> [S]everal aspects of the structures of interrupted genes hint that the presence of introns (coupled with the ability to excise them at the RNA level) could have been used to advantage during the evolution of higher eucaryotic genomes.
>
> Since exons can correspond to separate structural or functional domains of polypeptide chains, mixing strategies may have been employed during evolution to create new proteins. Having individual protein-coding segments separated by lengthy intron sequences spreads a eucaryotic gene out over a much longer stretch of DNA than the comparable bacterial gene. This will simultaneously increase the rate of recombination between one gene and another and also lower the probability that the recombinant joint will fall within an exon and create some unacceptable aberrant structure in the new protein. Instead, recombination will most likely take

place between intron sequences, thereby generating new combinations of independently folded protein domains.

WATSON ET AL., 1987

Note that this is very much like our later description of the possible "structural" effect of introns in the early and middle part of GP runs. In their structural role, introns would serve to separate good blocks of code and to direct the crossover operator to portions of the genome where crossover will not disrupt the good blocks.

On the other hand, there is no biological analogy to the runaway, exponential growth of introns at the end of GP runs that is referred to as bloat. So while it may be valid to draw the analogy between GP introns that serve a structural role and biological introns, the analogy is strained at best when it comes to bloat.

Introns Affect the Survivability of the Organism Only Indirectly

As we discussed in Chapter 2, the existence and contents of biological introns are correlated with significant effects on both the amount and the quality of the proteins expressed by the gene in which the introns occur. This is true despite the fact that introns have no *direct* effect on protein manufacture. The mechanism of the indirect effect is not known. Nevertheless, we can conclude that biological introns do indirectly affect the survivability of the organism and its genetic material.

GP introns do not have any known effect on the survivability of the GP individual, direct or indirect. However, they probably do affect the survivability of the offspring of a GP individual quite profoundly (see below). Thus, GP introns do have an indirect effect on the survivability of the *genetic material* of an individual.

Conclusion

Notwithstanding that both biological and GP introns have an indirect effect on survivability, the nature of the effect is different. Accordingly, while there are similarities between biological introns and GP introns, the differences are substantial and must be taken into account when reasoning from one domain to the other.

7.6 Introns in GP Defined

In Chapters 2 and 5, we discussed how GP frequently does not distinguish between genotype and phenotype structures. In that case, the GP program may be regarded as the genotype and its *behavior* (that is, the state transitions it causes in the computer) may properly be regarded as the phenotype [Maynard-Smith, 1994] [Angeline, 1994]. On the other hand, many GP and GA systems feature an explicit distinction between the genotype structure and the phenotype structure. In that case, the phenotype's *structure* is, by analogy with biology, properly regarded as *part* of the phenotype. But the behavior of the entity would be part of the phenotype in this situation also.

Defining a GP intron becomes rather tricky when one must be consistent with the biological analogy upon which the theory of introns is based. Recall that the strict biological definition of an intron is a sequence of base pairs that is clipped out before the maturing of messenger RNA. More generally, since an organism may survive only by structure or behavior, an intron is distinctive primarily as a sequence of the DNA that affects the chances of the organism's survival only indirectly.

Introns Defined

From the above argument, two features may be used to define introns:

- An intron is a feature of the genotype that emerges from the process of the evolution of variable length structures; and
- An intron does not directly affect the survivability of the GP individual.

Under the above definition, the following S-expressions would be examples of introns:

(NOT (NOT X)),
(AND ... (OR X X)),
(+ ... (- X X)),
(+ X 0),
(* X 1),
(* ... (DIV X X)),
(MOVE-LEFT MOVE-RIGHT),
(IF (2 = 1) ... X),
(SET A A)

By this definition, we draw very heavily on Angeline's 1994 concepts, which also focused on the very interesting dual aspect of introns. Introns are emergent *and* they do not directly affect the fitness of the individual [Angeline, 1994]. In GP, this means that any code that emerges from a GP run and does not affect the fitness of the individual may properly be regarded as an intron.

By this definition, we also ignore the details from biology of how an intron manages not to be expressed (i.e., the intron is snipped out before the maturing of mRNA). In short, we regard the fact that biological introns are not expressed as proteins to be more important than the manner in which that occurs. The importance of this distinction will become clear in the next section.

Intron Equivalents

In Chapter 6, we noted that many researchers have used various techniques to insert "artificial" introns into their GP systems. While both emergent and artificially inserted introns may behave in similar ways, the distinction is important. In this book, when we use the term *intron*, we mean an *emergent* intron as defined above. When we refer to artificially inserted introns, we shall use the term *artificial intron equivalents*. Therefore, the insertion into the genome of explicitly defined introns by Nordin et al. and Angeline, of introns by Wineberg et al., and of write instructions that have no effect by Andre et al. may all be regarded as different ways of inserting artificial intron equivalents into the genome. Unlike introns, artificial intron equivalents do not emerge from the evolutionary process itself.

Global vs. Local Introns

One final distinction: we shall call an intron *global* if it is an intron for every valid input to the program, and we call it *local* if it acts as an intron only for the current fitness cases and not necessarily for other valid inputs.

7.7 Why GP Introns Emerge

By definition, introns have no effect on the fitness of a GP individual. We would not expect to see strong selection pressure to create any genomic structure that does not affect the fitness of an individual. So why do introns emerge? The short answer is that, while introns do not affect the fitness of the individual, they do affect the likelihood that the individual's descendants will survive. We shall refer to this new concept of fitness, which includes the survivability of an individual's offspring, as *effective fitness*.

The effective fitness of an individual is a function not only of how fit the individual is now but also of how fit the individual's children are likely to be. By this view, the ability of an individual to have high-fitness children (given the existing genetic operators) is as important to the continued propagation of the individual's genes through the population as is its ability to be selected for crossover or mutation in the first place. It does no good for an individual to have high fitness and to be selected for crossover or mutation if the children thereby produced are very low in fitness. Thus, we would expect individuals to compete with each other to be able to have high-fitness children.

Destructive Genetic Operators

The reader will recall that the normal effect of crossover is that the children are much less fit than the parents; see Chapter 6. So too is the normal effect of mutation. Any parent that can ameliorate either of these effects even a little will have an evolutionary advantage over other parents. And this is where introns come in.

Simply put, the theoretical and experimental evidence today supports the hypothesis that introns emerge principally in response to the frequently destructive effects of genetic operators [Nordin et al., 1995]

[Nordin et al., 1996] [Soule et al., 1996] [Soule and Foster, 1997b] [Rosca, 1997]. The better the parent can protect its children from being the results of destructive genetic operators, the better the *effective fitness* of the parent. Introns help parents do that.

Effective Fitness

This concept of *effective fitness* will be very important later. It is clearly a function of at least two factors:

1. The fitness of the parent. The fitter the parent, the more likely it is to be chosen for reproduction.

2. The likelihood that genetic operators will affect the fitness of the parent's children.

Introns emerge as a result of competition among parents with respect to the second item.

An Important Caveat

It is important to note what we are *not* saying. We do not contend that destructive crossover is always bad or that it is always good that children have fitness as good or better than their parents [Andre and Teller, 1996]. In fact, such a rule would reduce GP to simple hill climbing. Rather, we are saying that destructive genetic operators are the principal cause of introns. Therefore, one should not take our use of the term *constructive crossover* to mean that crossover should always be constructive. Nor should one understand from the term *destructive crossover* that it should never happen. Evolution is more complex than that. Thus, whether introns are beneficial, detrimental, or both in the GP evolution process has nothing to do with the question of why introns emerge.

7.8 Effective Fitness and Crossover

It is possible to derive a formula that describes the effective fitness of an individual. That formula predicts that there will be selection pressure for introns. Before deriving this formula, it is necessary to define the effective and absolute complexity of a program.

Absolute vs. Effective Complexity

By the *complexity* of a program or program block we mean the length or size of the program measured with a method that is natural for a particular representation. For a tree representation, this could be the number of nodes in the block. For the binary string representation, it could be, e.g., the number of bits. The *absolute length* or *absolute complexity* of a program or a block of code is the total size of the program or block. The *effective length* or *effective complexity* of a program or block of code is the length of the active parts of the code within the program or block, in contrast to the intron parts. That is, the active parts of the code are the elements that affect the fitness of an individual.

In deriving the effective fitness of the individual, we must be careful to distinguish between effective and absolute complexity. We may start deriving the effective fitness of an individual by formulating an equation that resembles the schema theorem [Holland, 1992] for the relationship between the entities described above.

The following definition will be useful:

A Few Definitions and a Statement

- Let C_j^e be the effective complexity of program j, and C_j^a its absolute complexity.
- Let p_c be the standard GP parameter for the probability of crossover at the individual level.
- The probability that a crossover in an *active block* of program j will lead to a worse fitness for the individual is the probability of destructive crossover, p_j^d. By definition p_j^d of an absolute intron[5] is zero.
- Let f_j be the fitness[6] of the individual and $\overline{f}^t$ be the average fitness of the population in the current generation.

Using fitness-proportionate selection[7] and block exchange crossover, for any program j, the average proportion P_j^{t+1} of this program in the next generation is:

$$P_j^{t+1} \approx P_j^t \cdot \frac{f_j}{\overline{f}^t} \cdot \left(1 - p_c \cdot \frac{C_j^e}{C_j^a} \cdot p_j^d\right) \quad (7.6)$$

In short, equation (7.6) states that the proportion of copies of a program in the next generation is the proportion produced by the selection operator minus the proportion of programs destroyed by crossover. Some of the individuals counted in P_j^{t+1} might be modified by a crossover in the absolute intron part, but they are included because they still show the same behavior at the phenotype level. The proportion P_j^{t+1} is a conservative measure because the individual j might be recreated by crossover with other individuals.[8]

Equation (7.6) can be rewritten as:

[5]An absolute intron is defined as neither having an effect on the output nor being affected by crossover.

[6]Notice that this is not standardized fitness used in GP. Here a better fitness gives a higher fitness value (GA).

[7]The reasoning is analogous for many other selection methods.

[8]The event of recreating individuals can be measured to be low except when applying a very high external parsimony pressure that forces the population to collapse into a population of short building blocks.

$$P_j^{t+1} \approx \left(\frac{f_j - p_c \cdot f_j \cdot p_j^d \cdot C_j^e / C_j^a}{\overline{f}^t} \right) \cdot P_j^t \tag{7.7}$$

We may interpret the crossover-related term as a direct subtraction from the fitness in an expression for reproduction through selection. In other words, reproduction by selection *and* crossover acts as reproduction by selection *only*, if the fitness is adjusted by the term:

$$p_c \cdot f_j \cdot \frac{C_j^e}{C_j^a} \cdot p_j^d \tag{7.8}$$

Effective Fitness, Formal

Term (7.8) can be regarded as a fitness term proportional to program complexity. Hence we define *effective fitness* f_j^e as:

$$f_j^e = f_j - p_c \cdot f_j \cdot \frac{C_j^e}{C_j^a} \cdot p_j^d \tag{7.9}$$

The effective fitness of a parent individual, therefore, measures how many children of that parent are likely to be chosen for reproduction in the next generation.[9] A parent can increase its effective fitness by lowering its effective complexity (that is, having its functional code become more parsimonious) or by increasing its absolute complexity or both. Either reduces the relative target area of functioning code that may be damaged by crossover. Either has the effect of increasing the probability that the children of that parent will inherit the good genes of the parent intact. In other words, the difference between effective fitness and actual fitness measures the extent to which the destructive effect of genetic operators is warping the real fitness function away from the fitness function desired by the researcher.

Rooted-Tree Schemata

Recently, Rosca has derived a version of the schema theorem for "rooted-tree schemas" that applies to GP. A *rooted-tree schema* is a subset of the set of program trees that match an identical tree fragment which includes the tree root. The *rooted-tree schema theorem* also predicts that an individual may increase the likelihood that its genetic material will survive intact in future generations by either increasing the fitness of the individual or increasing the size of the individual [Rosca, 1997].

7.9 Effective Fitness and Other Operators

The above equations could easily be refined to quantify the effect of the destructive quality of any genetic operator simply by replacing the term for the probability of destructive crossover with, say, a new

[9] This assumes $\overline{f}^t \approx \overline{f}^{t+1}$.

term for the probability of destructive mutation. Since the effect of mutation is normally quite destructive, this model would predict that some form of intron should emerge in response to mutation. In fact, that is exactly what happens.

A Mutation-Resistant Intron

An example of such an intron we found in great numbers in AIMGP when increasing the probability of mutation is:

$$Register1 = Register2 >> Register3 \qquad (7.10)$$

Shift-right (>>) is effectively a division by powers of 2. Here, mutations that change argument registers or the content of argument registers are less likely to have any effect on fitness. By way of contrast, an intron like

$$Register1 = Register1 + 0 \qquad (7.11)$$

more typical for a high crossover probability run would very likely be changed into effective code by a mutation. In short, a type 7.10 intron may be relatively more resistant to mutation than a type 7.11 intron. Interestingly, these new introns are also resistant to crossover.

Solutions Defend Themselves

We can suspect a general tendency of the GP system to protect itself against the attack of operators of *any* kind. When it is no longer probable that fitness will improve, it becomes more and more important for individuals to be protected against destructive effects of operators, regardless of whether these are crossover, mutation, or other kinds of operators.

Equation 7.9 could thus be generalized to include the effects of other operators as well:

$$f_j^e = f_j[1 - \frac{1}{C_j^a} \sum_r p_r \cdot p_j^{r,d} \cdot C_j^{r,e}] \qquad (7.12)$$

where r runs over all operators trying to change an individual, p_r is the probability of its application, $p_j^{r,d}$ is the probability of it being destructive, and $C_j^{r,e}$ is the effective complexity for operator r.

7.10 Why Introns Grow Exponentially

Introns almost always grow exponentially toward the end of a run. We believe that the reason is that introns can provide very effective *global protection* against destructive crossover. By that, we mean that the protection is global to the entire individual. This happens because at the end of a run, individuals are at or close to their best performance. It is difficult for them to improve their fitness by solving the problem better. Instead, their best strategy for survival changes.

Their strategy becomes to prevent destructive genetic operators from disrupting the good solutions already found.

No End to Intron Growth

One can reformulate equation 7.12 as follows:

$$f_j^e = f_j \cdot [1 - \sum_r p_j^{D,r} \cdot \left(1 - (C_j^{i,r}/C_j^a)\right)] \qquad (7.13)$$

where $p_j^{D,r}$ now lumps together the probability both of application and of destructiveness of an operator r, and $C_j^{i,r}$ is the corresponding intron complexity. When fitness (f_j) is already high, the probability of improving effective fitness by changing actual fitness is much lower than at the beginning of the run. But an individual can continue to increase its effective fitness, even late in a run, by increasing the number of introns ($C_j^{i,r}$) against r in its structure.

Further, there is no end to the predicted growth of introns. Because the number of introns in an individual is always less than the absolute size of an individual, the ratio (supressing index r) C_j^i/C_j^a is always less than one. So introns could grow infinitely and continue to have some effect on the effective fitness of an individual as long as $p^D > 0$.

Here are two pictures of how that works in practice. Figure 7.1 represents a good solution of functioning (non-intron) code that has no introns. If crossover and mutation are destructive with high probability, then the individual in Figure 7.1 is very unlikely to have highly fit children. In other words, its effective fitness is low.

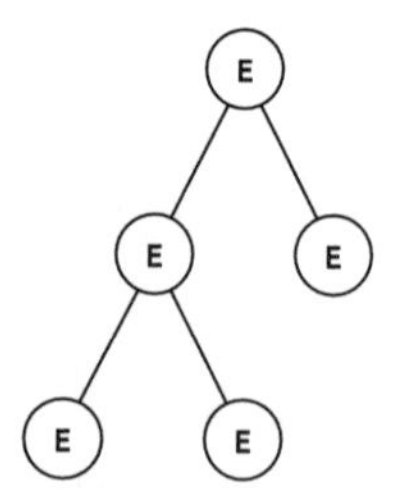

Figure 7.1
A group of functioning nodes. E indicates "effective" code.

By way of contrast, Figure 7.2 shows *the same* functioning code. But the individual in Figure 7.2 is attached to fourteen introns. Altogether, there are four crossover points that would result in breaking up the functioning code in the individual and fourteen crossover points that would not. This individual's effective fitness is, therefore, relatively higher than the effective fitness of the individual in Figure 7.1. That is, the individual in Figure 7.1 is likely to have children

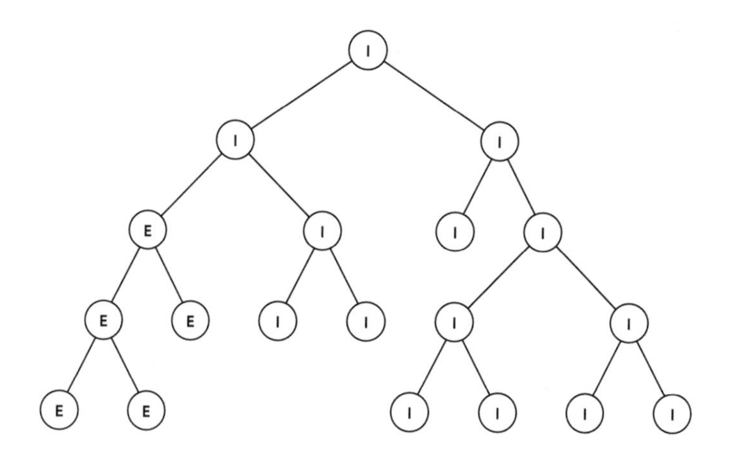

Figure 7.2
Same functioning nodes as in Figure 7.1. Introns (I) have been added.

that will not compete well with the children of Figure 7.2, even if their regular GP fitness is identical.

The Effect of Exponential Intron Growth

The effect of explosive intron growth is dramatic. Normally, crossover is destructive more than 75% of the time. But after introns occupy most of the population, destructive crossover is replaced almost completely with neutral crossover. Why? At that point, the individuals are just swapping introns with higher and higher probability each generation. Swapping code that has no effect between two individuals will have no effect. Hence neutral crossover comes to dominate a run after the explosive growth of introns.

A Prediction

We have presented a theory about why introns grow exponentially. The theory has a good theoretical basis and also some growing experimental support [Nordin et al., 1995] [Nordin et al., 1996] [Soule and Foster, 1997b]. One additional prediction of this theory is that exponential growth will come mostly or entirely from introns or groups of introns that are, in effect, terminals. This is because introns that occur in the middle of blocks of functioning code are unlikely to result in neutral crossover if crossover occurs in the introns – because there is functioning code on either side of the introns and it will be exchanged among individuals. Soule's research supports this prediction – bloat appears to grow at the end of GP trees [Soule and Foster, 1997b].

7.11 The Effects of Introns

Various effects have been attributed to introns [Angeline, 1994] [Nordin et al., 1995] [Andre and Teller, 1996] [Nordin et al., 1996] [Soule et al., 1996] [Wineberg and Oppacher, 1996]. Some have argued that introns may benefit evolution while others argue that introns almost always result in poor evolution and extended computation time. This field is very young and it may be that all of these studies, some seemingly inconsistent, just represent different points on the very complex surface that represents the effect of introns. Much more study is necessary before anything more than tentative conclusions may be stated. Accordingly, this section will be devoted more toward outlining the few effects of introns that may be clearly stated and, as to the other effects, trying to provide a clear outline of the unanswered issues.

We start by noting again that introns are an emergent phenomenon and probably exhibit a good deal of complexity and sensitivity to initial conditions. It is important, in outlining these issues, to keep several points clear:

- **Introns may have differing effects before and after exponential growth of introns begins.** After exponential growth occurs, the exponential effect surely overwhelms whatever effect the introns had previously, if any.

- **Different systems may generate different types of introns with different probabilities.** It may, therefore, be harder to generate introns in some GP systems than in others [McPhee and Miller, 1995].

- **The extent to which genetic operators are destructive in their effect is likely to be a very important initial condition in intron growth.** Equation 7.12 and Rosca's work [Rosca, 1997] predict that the maximum extent to which intron growth can modify effective fitness is equal or proportional to p^D (the probability of destructive crossover). Furthermore, the effect of any given change in the ratio C^i/C^a on effective fitness will be proportional to p^D. This underlines the importance of measuring and reporting on destructive, neutral, and constructive crossover figures when doing intron research (see Chapter 6).

- **Mutation and crossover may affect different types of introns differently.** How this works may depend on the system, the function set and the mix of genetic operators.

  - A function set that uses no conditionals can generate introns ($a = a + 0$). Such introns will be changed, with high probability, into functioning code by the mutation operator [Banzhaf et al., 1996]. Adding more mutation will, therefore, change the initial conditions of the stew.
  - On the other hand, adding conditional branching creates the possibility of such introns as `If 2 < 1 then X`. If X represents a subtree of any length, both X and each node of X would be introns. However, unlike the example in the prior paragraph, X type introns will be quite immune to mutation within X. Again the initial conditions are changed.

- Finally, it is important to distinguish between emergent introns and artifical intron equivalents. In most systems, the existence and growth of artifical intron equivalents is more or less free to the system – a gift from the programmer so to speak. This may well make artificial intron equivalents much more likely to engage in exponential growth than emergent introns.

With this in mind, here are some reflections on the possible benefits and drawback of introns.

7.11.1 Problems Caused by Introns

Researchers have identified possible problems caused by introns, including run stagnation, poor results, and a heavy drain on memory and CPU time.

Run Stagnation and Bloat

Run stagnation appears to be, in part, a result of bloat – the exponential growth of introns. This effect is not something for which the researcher should strive. Once the population is swapping only introns during crossover, no improvement is likely to come in the best or any other individual. All effective growth has ended.

One might argue that exponential intron growth is not really important because it merely reflects that the run is over. That is, the argument goes, individuals in the run cannot improve, given the problem and the state of the population. That introns explode at this point is therefore of no consequence. While this may be partly true, stagnation due to intron growth is probably the result of a *balance* between the probability of destructive crossover and the extent to which changes in effective fitness may be more easily found by finding better solutions or by adding introns (see equation 7.12). Of course, it is more difficult to find better solutions after quite some evolution had already occurred. But that does not mean that it is impossible or even improbable, just less likely. Thus, it is quite possible that the ease of improving effective fitness by adding introns just overwhelms the slower real evolution that occurs toward the end of a run. If this hypothesis is correct, then the exponential growth of introns would prevent the population from finding better solutions. This possibility, of course, becomes greater and greater as the destructive effects of crossover become greater.

One study suggests that effective evolution may be extended (and stagnation averted for at least a while) by increasing the mutation rate in a system where mutation converts introns to functional code with high probability. In the high mutation runs, the duration of the runs was extended significantly before stagnation, the number of introns in the population was reduced, and better overall solutions were found [Banzhaf et al., 1996].

Poor Results

Researchers [Andre and Teller, 1996] [Nordin et al., 1995] reporting poor best individual fitness results have attributed those results to introns. Andre et al. reported very bad results when they added a form of artificial intron equivalents to their system. They added nodes that were deliberately designed to do nothing but take up space in the code. Nordin et al. added explicitly defined introns (EDI) (another type of artificial intron equivalents), which allowed the probability of crossover to be changed at each potential crossover point. We reported good results up to a point but when it became too easy for the system to increase the EDI values, evolution tended to find local minima very quickly.

It is difficult to assess how general either of these results is. One possible explanation for the results is suggested by the argument above regarding balance. Both studies made it very easy to add introns in the experiments that showed bad results. The easier it is to add introns, the sooner one would expect the balance between improving fitness, on the one hand, and adding introns, on the other, to tip in favor of adding introns and thus cause exponential growth. By making it so easy to add introns, exponential growth may have been accelerated, causing evolution to stagnate earlier. In both studies, making it harder for the system to generate introns seemed to improve performance.

Some support for this tentative conclusion is lent by another study that used explicitly defined introns similar to ours but evolved them differently. Angeline [Angeline, 1996] inserted EDIs into a tree-based system. He made it quite difficult for the introns to change values rapidly by treating the EDI vector as an evolvable evolutionary programming vector. Thus, growth was regulated by Gaussian distributions around the earlier EDI vectors. This approach led to excellent results.

Computational Burden

The computation burden of introns is undisputed. Introns occupy memory and take up valuable CPU time. They cause systems to page and evolution to grind to a halt.

7.11.2 Possible Beneficial Effects of Introns

Three different benefits may be attributable to introns.

Compression and Parsimony

Although introns tend to create large solutions filled with code that has no function, they may tend to promote parsimony in the real code – that is, the code that computes a solution. As noted elsewhere, short effective complexity is probably correlated with general and robust solutions. Ironically, under some circumstances, theory suggests that introns may actually participate in a process that promotes parsimonious effective solutions and that this is an emergent property of GP [Nordin and Banzhaf, 1995a].

Recall equation 7.12. It may be rewritten as follows to make it easier to see the relationship between parsimony and introns:

$$f^e = f \cdot [1 - p^D \cdot (C^e / C^a)] \qquad (7.14)$$

supressing indices j and r.

If C^e is equal to C^a, there are no introns in the individual. In that case, the influence of destructive crossover on the effective fitness of an individual is at its maximum. On the other hand, if there are introns in the individual, a reduction in C^e increases its effective fitness. When C^e is smaller, the solution is more parsimonious [Nordin and Banzhaf, 1995a]. The conditions that would tend to promote such a factor are:

- A high probability of destructive crossover.
- Some introns in the population (almost never a problem).
- A system that makes it relatively easier to reduce the amount of effective code than to add more introns.

Soule's work tends to support the prediction of equation 7.14. When no destructive crossover is allowed, the amount of effective code (code that has an effect on the behavior of the GP individual) is significantly higher than when normal, destructive crossover is permitted. The effect persists in virtually all generations and grows as the run continues [Soule and Foster, 1997b]. Thus, the prediction that effective fitness will tend to compress the effective code of GP individuals and that the effect will be more pronounced as the destructiveness of crossover increases now has significant experimental support.

Structural Protection Against Crossover

Researchers have also suggested that introns may provide a sort of structural protection against crossover to GP building blocks during the earlier stages of evolution. This would occur if blocks of introns developed in a way so that they *separated* blocks of good functional code. Thus, the intron blocks would tend to attract crossover to the introns, making it more likely that crossover would be swapping good functional blocks of code instead of breaking them up.[10] This is very similar to Watson et al.'s view of biological introns in the quotation above.

Global Protection Against Crossover and Bloat

It is important to distinguish between this *structural* effect of introns and the *global* effect (explosive intron growth effect) discussed above. The two types of intron effects should have very different results and should look very different.

- The global effect usually has a very bad effect on GP runs. It protects the *entire individual* from the destructive effects of crossover. It is probably implemented when crossover swaps groups of introns that are, effectively, terminals.

- The structural effect would, on the other hand, tend to protect *blocks of effective code* from the destructive effects of crossover – not the entire individual. Unlike global protection, which almost always has negative effects, structural protection could be very beneficial if it allowed building blocks to emerge despite the destructive effects of crossover and mutation.

This structural effect would look considerably different than the global effects of introns discussed above. Structural protection would be characterized by groups of introns *between* groups of functional code instead of introns that were, effectively, terminals. In fact, in linear genome experiments, the authors have located a tendency for introns and functional code to group together. It can be observed by looking at the development of the so-called intron map which shows the distribution of effective and intron code in a genome.

Intron Maps

Figure 7.3 shows such an intron map. It records, over the generations, the feature of particular locations on the genome, whether they are introns or not. The figure was the result of a pattern recognition run with GP. In the early generations, the length of programs quickly increases up to the maximum allowed length of 250 lines of code. The further development leads to a stabilization of the feature at each location, which in the beginning had been switching back and forth between effective and intron code.

---

[10]The theoretical basis for this idea is formalized in [Nordin et al., 1995] [Nordin et al., 1996]. Angeline's results with explicitly defined introns tend to support this hypothesis [Angeline, 1996].

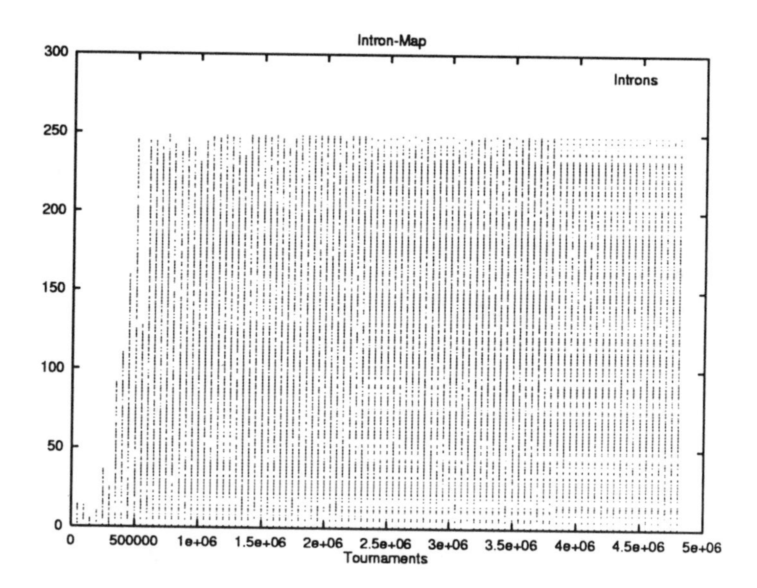

Figure 7.3
Development of an intron map over the course of a run. Introns are drawn in black; effective code is left white.

7.12 What to Do about Introns

The above equations and other experimental results suggest several directions for controlling and possibly harnessing introns.

7.12.1 Reduction of Destructive Effects

In Chapter 6 the destructive effect of crossover was discussed at some length along with recent attempts to improve the crossover operator. Equation 7.12 suggests that reducing the destructiveness of crossover will, itself, reduce the tendency of introns to emerge or at least postpone the time at which exponential growth begins [Soule and Foster, 1997b].

Therefore, all of the different techniques explored in Chapter 6 for improving the crossover operator – brood recombination, explicitly defined introns, intelligent crossover operators, and the like – are likely to have a measurable effect on GP bloat. We argued in that chapter that the essence of the problem with the GP crossover operator is that it fails to mimic the homologous nature of biological recombination. The problem of bloat may be viewed in a more general way as the absence of homology in GP crossover.

Of course, it would be easy to take reducing the destructiveness of crossover too far. We would not want to reduce the destructive effect of crossover to zero, or GP would become just a hill climber.

7.12.2 Parsimony

The effect of parsimony pressure is to attach a penalty to the length of programs. Thus, a longer solution will be automatically downgraded in fitness as compared to a short solution. Whereas this will prevent introns from growing exponentially, it depends on the strength of that penalty at what point in evolution introns become suppressed altogether.

Figure 7.4 shows an AIMGP run with parsimony pressure. Figure 7.5 shows a similar run, though without this pressure. Whereas initially both runs look the same, the run with parsimony pressure succeeds in suppressing introns, effectively prohibiting explosive intron growth.

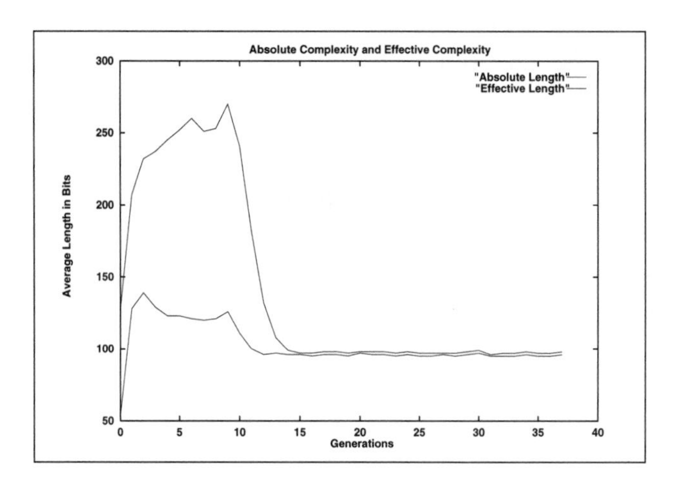

Figure 7.4
Length of program during evolution with parsimony pressure in AIMGP

7.12.3 Changing the Fitness Function

Changing the fitness function during a run may also help to suppress explosive intron growth. When the fitness function becomes variable, GP individuals might find ways to improve their fitness, despite the fact that they might have stagnated under a constant fitness function. If, for instance, one were to change a specific parameter in the fitness function for the sole purpose of keeping the GP system busy finding better fitness values, this would greatly reduce the probability of a run getting stuck in explosive intron growth. Fitness functions might just as well change gradually or in epochs; the latter technique is also used for other purposes. Even a fitness function that is constantly changing due to co-evolution will help avoid explosive intron growth.

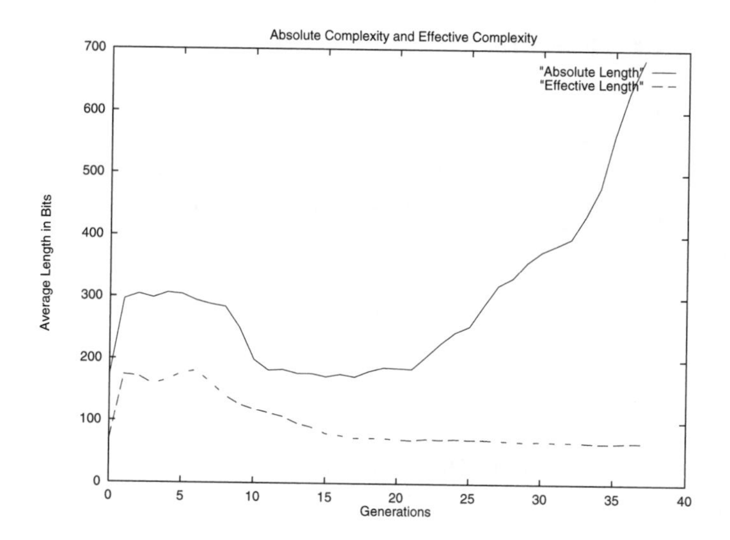

Figure 7.5
Length of program during evolution without parsimony pressure in AIMGP

Exercises

1. Give two examples that could be seen as GP evolving its own representation.
2. Give three examples of introns in GP.
3. What is an artificial intron equivalent?
4. What is effective fitness? Explain the difference between effective and absolute fitness.
5. Name a beneficial effect of an intron for an individual. What is a detrimental effect of an intron?
6. Is it possible to find code segments that are introns against crossover, but not against mutation?
7. Give three examples of how an individual can increase its effective fitness.
8. Describe two methods of preventing exponential intron growth.

Further Reading

D. Andre and A. Teller,
A Study in Program Response and the Negative Effects of Introns in Genetic Programming.
In J.R. Koza, D.E. Goldberg, D.B. Fogel, and R.L. Riolo (eds.),
Genetic Programming 1996:
Proceedings of the First Annual Conference.
MIT Press, Cambridge, MA, 1996.

P.J. Angeline,
Genetic Programming and Emergent Intelligence.
In K.E. Kinnear, Jr. (ed.),
Advances in Genetic Programming.
MIT Press, Cambridge, MA, 1994.

P.J. Angeline,
Two Self-Adaptive Crossover Operators for Genetic Programming.
In P.J. Angeline and K.E. Kinnear, Jr. (eds.),
Advances in Genetic Programming 2.
MIT Press, Cambridge, MA, 1996.

W. Banzhaf, F.D. Francone, and P. Nordin,
The effect of extensive use of the mutation operator on Generalization in Genetic Programming using Sparse Data Sets.
In H. Voigt, W. Ebeling, I. Rechenberg, and H. Schwefel (eds.),
Proc. PPSN IV,
Springer, Berlin, 1996.

P. Nordin, F.D. Francone, and W. Banzhaf,
Explicitly Defined Introns and Destructive Crossover in Genetic Programming.
In P.J. Angeline and K.E. Kinnear, Jr. (eds.),
Advances in Genetic Programming 2.
MIT Press, Cambridge, MA, 1996.

T. Soule, J.A. Foster, and J. Dickinson,
Code Growth in Genetic Programming.
In J.R. Koza, D.E. Goldberg, D.B. Fogel, and R.L. Riolo (eds.),
Genetic Programming 1996:
Proceedings of the First Annual Conference.
MIT Press, Cambridge, MA, 1996.

8 Analysis — Improving Genetic Programming with Statistics

Contents

The foregoing chapters established some theoretical grounding for genetic programming. Nevertheless, GP is 99% an *experimental* methodology. Like natural evolution, GP produces complex and emergent phenomena during training (see Chapter 7) and, as in natural evolution, predicting the results of changes in a GP system is difficult. In other words, the gap between theory and practice is wide indeed.

To illustrate this point, assume that GP run 1 produces a good result and continues for 200 generations before intron explosion occurs. For GP run 2, only the random seed is changed. All other parameters are the same as for GP run 1. But GP run 2 could easily produce a bad result and terminate after only 15 generations because of intron explosion. Such wide variances in results are typical in genetic programming.

As a result, measuring what is going on before, during, and after a GP run and measuring its significance is important for at least three reasons:

1. **Dynamic Run Control**

   One would like to be able to use the information to control the run itself.

2. **Data Preprocessing**

   It is good to present the data to the GP system in a way that maximizes the chance of testable, useful predictions.

3. **Significance or Meaning of Run**

   It is important to interpret the results of the run for meaning and statistical significance – that is, does an evolved GP individual have any statistically valid predictive value?

Online Analysis Tools

This chapter refers to measurements made during a GP run as "online measurements." Online measurements of fitness have been with GP since its beginning [Koza, 1992d]. In 1995, researchers started systematically to measure other aspects of GP runs online, such as introns, effective size, and crossover results.

Offline Analysis Tools

"Offline" analysis tools are measurements and statistical analysis performed before and after GP runs. Typical offline tools include data preprocessing and statistical analysis of whether a GP run is "generalizing" well to data it has not yet seen.

This chapter begins with some basic statistical concepts such as populations and samples, and several elementary statistical tests that are widely applicable in GP. Then, we look at ways to use both online and offline tools in analyzing and improving GP runs.

8.1 Statistical Tools for GP

8.1.1 Basic Statistics Concepts

Statistical Population

By *statistical population* we mean the entire group of instances (measured and unmeasured) about which one wishes to draw a conclusion. If, for example, one wished to know how many men living in the city of Dortmund, Germany, weighed more than 100 kg, the statistical population to examine would be all men living in Dortmund.[1]

Sample

A *sample* is a subset of a statistical population. To comprise a sample, the subset must be drawn randomly from the statistical population and the draws must be independent of each other (see Chapter 3).

Many populations cannot be measured *as a whole*. Often, the best we can do is to select a sample from the population (say, 1000 of the men who live in Dortmund) and measure their heights. We would hope that the distribution of men's heights in that sample is representative and would let us estimate the distribution of heights in the entire male population of Dortmund.

Statistics as a discipline does not give black and white answers. It gives probabilities that certain measurements or findings will be replicated on subsequent samplings from the population. For instance, suppose that the 1000-man sample from Dortmund had a mean weight of 80 kg with a standard deviation of 7 kg. Statisticians can estimate the probability that a second sample of 1000 men from Dortmund will weigh, on average, 110 kg. The probability would be very low. But statistics could not say that this event could not occur.

Statistical Significance Level

The *statistical significance level* is a percentage value, chosen for judging the value of a measurement. The higher the level chosen, the more stringent the test. Statistical significance levels chosen are usually values from 95% to 99%.

8.1.2 Basic Tools for Genetic Programming

Generically Applicable Tools

There are many statistical tools available to the GP researcher. This section is a brief survey of basic tools that apply generically to measurements taken from samples drawn from populations. They should, therefore, be considered whenever a measurement is made of a sample drawn from a population in GP.

[1] "Statistical population" is offbeat terminology – statistics texts would refer to all men living in Dortmund as the "population." However, the term *population* already has a special meaning in GP. Therefore, we will use the term *statistical population* to refer to what ordinary statistics literature would call a "population."

How to calculate the statistics referred to in this section may be easily found in any basic text or handbook on statistics.

Confidence Intervals

A confidence interval is a range around a measured occurrence of an event (expressed as a percentage) in which the statistician estimates a specified portion of future measurements of that same event. The confidence interval expresses how much weight a particular proportion is entitled to. A narrow confidence interval means that the measurement is accurate. A wide one means the opposite. Normally, confidence intervals are measured at the 95% and 99% statistical significance levels.

Confidence Intervals in the Real World

What follows is a real-world example of the use of confidence intervals to determine the feasibility of redesigning a GP system. The importance of measuring GP introns has been discussed elsewhere (see Chapter 7). Let us assume a GP population (not a statistical population) size of 3000 programs. To measure the percentage of nodes in the GP population that are introns, the *statistical* population is comprised of every node of every individual in the entire GP population. If the average individual is 200 nodes in size, the statistical population size is large – 600 000 nodes.

Sometimes it is possible to take the measure of the *entire* population. In fact, one recent study did measure the average number of introns per individual for the entire statistical population for every generation. Doing it that way was *very* time consuming [Francone et al., 1996].

After completing that study, we began to look for ways to speed up intron checking. One suggestion was to take a sample of, perhaps, 1000 nodes (out of the 600 000 nodes) and measure introns in the sample. This sample could be used to estimate the percentage of introns for the entire population – reducing the time used for intron checking by a factor of 600.

The payoff from this approach is clear. But what would be lost? Suppose that 600 of the 1000 sampled nodes were introns. Does this mean that 60% of the entire 600 000 node population are comprised of introns? No, it does not. The 95% confidence interval for this measured value of 60% is actually between 57% and 63%. This means, in 95% of all 1000-node samples drawn from the same GP population, between 57% and 63% of the nodes will likely be introns. Viewed another way, we could say that the probability is less than 5% that a 1000-node sample drawn from the same GP population would have fewer than 570 introns or more than 630 introns. In a run of two-hundred generations, however, one would expect at least ten generations with anomalous results just from sampling error. So we decided to run intron checks on about 2000 randomly gathered nodes per generation. This tightened up the confidence interval to between 59% and 61%.

Confidence Intervals and Intuition

In GP work, calculating confidence intervals on proportions is important – intuition about proportions and how significant they ought to be are frequently wrong. What would happen to confidence intervals if the sample size of 1000 fell to a more typical GP level of thirty samples? Of those thirty nodes, suppose that 60% (18 nodes) were introns. Would we be as confident in this 60% measure as we were when the sample size was 1000? Of course not, and the confidence interval reflects that. The 95% confidence interval for a sample comprised of thirty nodes is between 40% and 78% – a 38% swing. To obtain a 99% confidence interval from the thirty sample nodes, we would have to extend the range from 34% to 82%. The lesson in this is that proportions should *always* be tested to make sure the confidence intervals are acceptable for the purpose for which the measurement is being used.

Correlation Measures

There are several different ways to test whether two variables move together – that is, whether changes in one are related to changes in the other. This chapter will look at two of them – correlation coefficients and multiple regression.[2]

Correlation Coefficient

Correlation coefficients may be calculated between any two data series of the same length. A correlation coefficient of 0.8 means 80% of the variation in one variable may be explained by variations in the other variable. If the correlation coefficient is positive, it means increasing values of the first variable are related to increasing values of the second variable. If negative, it means the reverse.

Student's t-Test

A correlation coefficient comes with a related statistic, called the *student's t-test.* Generally, a *t*-test of two or better means that the two variables are related at the 95% confidence level or better.

Use of Correlation Analysis in GP

Assume a project involves many runs where the mutation rate is changed from 5% to 20% to 50%. The runs are finished. Did runs using higher mutation rates produce better or worse performing individuals? Calculating the correlation coefficient and *t*-test between mutation rate and performance would be a good way to start answering this question. If the value of the *t*-test exceeded 2.0, then the

[2] Correlation analysis does not test whether changes in variable 1 *cause* changes in variable 2, only whether the changes happen together. Statistics does *not* establish causation.

correlation between mutation and performance is significant at the 95% level.

Multiple Regression

Multiple regression is a more sophisticated technique than simple correlation coefficients. It can uncover relationships that the simpler technique misses. That is because multiple regression is able to measure the effects of several variables simultaneously on another variable. Like correlation analysis, multiple regression is linear and yields both a coefficient and a t score for each independent variable.

If the researchers in the mutation/performance study referred to above had not found a statistically significant relationship between mutation and performance using the correlation coefficient, they might have to deal with one additional fact. In their runs, they did not just vary the mutation rate, they also varied the parsimony rate. The effect of changes in the parsimony rate could well have obscured the effect of the changes in the mutation rate. Multiple regression might have helped in this case because it would have allowed to hold the parsimony rate constant and test for just the effect of the mutation rate.

Testing Propositions

F scores are frequently useful to determine whether a proposition is not false. Assume a project involving measurements of the value of best individual hits in thirty separate GP runs. That thirty-run sample will have a mean (call it 75%) and the individual samples will vary around that mean (some runs will have hits of 78% and others 72%).

F-Test

An F-test analyzes the mean and the variance of the sample and tests the proposition that the mean of the thirty samples is not zero. The F-test is expressed in probabilities. Therefore an F-test of 0.05 would mean the probability that the mean of the thirty samples is zero is only 5%. (See [Andre et al., 1996a] for a successful application of the F-test to establish that a tiny but very persistent difference in results was statistically significant.)

More complex F-tests are available. For example, one can test the proposition that the means of two different samples are different by more than a specified interval. This is a very useful measure in establishing the relative performance of different machine learning systems (benchmarking). In a typical benchmarking study, each system produces a variety of runs and results on the same data.

A recent study ran a GP system and a neural network on the same test data. The GP system did a little better than the neural network across the board but the difference was not large. For example, for the best 10% of runs on one data set, GP averaged 72% correct and the network averaged 68% [Francone et al., 1996]. Is that difference statistically significant? Fortunately, this was a big study and there were a lot of runs. The F-test for a hypothesized difference between two means assuming unequal variances is very useful in this situation. It gave a 95% confidence level that the difference in means cited above was at least a 1% difference.

Caveat

All of the above tests can provide excellent feedback about the weight that should be given to experimental results. Nevertheless, statistics is an analytic science. A number of assumptions had to be made to derive t-tests, F-tests, and the like. For example, most statistical measures assume that the distribution of the sample on which the test is being performed is approximately normal (see Chapter 3). In practice, the tests work pretty well even if the distribution is fairly distant from a Gaussian. But it is always a good idea to take a look at a histogram of the data one is analyzing to make sure it does not grossly violate this assumption of normality. If it does, the value of any of the tests described above is dubious.

8.2 Offline Preprocessing and Analysis

Before a GP run, there is only data. It is usually comprised of different instances of different data series. The task of the researcher is twofold:

1. To select which data series and data instances should be fed to the GP system as input(s);[3] and

2. To determine which, if any, transformations should be applied to the data before it is fed to the GP system.

Preprocessing and analysis of data *before* a run plays a crucial role in the machine learning literature, particularly the neural network literature.

Preprocessing and analysis comes in three forms:

1. Preprocessing to meet the input representation constraints of the machine learning system;

2. Preprocessing to extract useful information from the data to enable the machine learning system to learn; and

3. Analyzing the data to select a training set.

[3] This is a largely unexplored domain in GP. The little research that exists on selecting inputs indicates that GP does not do well when fed a lot of irrelevant inputs [Koza, 1992d].

The next three subsections discuss the integration of such techniques into GP.

8.2.1 Feature Representation Constraints

Much of the preprocessing of data for neural networks or other machine learning methods is required by the constraints on feature (input) representation in those methods (see Chapter 1). For example, most neural networks can accept inputs only in the range of -1 to 1. Boolean systems accept inputs of only `0` or `1`, `true` or `false`. So it is usually necessary to center and normalize neural network inputs and to transform inputs for a Boolean system to be either `0` or `1`.

By way of contrast, GP has great freedom of representation of the features in the learning domain. As a result, it can accept inputs in about any form of data that can be handled by the computer language in which the GP system is written and, of course, over a very wide range. It has been suggested that it may be useful to have the different inputs cover roughly the same ranges and that the constants allowed in the GP system cover that range also. The authors are not aware of any experimental proof of this concept but it seems a good suggestion to follow. However, unlike in most machine learning systems, this is not a major issue in GP.

8.2.2 Feature Extraction

One challenging goal among GP researchers is to be able to solve problems using no prior knowledge of the learning domain. Frequently, however, we can obtain prior knowledge of the domain by using well-established statistical techniques. These techniques may extract useful types of information from the raw data or filter out noise.

Some of these techniques are simple. For example, one might look at a histogram of an input data series and determine that outlying points in the distribution should be truncated back to three standard deviations from the mean of the series. Other techniques are much more complex but nevertheless very useful. This chapter will touch on only two techniques: principal components analysis and extraction of periodic information in time series data.

Principal Components Analysis

Real-world problems frequently have inputs that contain a large quantity of redundant information. One series of input components might frequently be correlated with another series of input components to a high degree. If both series are used as inputs to a machine learning system, they contain redundant information. There are two reasons why this is relevant:

1. GP starts in a hole. It has to learn how to undo the redundancy and to find the useful information.

2. Redundancy of this kind creates unnecessarily large input sets.

One way to attack this problem would be to calculate the correlation coefficient for each pair of potential input variables. In fact, this is a good idea before a GP run in any event. But suppose the correlation between two of the variables is high. Do we throw one out? Do we keep them both?[4]

Principal components analysis (PCA) extracts the useful variation from several partially correlated data series and condenses that information into fewer but completely uncorrelated data series. The series extracted by PCA are called the *components* or *principal components.* Each of the principal components is numbered (the first principal component, the second ... and so forth). The actual technique for calculating principal components is beyond the scope of this chapter. What follows is a visual metaphor for what PCA is doing when it extracts components.

PCA calculates the first principal component by rotating an axis through the n-dimensional space defined by the input data series. PCA chooses this axis so that it accounts for the maximum amount of variation in the existing data set. The first principal component is then the projection of each data element onto that axis.

PCA calculates the additional principal components in a similar manner except that the variation explained by previous components is not compensated for in positioning subsequent axes. It should be apparent that each new principal component must be measured on an axis that is orthogonal to the axes of the previous principal component(s). One very nice feature of PCA is that it reduces the number of inputs to the GP system substantially while keeping most of the information contained in the original data series. In fact, it is not normally necessary to extract more than two or three components to get the vast bulk of the variance. This is true even when there are many data series.

However, PCA is not an automatic process nor is it a panacea for all input problems, for at least three reasons:

[4]Actually, where there are only two highly correlated variables, it often helps to use their sum and their difference as inputs rather than the variables themselves. If there are more than two, it is necessary to use principal components analysis.

1. PCA involves judgment calls about how many components to use.
2. Weighting inputs to the PCA system to emphasize "important" variance and to deemphasize "unimportant" variance can help the performance of the PCA tool. But assigning importance to variance is a task GP itself ought to be doing. For GP purposes, the manual assignment of weights seems rather self-defeating unless one has specific domain knowledge to justify the decision.
3. When, for instance, twelve components are reduced to only three components, *some* information will be lost. It could be the information that the GP system needed.

Notwithstanding these drawbacks, PCA can be an effective tool and, used properly, helps far more often than it hurts.

Extraction of Periodic Information in Time Series Data

Engineers and statisticians have spent decades developing techniques to analyze time series. Some of that work has made its way into GP [Oakley, 1994a]. The purpose of this section is to give the broadest overview of the types of work that have been done in these areas to spur further research by the interested reader.

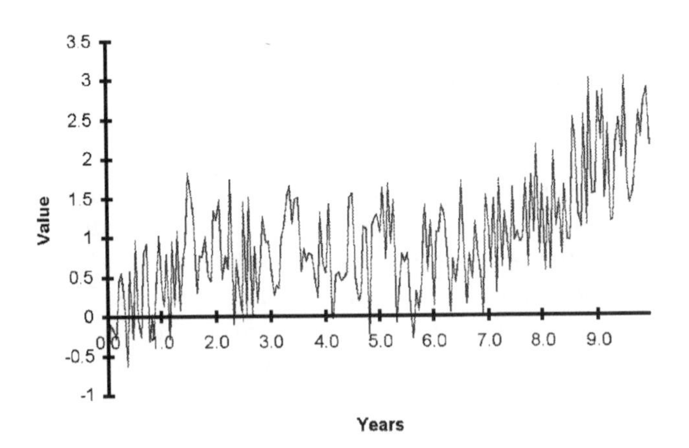

Figure 8.1
A time series shown in the time domain

To begin with, there are simple techniques to extract different types of information from time series data such as simple or exponential moving averages (SMAs or EMAs). For example, Figure 8.1 is a time series plotted over ten years. To the eye, it has some periodic components but there appears to be a lot of noise.

Figure 8.2 is the same time series as in Figure 8.1 except that two different simple moving averages have been applied to the series. Note that the simple moving averages serve as a sort of low pass filter, allowing only low frequencies through. This tends to filter out what might be noise. In many instances, SMA-type preprocessing greatly assists machine learning systems because they do not have to learn how to filter out the noise.[5]

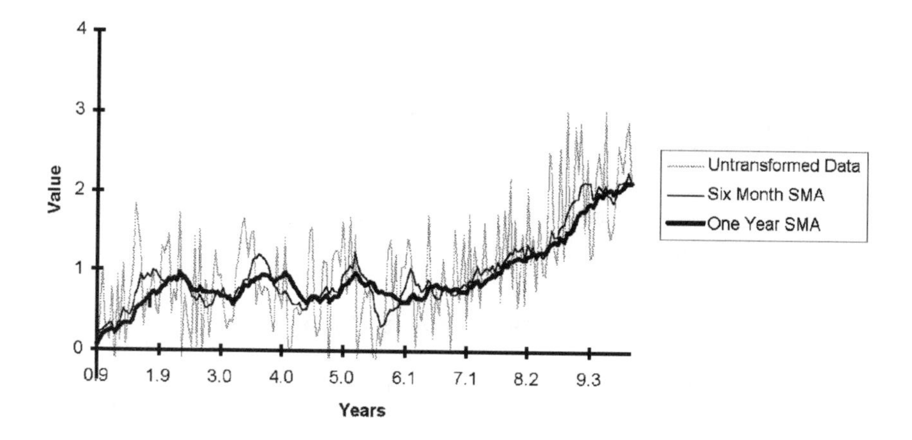

Figure 8.2
The time series from 8.1 with a simple moving average

This brief treatment of SMAs is far too simple. The price of the low pass filter is that the smoothed data series lags behind changes in the real world. The greater the smoothing, the greater the lag. Further, if the actual data exhibits genuine periodicity, this means that the smoothed data may be out of phase with the actual data unless exactly the right window was chosen for the SMA. There is a rich body of literature in the trading technical analysis world dealing with this issue and, in particular, how to identify the dominant cycle and how to adjust the filters so that the filtered data end up in phase with the real-world data [Ehlers, 1992].

Other Preprocessing Techniques

There are many other techniques to extract periodic components or otherwise preprocess time series data such as detrending and cen-

[5]Should there be concern that the SMA is filtering out valuable high frequency information, the residuals may be calculated (that is, the difference between the original series and the SMA). Those residuals may be fed to the GP system as a separate data source.

tering, differencing, digital filtering, the maximum entropy spectrum technique, quadrature mirror filters, and wavelets.[6]

Let us look at discrete Fourier transforms and the power spectrum as an example of the many preprocessing techniques that are available for real-world time series data.

The Discrete Fourier Transform

Figure 8.1 is a time series plotted in the time domain. This means that the series in Figure 8.1 is represented as a function of time and value. It is possible to transform the series into other domains. A popular transform is the discrete Fourier transform. Once the Fourier transform is performed, the real and imaginary parts of the transform are mapped to a power spectrum. This converts the series from the time domain to a frequency/power domain.

How is this useful? If Figure 8.1 were a sound wave, feeding the results of the discrete Fourier transform to a GP system might help the GP system to better extract useful information from the wave. In effect, the GP system would not have to go to the trouble of learning how to extract the periodic information. A much simpler statistical device would be doing that for the GP system.

8.2.3 Analysis of Input Data

Selecting a training set often presents two problems:

1. How to choose among input series
2. How to choose training instances

Choosing among Input Series

The GP researcher is frequently faced with many data series but has no idea (other than intuition) which of the data series are relevant. This problem arises frequently in real-world problems – especially in sound and image processing problems, time series problems, and data mining problems. How to choose input series is one of the great unexplored areas in GP. What little research there is suggests that, when GP is fed too many irrelevant inputs, its performance is severely compromised [Koza, 1992d].

Some Suggestions

In the long term, a meta-learning approach is probably the best solution. But for most GP systems, serious meta-learning will have to wait for faster computers. Here are some suggestions until then:

- Run correlation coefficients between each potential input and the output. This might help narrow what inputs to use. Obviously, inputs with higher correlations should be used, but it

[6] All of these subjects are treated well and in detail in [Masters, 1995b].

would be unwise to stop there. The correlation coefficient will frequently miss many of the important variables.

- Run correlation coefficients between each potential input. For those that are correlated, group them together, do a principal components analysis, as suggested above, and use the resulting components as inputs.

- Try different runs with different combinations of variables. Keep data on which runs do the best. Select the variables that are most often associated with good runs. Do more runs with the possibly good variables. Keep repeating this until the system produces acceptable results. In effect, this is performing meta-learning manually with respect to variable selection.[7]

Choosing Training Instances

Data analysis may also be of use in selecting a training set when there is an abundance of data. This is a different issue than choosing among data series. This issue assumes that the data series have been chosen. Now there are n examples of the training domain using those data series. Which of the examples will be used to train the GP system? The reader will, of course, recognize this as another incarnation of the problem of getting a representative sample from a population.

Data Mining

In data mining the GP system combs through a large, frequently mainframe or server database, and searches for patterns in the data. Increasingly, machine learning applications, including GP, are being used for data mining. The important characteristic here of data mining is that there are usually many more training instances available than a GP system could possibly digest. Large companies often have millions or more records of data. The problem is, how to select the particular training instances to make up the data set, given a gross overabundance of data?

Obviously, there will be problem-specific selection issues. But once the problem-specific criteria have been fulfilled, it is still very

[7]This suggestion illustrates the importance of calculating statistical significance levels. If input variable x appears in two runs that do well, that is not so impressive. If it appears in 800 runs that do well, it is. Somewhere in between 2 and 800 is a number of runs that will give you 95% confidence that having variable x in a run is related to having a good output on the run. Knowing what that number of runs is can save you from chasing a relationship that does not exist (if you draw conclusions based on too few runs). It can also save you much calculation time confirming a relationship for which you already have enough evidence.

important that the training set be representative of the overall data set. There are several possible approaches:

- Select a random sample of training instances from the million-plus database. But that raises the problem we have seen before, variance caused by sampling error. A random sample can be quite unrepresentative of the overall training set. This chapter has been filled with instances where statisticians would expect very substantial variation among different random samples from the same population.

- Large data sets for data mining usually have many input variables in addition to many records. It would take some time, but the mean and variance of each of those records could be calculated. From that, it should be possible to define an acceptable level of variance for each such variable. Either by calculation or by repeated sampling and measurement of different size samples, one could calculate the approximate sample size that produces a sufficiently representative training set. If the sample size chosen is not so large that GP training would be unduly slow, this is a workable option.

- One could also look at the distributions of the input variables in the overall data set and make sure that the "random sample" that is picked matches those distributions closely.

- The above approach is workable but unsatisfying. Perhaps this is a situation for stochastic sampling (see Chapter 10). In that method, the researcher never does pick a training set. Rather, the GP system is programmed to pick a new small training set regularly, train on it for a brief period, and then pick another new, small training set.

The above are just a few of the methods in which statistical analysis and data preprocessing may be used before a GP run even starts. The next sections will show what we can do with the results of a GP run and demonstrate how to measure what is going on *during* a GP run.

8.3 Offline Postprocessing

8.3.1 Measurement of Processing Effort

Koza has designed a method for GP performance measuring with respect to the processing effort needed for finding an individual –

with a certain probability – that satisfies the success predicate of a given problem [Koza, 1992d]. The success predicate sometimes is a part of the termination criterion of a GP run. It is a condition that must be met by an individual in order to be acceptable as a solution.

The processing effort is identified with the number of individuals that have to be processed in order to find – with a certain probability – a solution. Note that this measure assumes a constant processing amount for each individual, which is a simplification of reality. In general, different individuals will have different structures and this implies different processing efforts. For example, an individual having a loop may require a significantly higher processing effort for its evaluation than an individual that does not use loops.

Because of the simplification, we can concentrate on the population size M and the number of generations evolved in a certain run as the two important parameters.

Instantaneous Probability

Koza's method starts with determining the instantaneous probability $I(M, i)$ that a certain run with M individuals generates a solution in generation i. One can obtain the instantaneous probability for a certain problem by performing a number of independent runs with each of these runs using the same parameters M and i.

Success Probability

If $I(M, i)$ has been determined for all i between the initial generation 0 and a certain final generation, the success probability $P(M, i)$ for the generations up to i can be computed. That is, $p = P(M, i)$ is a cumulative measure of success giving the probability that one obtains a solution for the given problem if one performs a run over i generations.

With this function measured, the following question can be answered: given a certain population size M and a certain generation number i up to which we want to have a run perform at most, what is the probability of finding a solution, at least once, if we do R independent runs with each run having the parameters M and i?

To answer this question, consider that the probability of *not finding* a solution in the first run is $1 - P(M, i)$. Since the runs are independent of each other, the probability of again not finding the solution in the second run is $(1 - P(M, i))^2$. In general, the probability of not finding the solution in run R is $(1 - P(M, i))^R$. Thus, the probability of *finding* a solution by generation i, using R runs, is:

$$z = 1 - (1 - P(M, i))^R \tag{8.1}$$

Our question was: what is the probability of finding a solution, at least once, if we do R runs? We can now turn this question around and ask: how many runs do we need to solve a problem with a certain probability z? Solving the above equation for R gives

$$R = \frac{\log(1-z)}{\log(1-P(M,i))} \tag{8.2}$$

R must be rounded up to the next higher integer to yield the answer.

If mutation is one of the genetic operators, then the success probability $P(M,i)$ rises the more generations a run performs. For given values of M and z, this means that the evolution of more generations per run requires fewer runs to be done in order to find a solution with probability z, while, on the other hand, the evolution of fewer generations per run requires more runs to be done.

Which generation number requires a certain number of runs such that the overall number of processed individuals – the processing effort – is minimal? This question can only be answered in a problem-specific way since the instantaneous probability is empirically obtained by doing runs on the problem at hand.

8.3.2 Trait Mining

Usually, a program induced by genetic programming is not at all a piece of code that software engineers would rate as high quality with respect to understandability by a human. It often is very monolithic, contains problem-relevant but redundant code and – even worse – a lot of code irrelevant to the problem at hand. Tackett approaches this problem with a method he calls trait mining [Tackett, 1995]. Redundant and irrelevant code bloats the program and often very effectively disguises the salient expressions – the expressions that actually help in solving the problem. Trait mining helps to identify those expressions.

Knowledge emerges stepwise during a genetic programming run. Initially, the population-inherent knowledge about the problem is very small, because programs are random structures. Soon, at least in a successful genetic programming run, the behavior of more and more programs begins to reflect the problem, which means the population gradually acquires more knowledge about the problem.

However, due to their size and often tricky and counter-intuitive structures, the evolved programs mostly are hard for humans to read and understand. A naive approach to avoiding this unpleasant situation and alleviating program analysis is to restrict the size and complexity of the programs. On the other hand, program size and complexity may be needed for solving a problem.

Gene Banking

For this reason, trait mining does not impose restrictions on these program attributes in order to identify salient expressions. It rather keeps book on all expressions evolved so far during a genetic programming run. This approach has been called gene banking. Obviously, due to the large number of such expressions, gene banking consumes a significant amount of CPU time and memory. Tackett claims that an integrated hashing-based mechanism keeps this consumption within acceptable limits.

Finally, when a genetic programming run has ended, trait mining allows for the evaluation of expressions with respect to how salient they are.

8.4 Analysis and Measurement of Online Data

8.4.1 Online Data Analysis

The first question in online data analysis is: why would one want to do it? The answer is twofold and simple. Doing so is (i) fascinating and (ii) often immensely useful. GP runs usually start in a random state and go through a period of rapid change during which the fitness of the population grows. Eventually GP runs reach a point where change ends – even though simulated evolution continues.

Online tools monitor the transition from randomness to stability. Not only do they highlight how the transition takes place, they also raise the possibility of being able to control GP runs through feedback from the online measurements.

In the next subsections, we will first address measurement issues for online data. Then we will describe some of the online data tools that are already being used. Finally, we will discuss the use of online measurements for run control.

8.4.2 Measurement of Online Data

This chapter has already addressed some measurement issues for online measurement tools. Recall the earlier discussion of intron checking, a relatively recent addition to the arsenal of online analysis tools. Because intron checking is so computationally intensive, the reader will recall that a statistical approach to intron checking was recommended. A statistical approach to measuring any online data is possible. Whether it would be useful depends on the tradeoff between CPU cycles and the uncertainty caused by confidence intervals.

Generational Online Measurement

Many of the typical GP online measurements characterize the population as a whole – average fitness, percentage of the population that is comprised of introns, and so forth. In the generational model, each new generation is a distinct new group of individuals and it is created all at once (see Chapter 5). It therefore makes good sense to measure the online statistics once per generation at that time. This is the typical practice in GP.

Steady-State Online Measurement

Steady-state models, on the other hand, do not have distinct generations (see Chapter 5). Steady-state GP maintains the illusion of generations by stopping once every P fitness evaluations (where P is the population size) and measuring their online statistics. This convention has proved effective and is used almost universally in steady-state GP systems. Here, we will follow this convention. The reader should recall, however, that a generation is no more than an agreed convention in steady-state models and any online statistic could be calculated so as to reflect the reality of a steady-state system more precisely.[8]

8.4.3 Survey of Available Online Tools

Fitness

Understandbly, fitness was the earliest online measurement in widespread use [Koza, 1992d]. The most frequently used fitness measures are:

- Fitness of the presently best individual;
- Average fitness of the entire population;
- Variance of the fitness of the entire population.

The last two statistics characterize the population as a whole and are shown in Figure 8.3 for the same run of a pattern recognition problem. Both average and variance first decrease quickly, then increase again on finding a simple approximative solution which is not optimal. Later, average fitness decreases continuously, whereas variance stays at a certain level.

Diversity

Diversity is another measure of the state of the population. Genetic diversity is a necessary condition for the fast detection of a high-fitness individual and for a fast adaptation of the population to a

[8] One way to maintain online statistics in a steady-state system would be as EMAs. Each time a new fitness value was calculated, it would be added to the existing EMA for average fitness. If the illusion of generations is to be maintained, the width parameter of the EMA could be set so that the window of the average is approximately one generation. More interesting, however, would be to set the width to accommodate only a short window of data. Or possibly to maintain several different windows simultaneously.

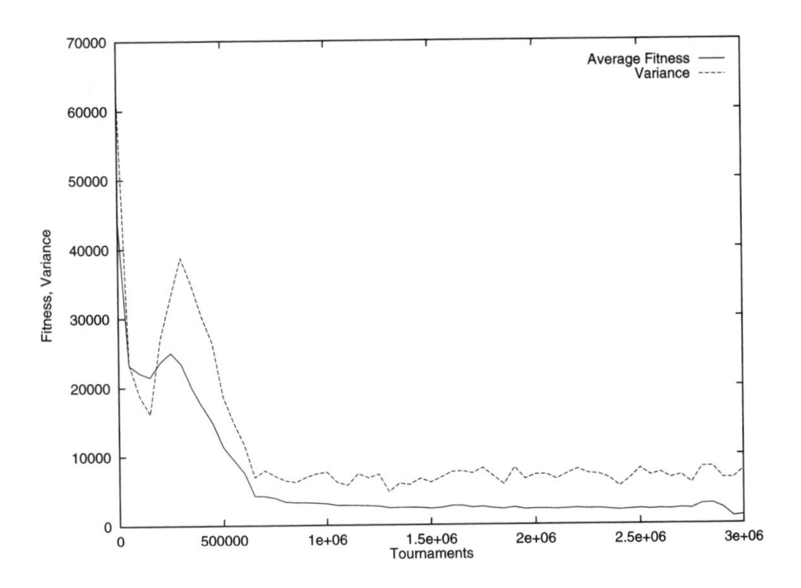

Figure 8.3
Average fitness and variance of fitness for a pattern recognition problem in AIMGP

changing environment. To maintain genetic diversity during runs, it is useful to measure diversity and changes in diversity. There are two overall approaches to measuring diversity:

- **Genotypic Diversity**
  Measuring the structural differences between genotypes; and

- **Phenotypic Diversity**
  Measuring differences in behavioral aspects of the phenotype that are believed to be related to diversity.

Measuring Genotypic Diversity

Genotypic diversity is diversity among the actual structures in the population – the trees, the graphs, or the linear genomes. Genotypic diversity might be an online statistic of the state of the GP population that is orthogonal to the online fitness measurements. This would be valuable indeed. To construct such a measure, it would be ideal if no quality (fitness) information were contained in the diversity measures. Viewed this way, diversity should be based on a comparison of the structure of the individuals only.

Edit Distance

One good measure of diversity that fits this requirement is to use "edit distances" between programs. The edit distance of two programs could be calculated as the number of elementary substitution operations necessary to traverse the search space from one program to another. The definition of the edit distance δ, which goes back to [Levenshtein, 1966], states that $\delta(g, h)$ of two genotypes g, h is the minimal number of applications of given elementary edit operations needed to transform g into h.

Edit Distance for Fixed Length Genomes

On the genospace of fixed length binary strings with single-bit flipping as the only edit operation, for instance, the Hamming distance, that is the minimal number of single-bit flips needed to transform one string into another, is a type of edit distance often used. As an example, consider the strings $g = 100001$ and $h = 011111$. In order to transform g into h, we have to apply a single-bit flip at least 5 times. Thus, we have $\delta(g, h) = 5$.

Edit Distance for Tree Genomes

In a tree space G, on the other hand, basic edit operations could be the insertion and deletion of nodes. Let the notation be $del(l)$ and $add(l)$, where l is a node label. Since we have a bijection between G and prefix strings, we can write each tree as a string the characters of which are leaf labels. For example, consider $g = +a/bc$ and $h = +*abc$. The sequence

$$del(b), del(c), del(/), add(c), del(a), add(*), add(a), add(b)$$

is a minimal sequence leading from g to h. Thus, we have $\delta(g, h) = 8$.

If we imagine the space of all genotypes as a high-dimensional space [Keller and Banzhaf, 1994], a genotype – that is a structure – takes exactly one position. Intuitively, there is high genetic diversity in a population if the individuals represent many different genotypes that are "equally distributed" over this entire space.

Edit Distance and Introns

The edit distance approach to diversity is based on the assumption that edit distance actually measures diversity. It is not clear that this is the case. For example, two individuals with identical working code would be far apart in edit distance if one of the individuals had an intron structure comprised of hundreds of nodes. Such structures are not uncommon. Although there would be great apparent diversity caused by the introns, in fact the only contribution of the introns would be to change the probability of crossover in the working code, that is, the *effective fitness* of the working code [Nordin et al., 1996]. Arguably, the difference between the *effective fitness* of the two individuals constitutes diversity. Because effective fitness increases with the number of introns, the edit distance would be related to the difference in effective fitness. It is especially unclear that the edit distance necessary to factor in introns is in any way equivalent to the edit distance to get from one piece of working code to another.

Because it has been so hard to define genotypic diversity and because measuring it is computationally intensive, edit distance as a measure of structural diversity is not widely used.

GP researchers have also measured diversity by measuring variance in the performance of the phenotype – a good example is measurement of variance of fitness in the population. Such measurements compute some aspect of phenotypic behavior. While such measures are clearly not orthogonal to fitness, they are computationally tractable. Several methods have been used to calculate phenotypic diversity:

- **Fitness Variance**
  Experimental use of fitness variance suggests that it measures some quantity of the state of the population that is different from fitness. Changes in such measures of diversity within the population may presage major leaps in the quality of the best individuals in the population [Rosca, 1995b] (see Figure 8.3). Thus, fitness variance provides a look into the diversity issue but it is far from the ideal of being orthogonal to the fitness vector.

- **Fitness Histograms**
  Fitness histograms provide a more subtle look into the diversity of the population. Both Rosca and Koza have experimented with this measure. Figure 8.4 shows an example of a fitness histogram used by [Rosca, 1995b].

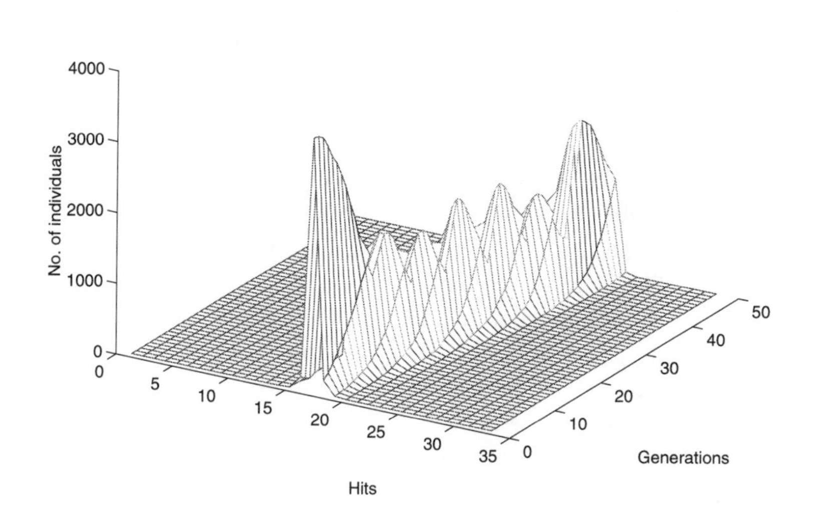

Figure 8.4
Fitness histogram. The number of hits (fulfilled fitness cases) increases over the course of a run. (Reproduced with permission from [Rosca, 1995b].)

- **Entropy**
  In accordance with developments in the GA community, entropic measures of diversity are also proliferating. Rosca points out that diversity in populations of program individuals can be

represented by a measure

$$E(P) = -\sum_k p_k \; log \; p_k \tag{8.3}$$

where p_k is the proportion of a population P that is in category or partition k with reference to a specific feature, such as fitness.

Regardless of what measurement is used, all phenotypic fitness measures assume, implicitly, that fitness differences do in fact reflect differences in the genotypes of individuals. While this assumption is intuitively appealing, neutral variants having huge differences in their structure often behave in virtually the same manner, and thus evaluate to the same fitness. It is not clear how to handle this problem.

What is the purpose of observing diversity, either of the phenotypes or the genotypes? The reason one would want information about the status of diversity of the population is simply that it helps estimate the chances that continuing the run would have some prospect of discovering a solution to the problem. High diversity would give an indication that it is profitable to extend the run; low diversity would indicate the opposite. Thus, a re-start with a newly seeded population might in this situation be the better choice.

Measuring Operator Effects

Each variation operator in GP has its own effect on the population and on the emergent dynamic of the run. By observing these effects, one can try to see which operators are useful in a certain stage of the evolution and which are not.

To distinguish between useful and useless operators, one has to isolate their effects radically. If a generational selection scheme is used, it will in general not be possible to discern the effect of different operators. Therefore, a tournament or (μ, λ) selection scheme should be used to check immediately which effects have been caused by an operator. Also, each operator should be applied exclusively, i.e., no sequence of operators (apart from reproduction) should be allowed.

Crossover Effects

The most useful measurement of operator effects to date has been that of the effects of crossover on the fitness of parents relative to the fitness of their children. The goal is to be able to assign any operator event – here, crossover – into a category that tells us whether it was successful, neutral, or unsuccessful. Two basic approaches to measuring the effect of crossover have been discussed earlier in the book.

- The average fitness of both parents has been compared with the average fitness of both children [Nordin and Banzhaf, 1995a]

[Nordin et al., 1996]. If the average of the children is more than 2.5% better than the average of the parents, the crossover is counted as being constructive; if more than 2.5% worse, it is counted as being destructive; and if the average fitness for the children is within 2.5% of the parents, it is counted as neutral. The effect of this measurement technique is that both parents and both children are counted as one crossover event. This method gives three simple online measurements that can be followed during evolution.

- The fitness of children and parents are compared one by one [Francone et al., 1996] [Teller, 1996]. In this approach, one child is assigned to one parent. Therefore, each such pairing is counted as one crossover event.

Measurements of crossover effects have permitted at least two different systems to achieve run-time control over GP runs. Those successes will be described later in this chapter.

Let us look at a more formal way to study operator effects. Given the number of predecessors $I = 1, 2, ..., N$ with $I = 1$ for mutation, $I = 2$ for conventional crossover, $I = N$ for multirecombinant operators, and a number of offspring $J = 1, ..., M$ one can quantify the "progress" between individuals gained from applying operator k as

$$fit_{gain}(k) = (f_I - f_J)/f_I \tag{8.4}$$

where f_I, f_J are scalar measures of the participating predecessors and offspring, respectively. In the simplest case these measures could be generated by taking the arithmetic average

$$f_I = \frac{1}{I}\sum_{i=1}^{I} f_i \tag{8.5}$$

and

$$f_J = \frac{1}{J}\sum_{j=1}^{J} f_j \tag{8.6}$$

Improvements are signaled by positive values of fit_{gain}, whereas deteriorations are signaled by negative values (assuming fitness falls to 0 for the optimal value).

Intron Measurements

Introns are code in a GP individual that does not affect the output of the individual. As we have seen in Chapter 7, a large proportion of introns in a population often indicates that all effective evolution is over and that crossover is only swapping introns back and forth. The process of ascertaining the number of introns is called *intron counting*. It is important to note that all intron measurement methods are estimations.

Intron Counting in Tree Structures

Assume a binary tree with addition nodes where two values are transmitted up into a node via the input branches and one value is transmitted from the node further up in the tree via the result branch. The top node in Figure 8.5 receives two inputs, 6 and 0. The node itself adds the two inputs together and produces the sum as its output. The output value is obviously 6. Further, it is obvious that, so long as the right input is 0, the result will always equal the left input. Viewed in this manner the right tree, which produces 0 as an output, is irrelevant because it does not affect the output value of the function at all.

Of course, introns are code that does not affect output. But what does that mean? For an intron, the output of the node in Figure 8.5 is the same as the left-hand side input for all fitness cases. Put another way, if the output is the same as one of the inputs for all fitness cases, then we regard the irrelevant subtree together with the node under examination as an intron.

The algorithm for measuring tree-based introns is not very computationally expensive. Each node contains a flag indicating whether this node so far has had the same output as one of the inputs. This flag is then updated by the evaluation interpreter each time it passes the node – for each fitness case. When all fitness cases have been calculated, the tree is passed through again to sum up the intron contents by analyzing the flag values.

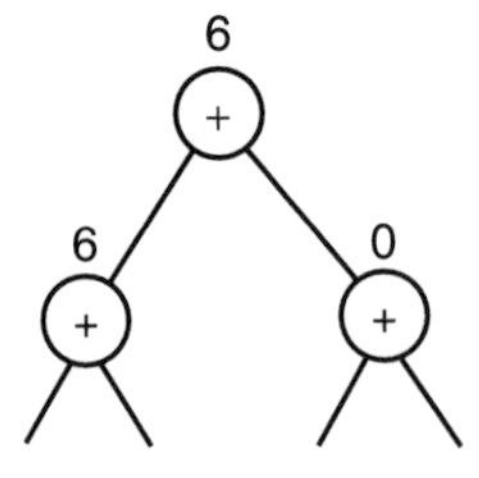

Figure 8.5
Intron measurement in tree-based GP

Intron Counting in Linear Genomes

Intron counting began with linear genomes in AIMGP. Not surprisingly, the technique is straightforward. The genome is a string of machine instructions because AIMGP evolves machine code directly. Other linear GP systems could evolve linear sequences of code (LoC). A no-op instruction (NOP), i.e., an operation that does nothing, can be inserted in place of each instruction/LoC in order to check whether this particular instruction/LoC is effective in the code or not. Thus, an intron counting method would read:

For each instruction/LoC in the individual do:

1. Run each fitness case through the individual and store the output values.
2. Replace the instruction/LoC with a NOP-instruction.
3. Rerun the fitness function on each fitness case.
4. If there was no change for any of the fitness cases after the NOP was exchanged *then* classify the instruction/LoC as an intron *else* classify it as an exon.
5. Remove the NOP-instruction/LoC and replace it with the original one.

When this procedure is completed the number of introns is summed together as the intron length of that individual. However, this is not the only method to measure introns. At the moment, researchers are actively examining various methods for intron removal.

Compression Measurement

In the previous chapter it was argued that the effective length of an individual often is reduced during evolution. This phenomenon is called compression. Compression can be estimated with a simple method. We measure the effective length by subtracting the length of the introns from the total length. Every time the fitness of the best individual changes, we store the value of the effective length. At termination of a run, we measure the difference between effective length of the best individual and the stored value. When fitness stagnates at the end of a run, we can see how much the effective size of an individual has decreased since it assumed its fitness value for the first time. It is also possible to store the number of generations over which this compression has taken place.

Node Use and Block Activation

Node use and *block activation* are two other important online analysis tools. Node use measures how often individual nodes from the terminal and function set are used in the present generation. The run-time overhead is small, since measuring just means incrementing a counter associated with a node.

The same measure can be applied to blocks of code, that is, to subtrees of arbitrary depth, although this approach will quickly become unfeasible for larger subtrees due to combinatorial explosion (see Section 8.3.2).

Block activation is defined as the number of times the root node of a block is executed. This requires an additional counter recording the number of times the node has been activated.

By looking at node use one can easily get an impression of how important a certain function or terminal is in a particular problem. A node might turn out never to be used over consecutive generations, indicating that it is mainly a burden in the search process and has no obvious advantage for successful programs. In a new run it could be discarded or substituted by a potentially better suited node.

The reader is advised, however, to be careful when using such an observational measure, because the observed frequencies could be misleading. Introns often comprise a large part of evolved programs, and measuring node use for introns would indicate high usage where no use is made of the node for behavioral purposes. It follows that it would be better to measure node use on effective code only. This way, one could at least ignore those parts of the code that do not contribute to fitness.

Rosca [Rosca, 1995a] introduced another term, *salient block*, as a means to indicate those parts of the code which influence fitness evaluation. A good modularization method, in his proposal, should identify those blocks. Rosca proposes to consider block activation as one piece of information necessary to identify salient blocks. Block activation, however, can usually be measured only for trees with a small depth. For details of the method, see [Rosca and Ballard, 1994a].

Real-Time Run Control Using Online Measurements

Two systems have successfully used online measurements to effect online control during a GP run.

❑ **PADO**
Chapter 6 describes how Teller conducted meta-learning during PADO runs. The meta-learning module changed the crossover operator itself during the run. Like its fitness function, the meta-learning module used online measurements of the effect of crossover.

❑ **AIMGP**
We performed a very large GP study involving 720 runs. A running total was maintained for each generation of the percentage of total crossover events that constituted destructive crossover. Destructive crossover usually exceeds 75% of total crossover events. Previous studies [Nordin et al., 1996] established that, when destructive crossover fell below 10% of total

crossover events, intron explosion had occurred and all effective evolution was over.

As a result of this previous study, a run-time termination criterion was adopted: terminate the run after 200 generations or when destructive crossover falls below 10% of total crossover events, whichever comes first. The result of using this termination criterion was that the run time for all 720 runs was reduced by 50%. In short, this run-time termination criterion effectively doubled the speed of the system [Francone et al., 1996].

8.5 Generalization and Induction

Once upon a time, there was a little girl named Emma. Emma had never eaten a banana, nor had she been on a train. One day she went for a journey from New York to Pittsburgh by train. To relieve Emma's anxiety, her mother gave her a large bag of bananas. At Emma's first bite of a banana, the train plunged into a tunnel. At the second bite, the train broke into daylight again. At the third bite, Lo! into a tunnel; the fourth bite, La! into daylight again. And so on all the way to Pittsburgh and to the bottom of her bags of bananas. Our bright little Emma told her grandpa: "Every odd bite of a banana makes you blind; every even bite puts things right again."

After LI and HANSON

This story is an example of the principles and risks of *induction.* We use GP to find a generally valid set of rules or patterns from a set of observations. Chapter 1 described how machine learning systems do their learning on a *training set.* For Emma, her training set is her train ride. GP, machine learning, and Emma all have tried to do something more – they have tried to come up with general rules that will work on data *they did not learn upon.* Chapter 1 referred to this type of data as a *testing set.*

Generalization as a Sampling Problem

Generalization occurs when learning that occurs on the training data remains, to some extent, valid on test data. Viewed another way, generalization is a problem of drawing a sample from a population and making predictions about other samples from the same population. The noise from one sample to the next is probably different, perhaps a lot different. In other words, sampling error can be a big problem. Recall that sampling error can cause the 95% confidence interval on thirty samples to extend all the way from 40% to 78%.

Overfitting

But there is another, perhaps equally difficult problem commonly referred to as overfitting. Ideally, GP would learn the true relationship between the inputs and the outputs and would ignore the noise.

But GP is constantly pushed by its fitness function to lower the error on the training data. After GP has modeled the true relationship, GP individuals can and do continue to improve their fitness by learning the noise unique to the sample that comprises the training data – that is, by overfitting the training data. An overfit individual will often perform poorly on the test set.

Let's go back to poor Emma. Of course, her learning is not likely to generalize the next time she eats a banana in, say, her kitchen. She has learned the noise in her training set. But can this be a problem in actual machine learning and GP? It can. Overfitting has plagued the neural network world and some GP researchers have noted its existence in their runs. Three factors may play a part in causing overfitting:

1. **Complexity of the learned solution**
   In Section 3.3 we discussed the complexity of computer programs. The simpler the solution, the higher the probability that it will generalize well. More has to be said about this below.

2. **Amount of time spent training**
   In neural networks, it is important to stop training before the network overfits the data. A large part of neural network literature deals with this problem. It is not clear yet how this factor fits into GP training because the end of training can be a very complex dynamic (see Chapter 7).

3. **Size of the training set**
   The smaller the training set (i.e., the smaller the sample size from the population), the less reliable the predictions made from the training set will be.

The next section presents an example of overfitting caused by the first and second factors above.

8.5.1 An Example of Overfitting and Poor Generalization

Here is an example of how extensive training and complex solutions can cause overfitting of noisy data. The task is to learn the simple quadratic function:

$$y = x^2 \tag{8.7}$$

in the interval between $x = -1$ and $x = +1$.

A simple sine function formed the atoms of our learning system:

$$y = a \ \sin(bx + c) + K \tag{8.8}$$

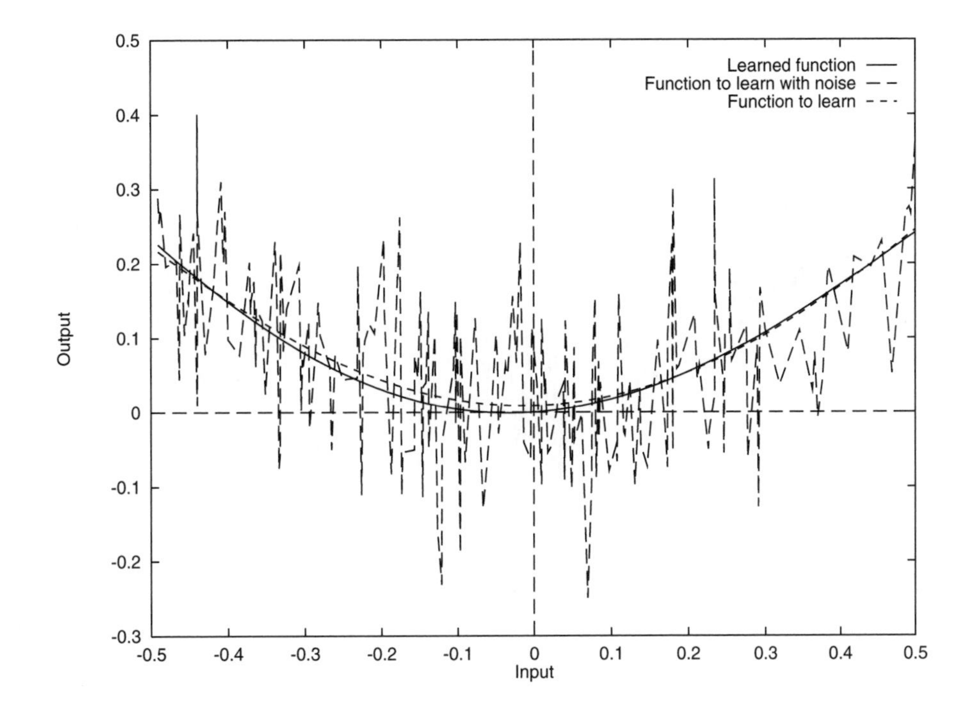

Figure 8.6
Results of learning with 1 degree of freedom

These were subjected to a hill climbing algorithm. We varied the complexity of the solution by adding more sine terms. For example:

$$y = a\ \sin(bx + c) + d\ \sin(ex + f) + K \tag{8.9}$$

Let us refer to a solution containing *one* sine expression as having one degree of freedom, a solution containing *two* as having two degrees of freedom, and so forth. While this is not a formal use of the term degrees of freedom, we use it here for convenience.

Noise was added to the function's 200 sample values between -0.5 and 0.5. The noise was distributed normally around 0 with a standard deviation of 0.1. Newtonian descent was then used to find solutions to the data with $1, 2, 4, 8$, and 16 degrees of freedom. The algorithm was run as long as any improvement was found by the algorithm (this was to encourage overfitting by training for too long).

Figures 8.6, 8.7, and 8.8 show a dramatic change in the nature of the learning that occurs as the degrees of freedom increase. With only one degree of freedom, the real function is found in the noise very accurately. But as the degrees of freedom increase, the learning began to fit the noise. At sixteen degrees of freedom, the system is modeling the noise rather effectively.

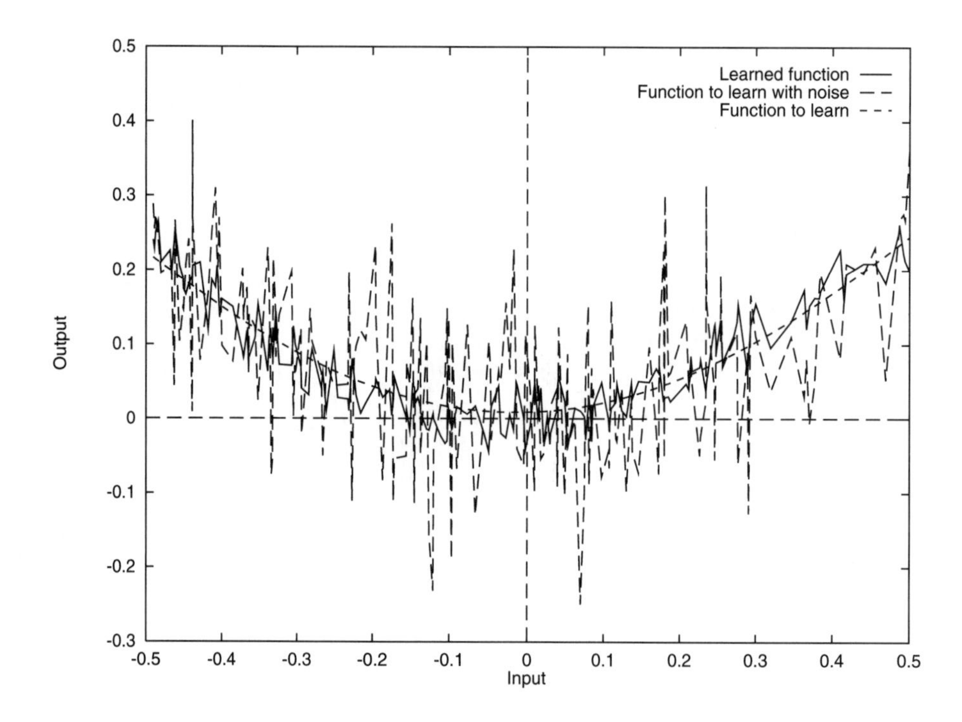

Figure 8.7
Results of learning with 4 degrees

When the complex solutions were tested on new values of x between -1 and $+1$ with *different* random noise,[9] the story was completely different. The *in-sample testing error* (the error on points between -0.5 and $+0.5$) increased by 31% as the number of degrees of freedom rose from one to sixteen. The out-of-sample testing error (the error from -1 to $+1$) more than doubled. The *simpler* solutions generalized *better*.

Table 8.1 shows that the more complex solutions modeled the *training data* better than did the simple solutions.

Table 8.1
Comparison of learning and generalization performance for varying degrees of freedom. ISE *(in-sample error):* $-0.5 \leq x \leq +0.5$*;* OSE *(out-of-sample error):* $-1.0 \leq x \leq +1.0$

Deg. of freedom	Training error	ISE	OSE
1	1.00	1.00	1.00
2	1.00	1.01	0.98
4	0.95	1.08	0.95
8	0.91	1.19	2.04
16	0.83	1.31	2.09

[9]The noise was generated with precisely the same mean and standard deviation; the only difference was the random seed.

This example demonstrates clearly that GP researchers must pay careful attention to generalization issues any time they use GP to make predictions on data other than training data.

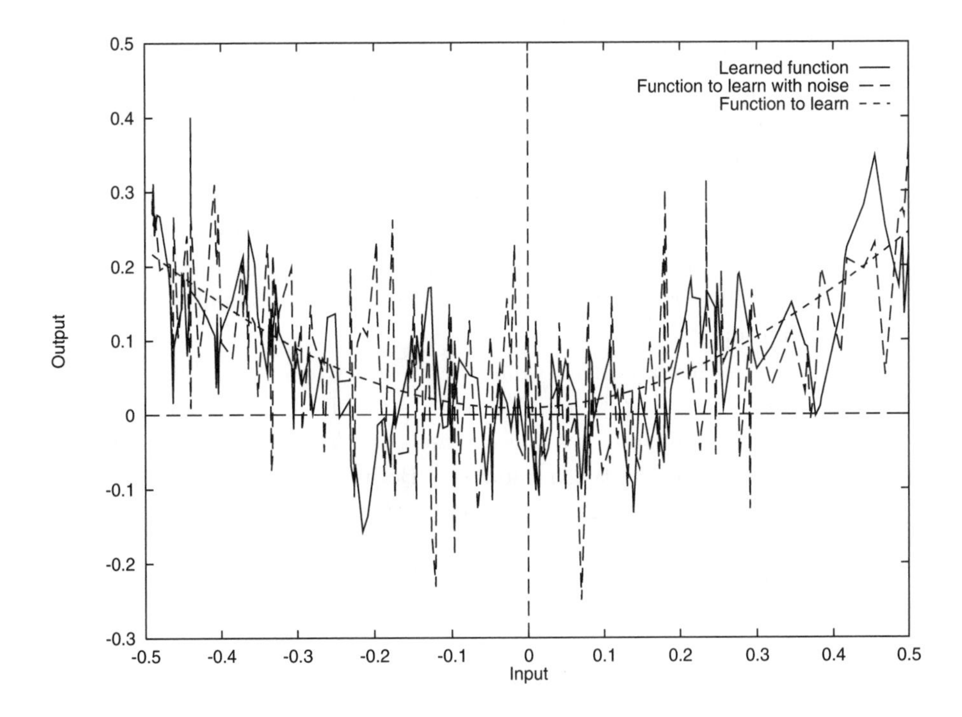

Figure 8.8
Results of learning with 16 degrees

8.5.2 Dealing with Generalization Issues

Overfitting and sampling error are the principal problems in machine learning generalization. There are several methods of addressing these issues statistically. The problem with these methods is that most GP systems are very computationally intensive. Performing enough runs to produce a statistically significant result is sometimes very difficult. So the GP researcher frequently opts for no statistical validation or the first type of validation to be discussed. Regardless of what is possible under particular circumstances, there is a range of options for dealing with generalization issues.

Training and Test Set

The traditional way for GP to test generalization is to split the data into a training and a test set. What this procedure really amounts to, however, is having *two* samples from *one* population. After training, the best individual on the training data is run on the test set. Because the best individual has never before seen any test data, its performance is then used as a measure of how well that individual will generalize on further samples drawn from the overall statistical population.

Though this approach is a good starting point to check generalization, it has some drawbacks. To begin with, what criterion should be used to choose the best individual on the training set? The individual with the highest fitness on the training set may be precisely the individual to avoid because it might overfit the training data. One way out of this problem would be to have a good stopping criterion. If we knew just when to stop the run, that would be a powerful solution.[10]

Another approach would be to test other individuals from the run on the test set to find a better generalizer. Based on experience, this method produces decent results as long as the individual also performs well on the training set. Because the test set becomes now part of the training set, however, this method has a serious problem: The test set ceases to be an independent set used only for measuring performance. Therefore, it is called the validation set and a new test set has to be introduced.[11]

Adding a New Test Set

Adding a third set of data that the system has never previously seen for performance-measuring purposes should be encouraged. But a single independent *test set* has problems all of its own. Again, variance and sampling error are important quantities to consider.

The Variance of the Mean Prediction Error

An example will help. Suppose a test set is comprised of 100 instances and the best individual has a mean error of only 1.2% on the test set. We should not yet be satisfied, and a few tests should be performed first.

The variance of the predictor on the test set is one essential test. The mean error of 1.2% hides whether the individual errors are all close to 1.2% or whether they frequently exceed 5% or −5%. A −5% error could be a disaster in many predictive models. It is possible to estimate the standard deviation of the population for the mean of the prediction error. Let μ be the mean of the absolute errors for the predictor. Let μ_i be the i-th absolute error. Let n be the sample size. An estimate of the standard deviation of the errors from the

[10]The increasing power of the online indicators lends hope that this may be possible some day.

[11]In the literature, test and validation set are sometimes interchanged.

population can then be given as:

$$\sigma_\mu = \sqrt{\frac{1}{n^2 - 1} \sum_{i=1}^{n} (\mu_i - \mu)^2} \tag{8.10}$$

If the distribution of the errors appears normal and if $\sigma_\mu = 0.8$, then one could cautiously conclude that 95% of the predictions will have an error no greater than 1.6%.

Measuring the Variance of Statistics Other Than Mean Error

The above equation is not valid for statistics like percentiles or percentage of predictions where the error is greater than a certain percentage. For those predictions, one can add new sets of test data. Were it possible to draw 1000 test sets of 100 instances each and to repeatedly test the best individual on those 1000 test sets, the variance caused by the sampling error could be calculated easily and one could assess in this way whether the best individual was worth keeping. In data-mining applications, such a brute force approach may be feasible. But in most GP applications, data is dear and one has to make the most of limited data.

There are powerful statistical techniques not covered here that may be used to *estimate* the bias and variance of any statistic from sparse data. The techniques go by such names as the *Jackknife* and the *Bootstrap*. If testing a GP individual uses measures other than the mean error, these techniques can improve the researcher's confidence in the integrity of the results. Or they can reveal variance so wide as to render the predictions made by the GP individual useless. Either way, crucial information can be obtained.

Special Techniques for Classification

Classification problems are increasingly being confronted by GP systems as more GP researchers attack real-world problems. Many classification models are structured so that the output indicates whether an instance is a member of a class or not. In that case, Kendall's t score measures how well the system distinguishes instances from non-instances [Walker et al., 1995]. It is a good starting measure with which to judge the predictions of a classification system.

8.6 Conclusion

The common thread running through this chapter is *observe*, *measure*, and *test*. The tools to estimate the predictive value of GP models exist and are easily applied to GP. The tools to improve GP systems or to test theories about, say, the crossover operator, exist and can be applied easily. What is needed is a decision to use them.

Exercises

1. What is a statistical population?
2. What is a sample?
3. Describe two basic statistical tools for GP.
4. Describe a moving average.
5. Give an overview of how to choose among input series.
6. What is an online tool, and what is an offline tool?
7. Give four different useful fitness measurements.
8. Give two examples for diversity measures in populations of program individuals.
9. Describe a method for measuring introns in a tree-structured individual.
10. What is generalization? What is overfitting? How can overfitting be avoided?

Further Reading

B. Efron,
THE JACKKNIFE, THE BOOTSTRAP
AND OTHER RESAMPLING PLANS.
Society for Industrial and Applied Mathematics,
Philadelphia, PA, 1995.

B. Efron and R.J. Tibshirani,
AN INTRODUCTION TO THE BOOTSTRAP.
Chapman & Hall, London, 1993.

J.S.U. Hjorth,
COMPUTER INTENSIVE STATISTICAL METHODS VALIDATION,
MODEL SELECTION, AND BOOTSTRAP.
Chapman & Hall, London, 1994.

S.M. Weiss and C.A. Kulikowski,
COMPUTER SYSTEMS THAT LEARN.
Morgan Kaufmann, San Francisco, CA, 1991.

Part III

Advanced Topics in Genetic Programming

9 Different Varieties of Genetic Programming

Contents

Structures Used in GP

We have already seen how the GP arena has been populated with numerous different approaches to program evolution. It is astonishing how the simple theme can be varied without destroying the basic mechanism. In this chapter we will take a closer look at GP variants. Table 9.1 gives an indication of which kinds of structure have been used with GP. As we can see, some of these variants are very different and quite incompatible with each other, a strength rather than a weakness of the fundamental idea.

Structure name	Description	Source
S-expression	tree structure	[Koza, 1992d]
GMDH primitives	tree structure	[Iba et al., 1995b]
TB	linear postfix	[Cramer, 1985]
JB	linear prefix	[Cramer, 1985]
bits	linear genomes	[Banzhaf, 1993b]
bits	machine code instructions	[Nordin, 1994]
abstract date types	lists, queues, stacks	[Langdon, 1995b]
production rules	grammars	[Whigham and McKay, 1995]
production rules	graph structure	[Jacob, 1996a]
PADO	graph structure	[Teller and Veloso, 1995b]
cellular encoding	tree grammars	[Gruau, 1993]

Table 9.1
Some of the many different structures used for GP

The following sections present a selection of these variants, starting with a look at tree-based GP, currently in most frequent use in the GP community.

9.1 GP with Tree Genomes

The general working of GP on trees was introduced in Chapter 5.[1] Here we shall review the different types of operators that have been introduced over the years for manipulating tree structures. As mentioned earlier, the simplest operator is the *reproduction operator* which does nothing but copy the individual into the next generation. The next more complicated operator is the *mutation operator* which acts on a single tree at a time. Table 9.2 gives an overview of what sorts of mutation operators have been used with trees. Figures 9.1–9.6 summarize graphically the effect of certain mutation operators.

In many applications, the mutation operator is not applied directly to a reproduced individual, but is applied to the result of

[1]There are trees of different kinds. Mostly we shall use "expression" trees in our examples.

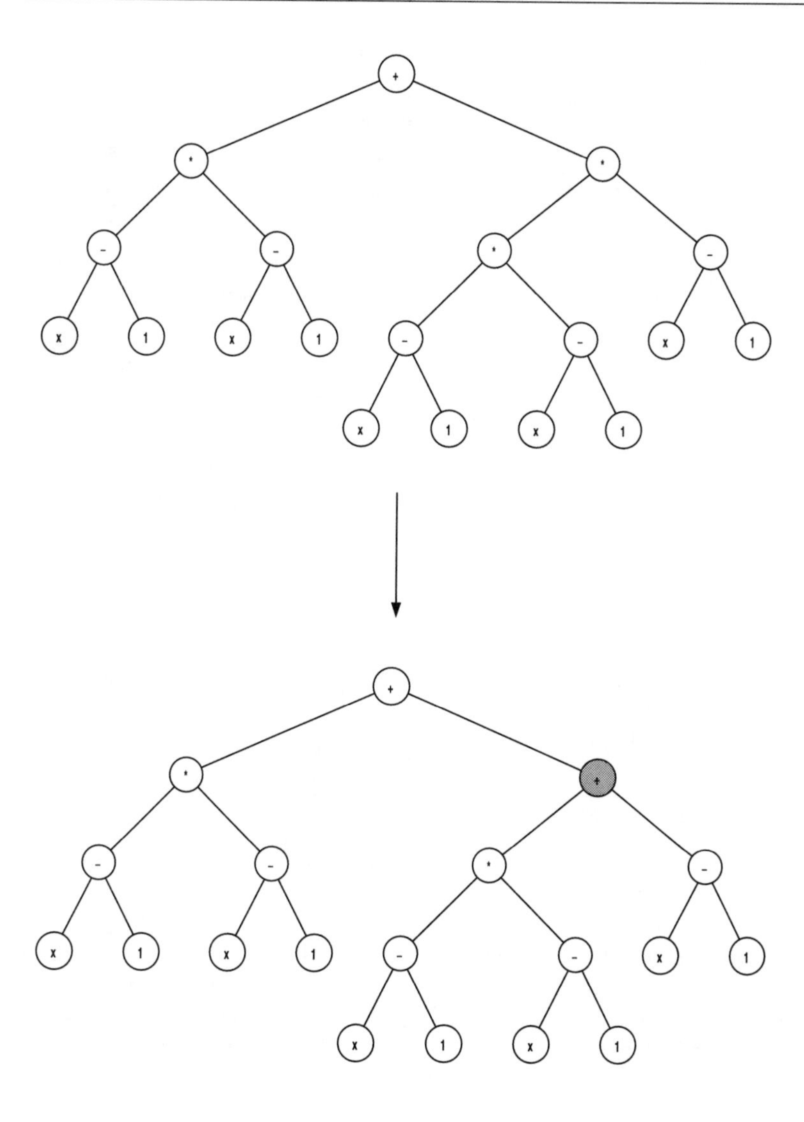

Figure 9.1
Different mutation operators used in tree-based GP: point mutation

another operator, canonically working on two trees: crossover. If crossover is used as the follow-up operator to reproduction, two individuals are copied into the next generation. Subsequently, one node is selected in each tree and the subtree below this node is cut out and transferred to the location of the node selected in the second tree, and vice versa. The function of crossover is to swap genetic material between trees.[2] As Table 9.3 shows, various crossover operators are applicable to trees and it is a matter of *a priori* choice and implementation which one to take in an actual run.

[2]Usage of the crossover operator to generate *one* new individual is also very common.

Table 9.2
Mutation operators applied in tree-based GP

Operator name	Description of effect
point mutation	single node exchanged against random node of same class
permutation	arguments of a node permuted
hoist	new individual generated from subtree
expansion mutation	terminal exchanged against random subtree
collapse subtree mutation	subtree exchanged against random terminal
subtree mutation	subtree exchanged against random subtree
gene duplication	subtree substituted for random terminal

Operator name	Description of effect
subtree exchange crossover	exchange subtrees between individuals
self crossover	exchange subtrees between individual and itself
module crossover	exchange modules between individuals
context-preserving crossover	exchange subtrees if
SCPC	coordinates match exactly
WCPC	coordinates match approximately

Table 9.3
Crossover operators applied within tree-based GP. CPC: context-preserving crossover

There should be a close correspondence between crossover and mutation operators, the former being applied to two individuals and swapping information between them and the latter being applied to one individual alone with additional random action. One can assume that a crossover operator can be constructed to correspond to each mutation operator, and vice versa. The fact that this has not been done systematically in the past says more about the status of systematization in GP than about the potential of those operators.

Figures 9.7–9.9 summarize graphically the effect of certain crossover operators on trees.

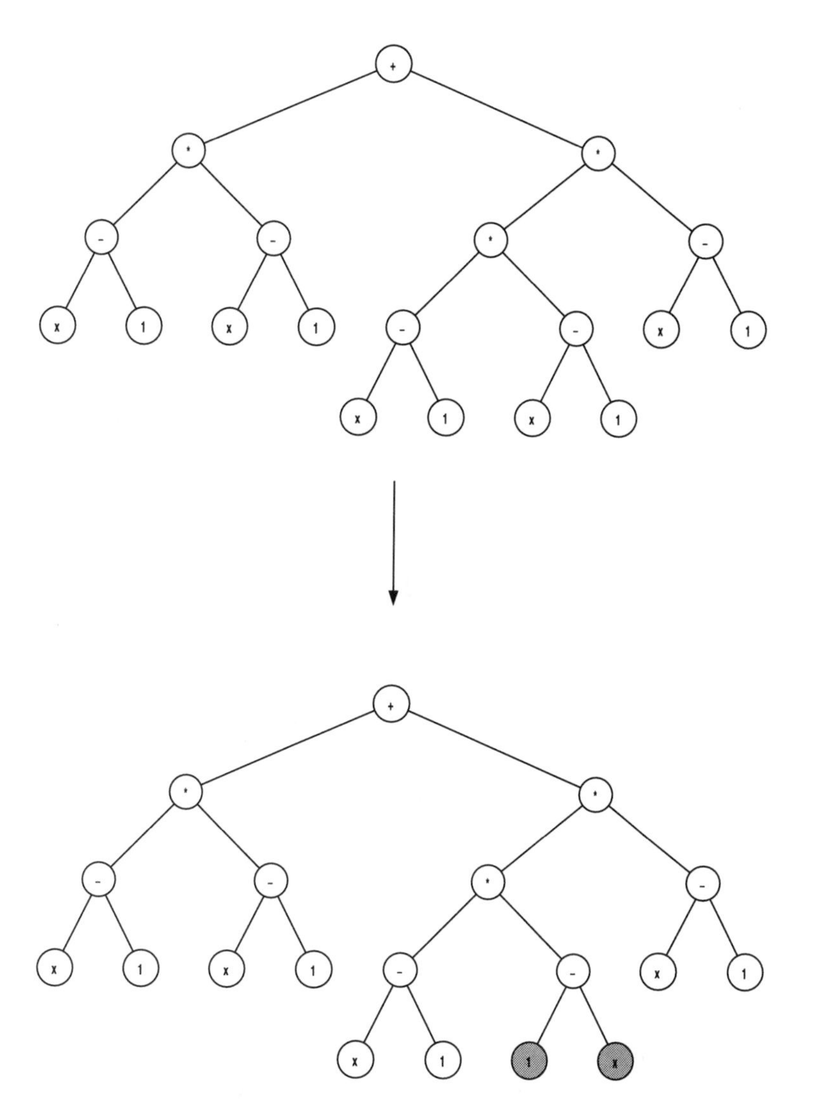

Figure 9.2
Different mutation operators used in tree-based GP: permutation

9.2 GP with Linear Genomes

Linear GP acts on linear genomes, like program code represented by bit strings or lines of code for register machines. Before we discuss a few variants of linear GP we should contrast the behavior of linear GP to that of tree-based GP.

When we consider the effects of applying different operators to tree individuals there is one aspect of the hierarchical tree representation that specifically attracts our attention: in a hierarchical representation there is a complicated interplay between the order of execution of the program and the influence of changes made at var-

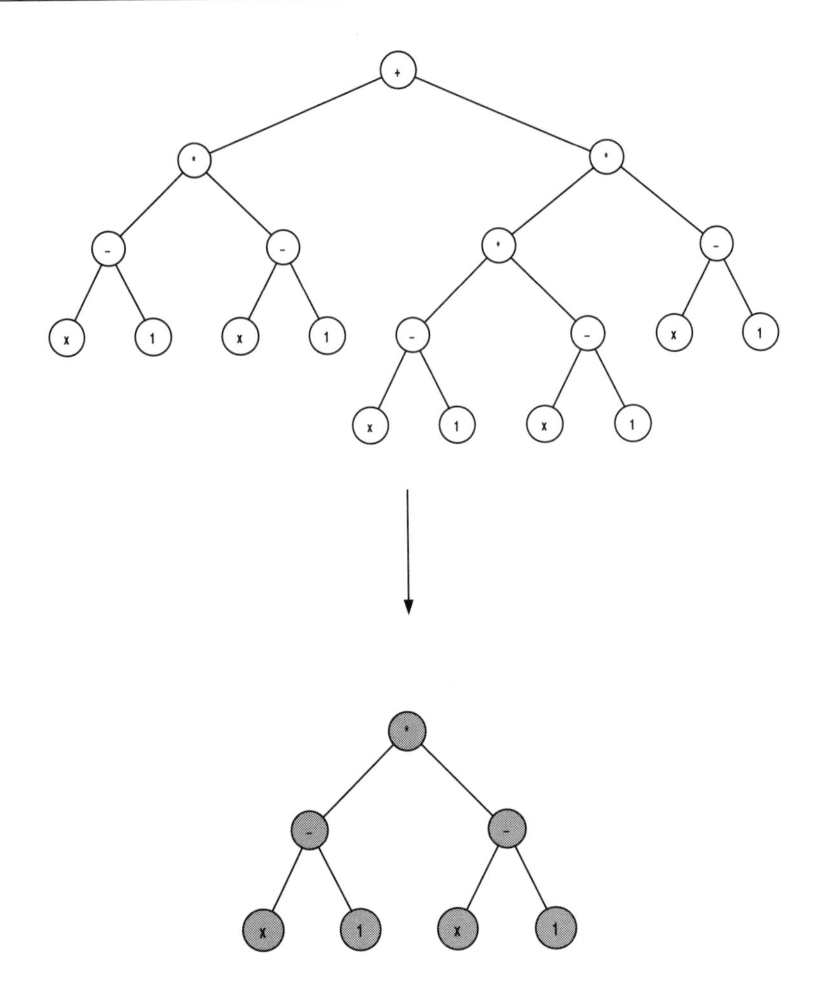

Figure 9.3
Different mutation operators used in tree-based GP: hoist

ious levels of the tree. Trees are usually executed in a depth-first manner, i.e., the left branches of nodes are evaluated before the right branches. Thus order of execution is from left to right. However, the influence of change is determined in a hierarchical representation by the level at which a particular node resides. An exchange at a deeper level in the program tree will naturally have fewer consequences than an exchange at a higher level.

For the following consideration we have to make two simplifying assumptions, they are, however, not very restrictive. First, we shall assume that our tree is constructed from binary functions. Second, we shall assume that trees are (nearly) balanced.

Operator Hit Rate

Definition 9.1 *The* **hit rate** *of an operator in relation to a feature of nodes is the probability by which one node or more nodes in the tree possessing this particular feature are selected for operation.*

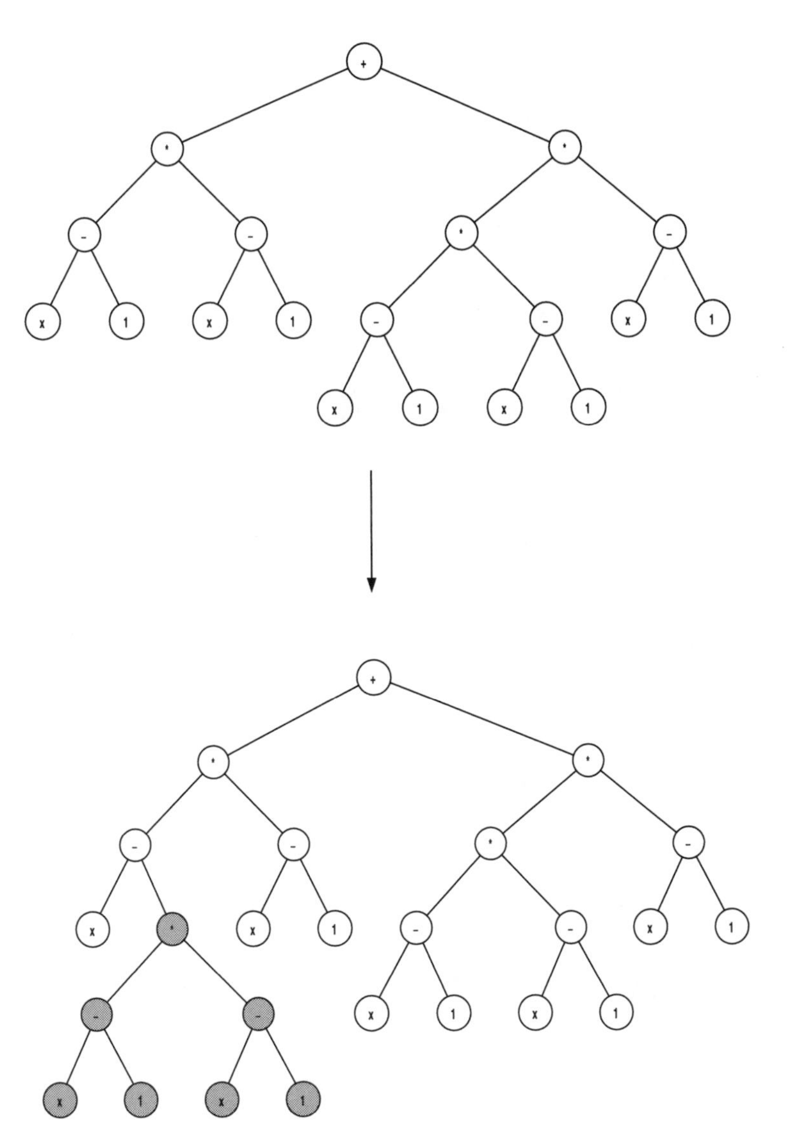

Figure 9.4
Different mutation operators used in tree-based GP: expansion mutation

Put plainly, the hit rate of nodes in the deepest level of a tree, i.e., nodes with the feature of being on the deepest level, is – assuming balanced binary trees – 1/2. It follows that all nodes in the rest of the tree possess a hit rate of 1/2 in total, with 1/4 going to the second deepest level, 1/8 to the next higher level, and so on. Hence, with overwhelming probability, nodes at the lowest level are hit most often for crossover or mutation operations.[3]

[3]Very often, therefore, precautions are taken in tree-based GP systems to avoid high hit rates at the lowest level of a tree by biasing the hit rate.

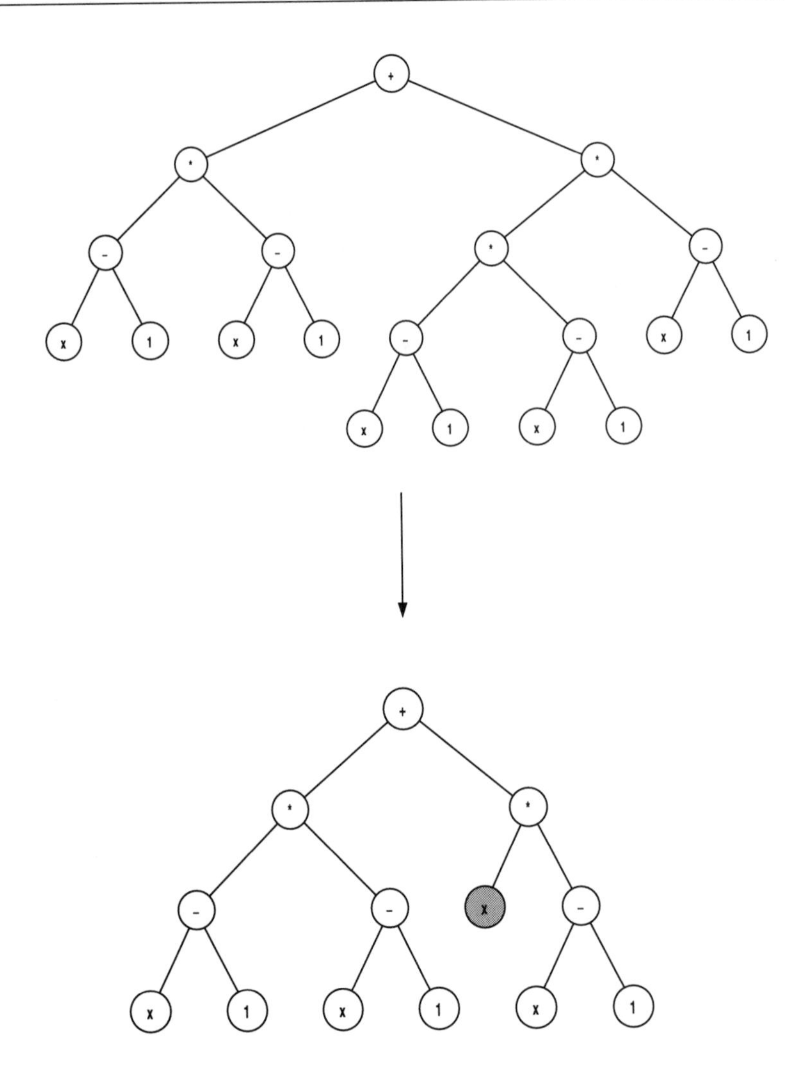

Figure 9.5
Different mutation operators used in tree-based GP: collapse subtree mutation

Together, the two facts mentioned above result in the conclusion that in tree-based GP, the most frequent changes are small. This is true for all operators uniformly selecting nodes from a tree.

The influence of change in a linear structure can be expected to follow the linear order in which the instructions are executed. An exchange later in the sequence will have fewer behavioral consequences than an exchange earlier on. The hit rate of operators, on the other hand, in relation to the position of a node or instruction in that sequence is equal for the entire linear genome. Hence, it can be expected that in GP with linear representations, changes of all sizes are equally frequent. This is true for all operators uniformly selecting nodes from a sequence.

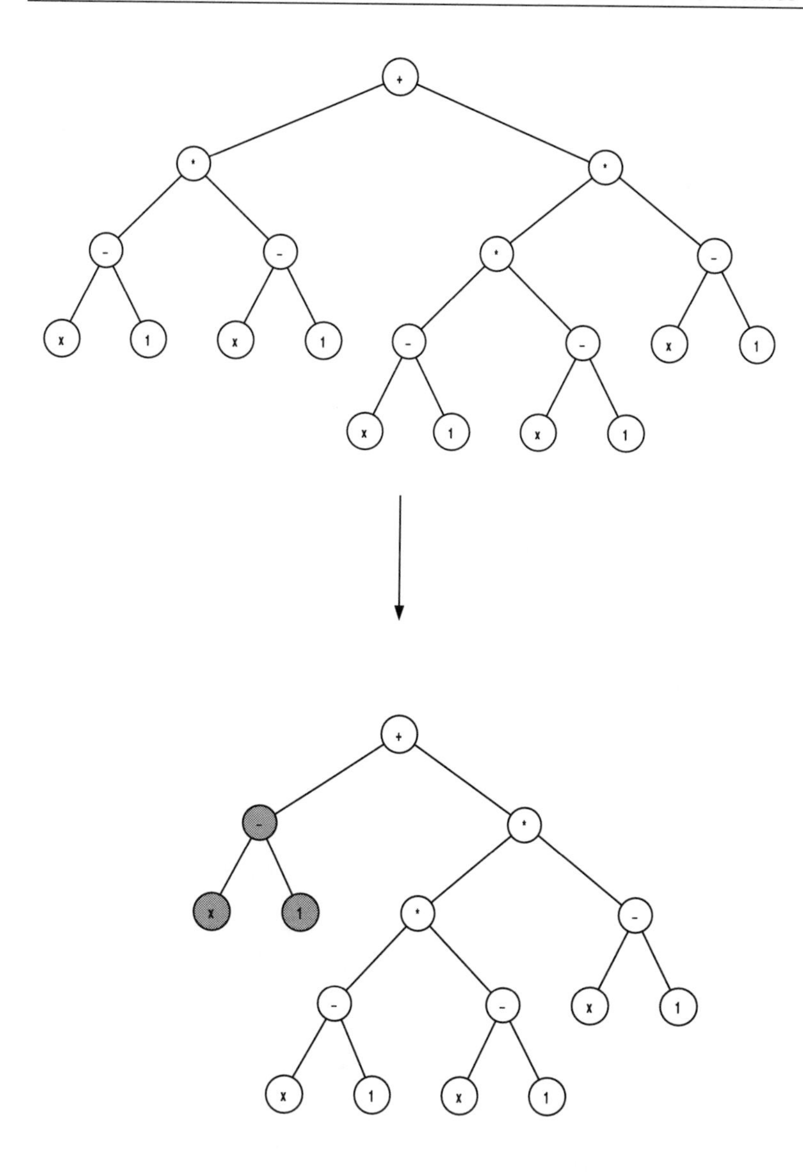

Figure 9.6
Different mutation operators used in tree-based GP: subtree mutation

Figure 9.10 demonstrates the behavior of tree-based GP for a simple regression problem. The amount of change (improvements as well as deteriorations) relative to the fitness of predecessors is recorded. Small changes are by far the most frequent ones.

Figure 9.11 shows a marked difference to Figure 9.10. Here, the same regression problem has been approached with machine language GP. Changes of all sizes are more frequent, which can be seen from the distribution of the aggregations of black dots.

We should keep in mind that results from this comparison are not yet conclusive, but it is an interesting research area to compare the behavior of different representations in genetic programming.

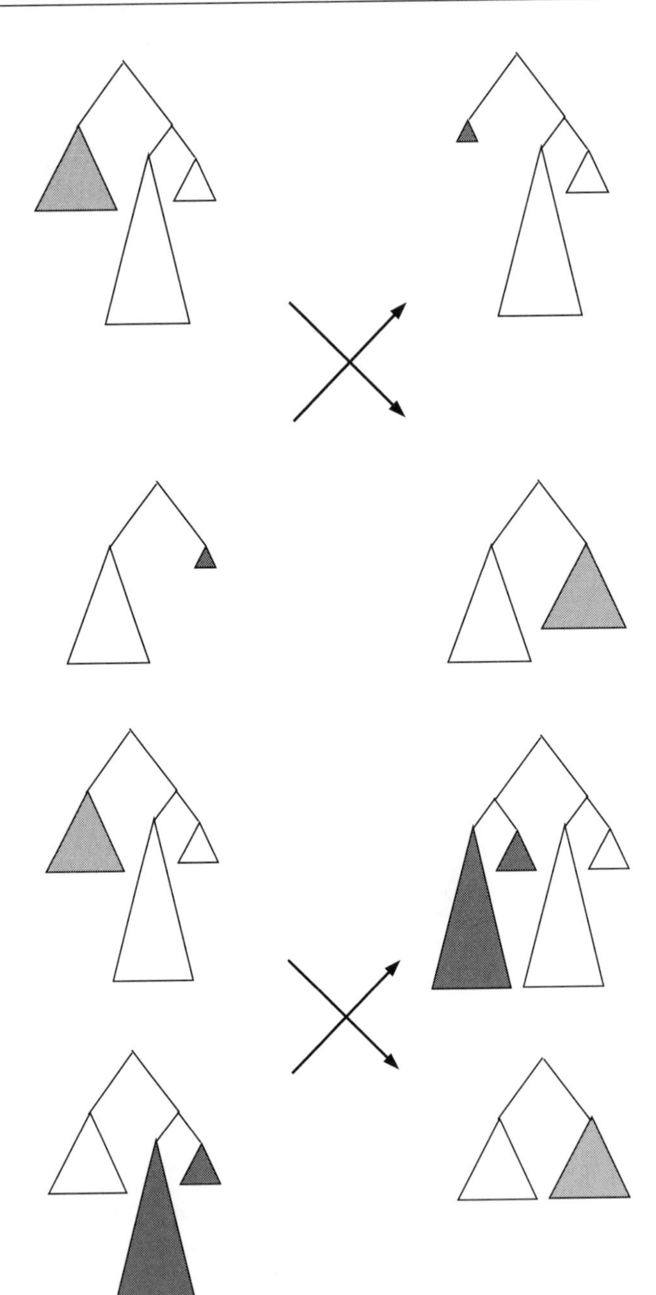

Figure 9.7
Different crossover operators used in tree-based GP: subtree exchange crossover

Figure 9.8
Different crossover operators used in tree-based GP: selfcrossover

9.2.1 Evolutionary Program Induction with Introns

Wineberg and Oppacher [Wineberg and Oppacher, 1994] have formulated an evolutionary programming method they call *EPI* (evolutionary program induction). The method is built on a canonical genetic algorithm. They use fixed length strings to code their indi-

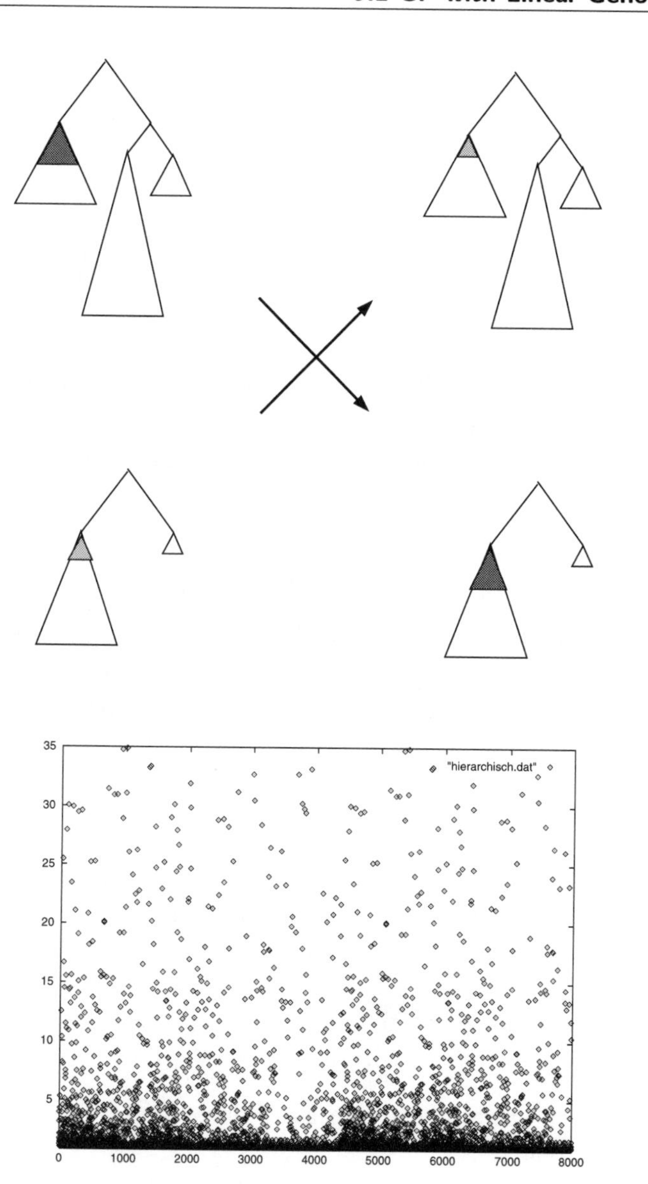

Figure 9.9
Different crossover operators used in tree-based GP: module crossover

Figure 9.10
Changes during an evolutionary run for a regression problem in tree-based GP. Each dot denotes a change relative to predecessors. Small changes are most frequent.

viduals and a GA-like crossover. However, their linear genomes code for trees that are identical to program trees in tree-based GP. The fixed length of individuals and their coding scheme imply a predefined maximal depth of trees in evolutionary program induction with introns. The coding is constructed to maintain a fixed structure within the chromosome that allows similar alleles to compete against each other at a locus during evolution. As a consequence, the genome normally will be filled with introns, which they argue is beneficial to the search process [Wineberg and Oppacher, 1996].

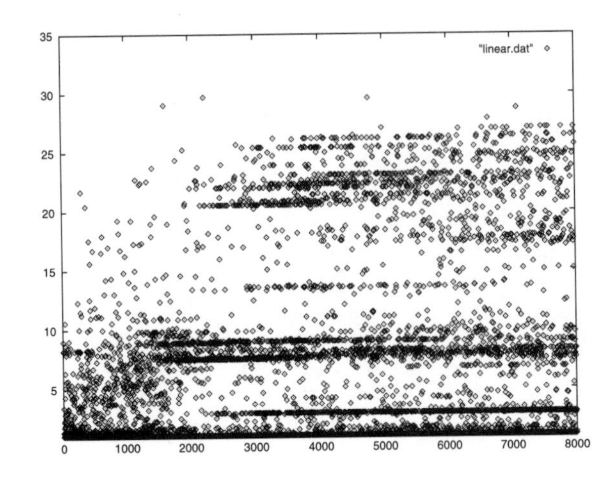

Figure 9.11
Changes during an evolutionary run for a regression problem in linear (AIMGP) representation. Each dot symbols a relative change to predecessors. Changes show levels that are most frequent.

9.2.2 Developmental Genetic Programming

Genotypes vs. Phenotypes

Developmental genetic programming (*DGP*)[4] is an extension of GP by a developmental step. In tree-based GP, the space of genotypes (search space) is usually identical to the space of phenotypes (solution space) and no distinction is made between genotypes and phenotypes: an individual is always identified with a computer program of syntactically correct structure which is interpreted or compiled for execution. Developmental genetic programming, on the other hand, maps binary sequences, *genotypes*, through a developmental process [Banzhaf, 1994] into separate phenotypes. These *phenotypes* are the working programs with the syntactic structure of an LALR(1) grammar.

Two abstract properties of how GP approaches a problem are evident:

1. GP optimizes a fitness measure that reflects the behavior of a program

2. The optimization problem is *hard-constrained*, since the set of feasible points in search space is identical to the solution space.

In the tree-based GP approach, the second property is reflected by the fact that it is a constrained search process: creation and variation are only allowed to yield feasible programs.

Feasible vs. Infeasible Structures

However, given a certain syntax, a function and terminal set, and a maximal sequence length, the set of infeasible programs is often extremely large compared to the set of feasible programs – as every

[4]We had earlier called this method *binary genetic programming* or *BGP* [Keller and Banzhaf, 1996] but have now decided DGP is more to the point.

programmer knows from painful personal experience. Despite this fact, infeasible programs may contain syntactically correct parts that could prove useful during evolution. A huge amount of genetic diversity might become unavailable to the search process if it could not make use of these parts. To make the diversity of infeasible programs accessible, the search process should be unconstrained, i.e., the search operators should be able to accept and to produce an arbitrary sequence. Prior to fitness evaluation, however, an infeasible program should be mapped into a feasible program. DGP implements this function by using a specifically designed *genotype–phenotype mapping.*

There is one further argument to add plausibility to these considerations: it is sometimes preferable to search in a space with more dimensions than the solution space. More dimensions enable the search to cut through dimensions instead of being forced to evolve along longer paths in solution space. One can expect that searching with shortcuts will increase the probability of finding better solutions dramatically.

As we already mentioned, genotypes in DGP take the simplest and most universal form of a binary representation. The representation of phenotypes (programs), on the other hand, is determined by the language used. Simple search operators, as in standard GAs, can be used in DGP.

The Genotype–Phenotype Mapping

Once the genotypic representation has been fixed, we can define the mapping between genotype and phenotype. At least two steps are needed to make the genotype–phenotype mapping work:

1. The high-dimensional genotype (binary sequence) must be *transcribed* into the low-dimensional phenotypic representation (raw symbol sequence).

2. In case the raw symbol sequence is infeasible, it must be *edited* into a feasible symbol sequence.

Transcription

For transcription, the obvious approach is to identify each symbol from the terminal and symbol set – which we could call token – with at least one binary subsequence which we shall call codon. If all codons have equal length, there is a mapping from the set of all n-bit codons into the set of all symbols. A binary sequence of m codons thus represents a raw sequence of m symbols.

For instance, if `a`, `b`, `+` are represented by 00, 01, 10, then the raw (and feasible) sequence `a+b` is given by 001001. In this example, we have a three-dimensional solution space if we consider all syntactically correct symbol sequences of length 3. The search space, however, has 6 dimensions.

Editing

The raw sequence produced by transcription is usually illegal, and an editing step is needed to map a raw sequence into an edited sequence. Here we consider only LALR(1) grammars [Aho, 1986], i.e., the DGP as discussed here can only evolve programs in languages defined by such a grammar. Many practically relevant languages like C are LALR(1) languages. Let us introduce the notion of a *legal symbol set*: a symbol that represents a syntax error at its position in a raw sequence will be called *illegal*. When, during parsing of a raw sequence, an illegal symbol s is detected, the corresponding legal symbol set is guaranteed to be computable in LALR(1) grammars. This set is computed and a subset of it, the *minimal-distance set*, is selected. It holds the candidates for replacing s. Depending on the actual mapping, there can be more than one such symbol. In order to resolve this ambiguity, the symbol with lowest integer value among all codons of closest symbols will be selected.

If this procedure is applied, editing will often produce symbol sequences that terminate unfinished. For instance, editing could result in the unfinished arithmetical expression $\sin(a) * \cos(b)+$. In order to handle this problem, we assign a *termination number* to each symbol. This number indicates how appropriate the symbol is for the shortest possible termination of the sequence. In the above expression, for instance, a variable would terminate the sequence, while an operator symbol like `sin` would call for at least three more symbols, (, variable,). The termination number of a symbol is simply the number of additionally needed symbols to terminate the expression. Thus, in the example, a variable-symbol like a has termination number 0, while a symbol like sin has termination number 3. Note that the number of symbols *actually* needed for termination when using a certain symbol can be larger than its termination number. For instance, there could be an open parenthesis prior to the end of the unfinished sequence, which had to be closed before termination. Such context-sensitive circumstances shall not be reflected in a termination number. In case there are several symbols with equal minimal termination number, that one is taken which is encoded by a codon with the lowest index among all codons of the symbols in question.

Translation

Editing yields a feasible symbol sequence which is subsequently completed by language-specific standard information such as a function header. Finally, the resulting program is interpreted or compiled and executed in order to evaluate its fitness.

Consider the example in Table 9.4. The genotype 000 001 011 is transcribed into a raw sequence $ab*$. Editing scans a as first and legal symbol. It then scans b as an illegal symbol with 001 as its codon in the genotype. $\{+, *\}$ is the legal symbol set. The symbol closest to b is $*$. Thus, b gets replaced by $*$, thereby terminating the removal of

Binary code	Token
000	a
001	b
010	+
011	*
100	a
101	b
110	+
111	*

Table 9.4
Example of redundant genetic code mapping codons into tokens

the syntax error. The partially edited raw sequence now equals $a**$. Editing continues, replacing the last symbol $*$ with b.

The edited sequence $a*b$ is passed on to translation, which adds, for instance, a function frame. The translated sequence might look like `double fnc(double a, double b){return a*b;}`. This sequence could now be integrated with other edited sequences, for example, into a main program that a C compiler could handle. Execution would then allow the fitness of the corresponding genotypes to be evaluated.

As can be seen, 000 001 011 gets mapped into the phenotype $a*b$. However, 010 011 101, 000 011 001, and 100 111 101 all get mapped into $a*b$ as well. When analyzing this phenomenon, it becomes clear immediately that both the redundancy of the genetic code and the editing mechanism are responsible for this effect.

Search Operators

The following unconstrained search operators are used in DGP. They perform operations on sets of codons and bits within codons.

- Creation
- Mutation
- Recombination

Creation generates individuals as random binary sequences that consist of m-bit codons. *Mutation* may work within codons or it may change bits of different codons at the same time. A *codon-limited* mutation corresponds to exploring few dimensions in the same subspace, while a *codon-unlimited* mutation may explore the search space in many dimensions. These mutation operators do not produce vast changes in the genotype because they do not replace complete syntactic units, as they would in tree-based GP. Thus, this type of mutation seems closer to natural mutation. *Crossover* is implemented in developmental genetic programming by a standard GA operator like one-

or two-point crossover. In particular, swapped subsequences may be of different length, thus producing offspring with different length.

Figure 9.12 outlines the central piece of the DGP algorithm. Addition of a genotype–phenotype mapping step could be considered for any GP algorithm. The genotype–phenotype mapping must always happen prior to fitness evaluation.

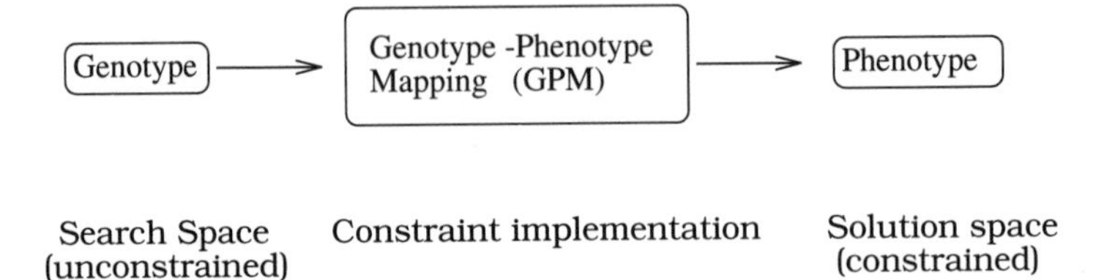

Figure 9.12
The central piece of the DGP algorithm, the genotype–phenotype mapping

9.2.3 An Example: Evolution in C

Major DGP Parameters

DGP can be used for the evolution of a program in an arbitrary LALR(1) target language. Assume DGP has been evolving FORTRAN programs for some problem, and now we want it to produce C programs. In principle, four system parameters must be changed. First, we exchange the FORTRAN parser for a C parser. Second, we substitute those target symbols in the genetic code that are specific of FORTRAN by the semantically corresponding C symbols. Third, the editing phase must supply C-specific phenotype parts, like C-style function headers. Fourth, any commercially available C compiler will be called prior to fitness evaluation. DGP will now happily evolve C programs for the given problem.

Sample Problem

Let us consider an example for the evolution of C programs along the lines of [Keller and Banzhaf, 1996]. Since C is the chosen target language, each phenotype will be a symbol sequence obeying C syntax. The test problem is a symbolic function regression on a four-dimensional parameter space. The function we would like to model is

$$f = \sin(m) \cdot \cos(v) \cdot \frac{1}{\sqrt{e^q}} + \tan(a) \tag{9.1}$$

All parameter values shall be real-valued. The domain of the test problem suggests variables, unary and binary arithmetic functions, and parenthesis operators as elements of the terminal and function sets. To protect against division by zero, we use a division function D that returns the reciprocal value of its single argument. If the argument equals zero, the function returns 1. We supply a protected square root function *sqrt* that returns the square root of the absolute

Protected Functions

value of its argument. Furthermore, an overflow-protected exponential function $exp(x)$ is provided that normally returns e^x, but in case the value of x causes an overflow, returns 1.

The code shown in Table 9.5 is used.

Binary code	Token	Abbr. token
0000	+	+
0001	*	*
0010	*	*
0011	D	D
0100	m	m
0101	v	v
0110	q	q
0111	a	a
1000	(	(
1001	)	)
1010	sin	S
1011	cos	C
1100	tan	T
1101	sqrt	R
1110	exp	E
1111	)	)

Table 9.5
Genetic code of the example function regression problem, including one-character abbreviations

The codons get mapped into a set of 16 symbols, featuring 14 different symbols of variables and operations. This genetic code is redundant with respect to the multiplication and the closing-parenthesis operator. This might have positive and negative effects on convergence. For instance, the redundancy of the multiplication operator could result in phenotypes that consist of many such operators. This can be advantageous since the problem function features two such operators. On the other hand, the redundancy concerning the closing-parenthesis operator could pose a handicap to evolution, since it enlarges the probability that a needed long subexpression never emerges.

The genotype length is fixed at 25 codons; that is, each genotype consists of 25 codons. Since there are 16 different codons in the genetic code, the search space contains 16^{25} or approximately 1.3E30 genotypes. When using the unrestricted mutation operator described above, DGP faces $25 \cdot 4 = 100$ degrees of freedom, since each codon consists of 4 bits. In other words, in addition to the relatively large size of the search space, this space is high-dimensional, with 100 dimensions.

Sometimes the actual length of phenotypes after repair can surpass 25 target symbols. Imagine, for instance, the – improbable – case of a phenotype like

```
D(D(D(D(D(D(D(D(D(D(D(D(D (a)))))))))))))
```

Prior to the space character within the symbol sequence, there are 25 symbols. Shortest possible termination of the phenotype still requires appending 15 more symbols, most of them being) to close open parenthesis levels.

Due to the real-valued four-dimensional parameter space, a fitness case consists of four real input values and one real output value. We supply a set of fitness cases with randomly chosen input values. Although the problem is significantly harder than many regression problems used to test GP systems, we increase the difficulty further by providing only ten fitness cases to the system. Note that GP systems will take advantage of this situation and develop a simpler model of the data than would have evolved with a large number of fitness cases.

An Example Result

Runs lasted for 50 generations at most, with a population size of 500 individuals. In one experimental run, the genotype

```
1100 0010 1000 0111 1001 0010 1101 1001 0111 1110
0000 1011 1001 1110 1001 1010 1101 0011 1100 1111
0101 1010 0110 1110 0001
```

evolved. What must happen to this genotype in order to evaluate the fitness of the corresponding individual? Transcription derives from it the raw symbol sequence (using the one-letter abbreviations)

```
T*(a)*R)aE+C)E)SRDT)vSqE*
```

Repairing transforms this illegal sequence into

```
{T((a)*R(a+m))+(S(D((v+q+D}
```

Since this sequence is unfinished, repairing terminates by completing the sequence into

```
{T((a)*R(a+m))+(S(D((v+q+D(m)))))}
```

Finally, editing produces

```
double ind(double m,double v,double q,double a)
{return T((a)*R(a+m))+(S(D((v+q+D(m)))));}
```

A C compiler takes over to generate an executable that is valid on the underlying hardware platform. This executable is the final phenotype encoded by the genotype. As a formula this reads

$$\tan(a * \sqrt{a+m}) + \sin(\frac{1}{v+q+\frac{1}{m}}) \tag{9.2}$$

This phenotype featured a normalized least-square error fitness of 0.99 (with 1.0 as perfect fitness), which is quite acceptable, considering the size of the search space and the use of mutation as the only search operator besides creation.

As a side effect, this example shows that GP can be used to find "simpler good models" (i.e., solutions) for a "reality" (i.e., a problem) described by fitness cases. While the perfect solution is an expression that uses the cosine and the exponential function, the almost perfect solution above does without these functions. This inherent potential of GP can be applied to problems where a certain degree of compromise between quality and complexity is acceptable.

9.2.4 Machine Language

Commercial computers are – at a low level of abstraction – register machines. A register machine executes a program by reading the numbers in its memory cells and interpreting them as operations between memory and registers. The registers form a small set of memory cells *internal* to the processor. A typical machine code instruction may look like `x=y+z`, which should be interpreted as "add the content of register y to the content of register z and store the result in register x." The instructions could also access and manipulate the RAM memory. The instructions `x=memory cell[y]` will interpret the content of register y as an address pointing to a memory cell whose value should be moved into register x. Instructions that realize jumps and comparisons enable the creation of conditional branches and loops.

A simple four-line program in machine language might look like this:

```
1: x=x-1
2: y=x*x
3: x=x*y
4: y=x+y
```

This program uses two registers, x, y, to represent a function, in this case the polynomial $f(x) = (x-1)^2 + (x-1)^3$. The input for the function is placed in register x and the output is what is left in register y when all (four) instructions have been executed. Register

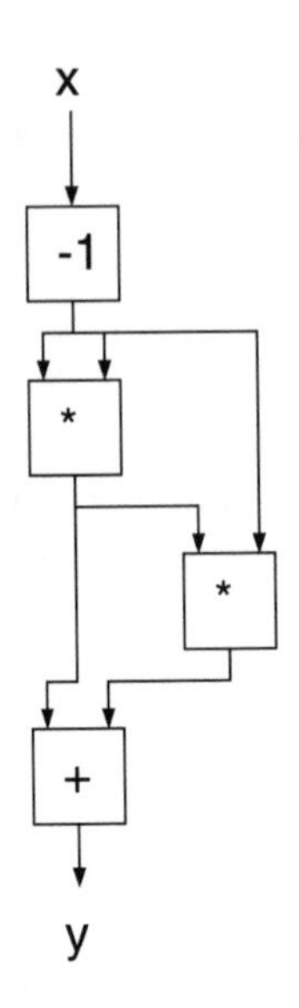

Figure 9.13
The dataflow graph of the $(x-1)^2+(x-1)^3$ polynomial

y is initially zero. Note that registers are variables which could be assigned at any point in the program. Register y, for instance, is used as temporary storage in instruction 2, before its final value is assigned in the last instruction. The program has more of a graph than a tree structure, where the register assignments represent edges in the graph. Figure 9.13 shows the dataflow graph of the function computation for $(x-1)^2+(x-1)^3$. We can see that the machine code program corresponds to this graph very closely. Compare this to an equivalent individual in a tree-based GP system such as in Figure 9.14.

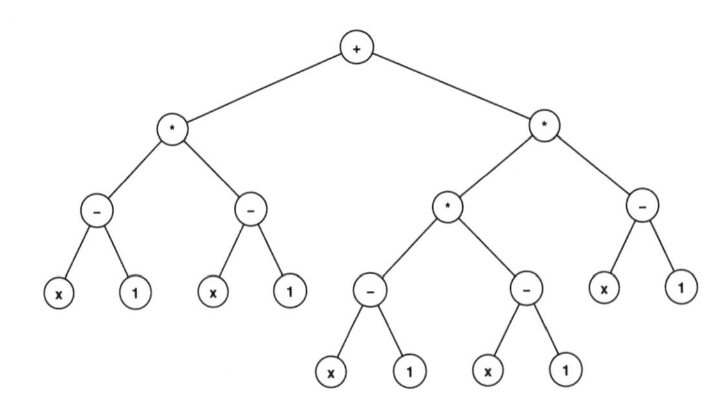

Figure 9.14
The representation of $(x-1)^2+(x-1)^3$ in a tree-based genome

A disadvantage of the more compact register machine representation might be that it may be more brittle because of the dependencies between registers. On the other hand, temporary storage of values in registers can be seen as a simple substitute for automatically defined functions (see Section 10.1), since the reuse of calculated values could under certain conditions replace the need to divide the programs into functions [Nordin, 1994].

Machine code programming is often used when there is a need for very efficient solutions – with applications having hard constraints on execution time or memory usage. In general, the reasons for using machine code in GP – as opposed to higher-level languages – are similar to those for programming in machine code by hand:

- The most efficient optimization can be done at the machine code level. This is the lowest level of a program and is also where the biggest gains are possible. Optimization could aim at speed or memory space or both. For instance, GP could be used to evolve short machine code subroutines with complex dependencies between registers and memory.

- Another reason for using machine language is that high-level tools might simply not be available for a target processor. This is often the case for special-purpose processors in embedded control applications.

- Machine code is often considered hard to learn, program, and master. Although this is a matter of taste, sometimes it could be more convenient to let the computer evolve small pieces of machine code programs itself rather than learning to master machine code programming.

Although it is possible to evolve machine code with a tree-based system (by building trees of machine instructions), there are additional reasons for using binary machine code directly:

- The GP algorithm can be made very fast by having the individual programs in the population in *binary machine code.* This method eliminates the interpreter in the evaluation of individuals. Instead, evaluation requires giving control of the processor to the machine code constituting the individual. This accelerates GP around 60 times compared to a fast interpreting system [Nordin and Banzhaf, 1995b]. As a result, this method becomes the fastest approach to GP.

- The system is also much more memory efficient than a tree-based GP system. The small system size is partly due to the fact that the definition of the language used is supplied by the CPU designer in hardware – there is no need to define the language and its interpretation in the system. Low memory consumption is also due to the large amount of work expended by CPU manufacturers to ensure that machine instruction sets are efficient and compact. Finally, manipulation of individuals as a linear array of integers is more memory efficient than using symbolic tree structures.

- An additional advantage is that memory consumption is stable during evolution with no need for garbage collection. This constitutes an important property in real-time applications.

In the rest of this section we will take a look at the methods used to realize machine code GP systems.

Evolving Machine Code with an Interpreting System

The most straightforward approach to GP with machine code would be to define a function and terminal set suitable for a machine code variant. The terminal set would be the registers and the constants while the function set would be the operators used. The problem with such an approach is that one cannot crossover arbitrary subtrees without violating the syntax of the machine language used. A solution is to have a strongly typed GP system where one can only crossover subtrees of the same type [Montana, 1994]. In strongly typed GP (STGP) one cannot crossover a subtree returning a Boolean with a subtree returning a float, to give one example. In a similar way, we would not crossover nodes that are registers with nodes that are operators. Figure 9.15 illustrates how the small machine code segment representation below could be represented in a strongly typed tree structure:

```
1: x=x-1
2: y=x*x
3: x=x*y
4: y=x+y
```

There are three different types of nodes in this figure. Terminal nodes represent constants or register values. Operator nodes calculate values from two of their operands and store the result in the register specified by their first operand. The operands are tied together with `I` nodes which do not represent a calculation but instead are used to give the operand nodes a defined execution order. In our example we execute the operand nodes in depth-first order.

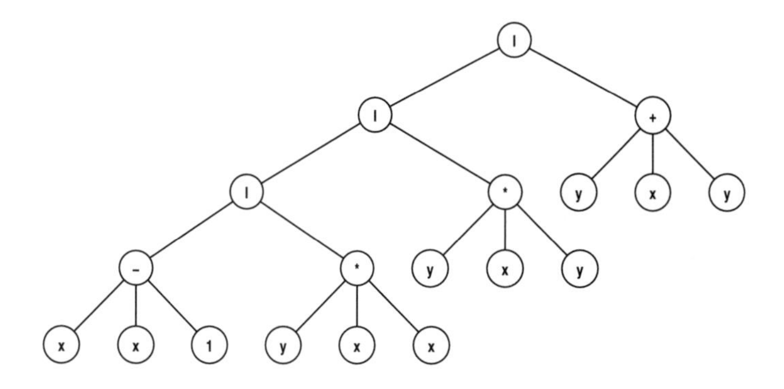

Figure 9.15
Representation of machine code language individual in a tree structure

Strongly typed GP is similar to grammar-restricted GP, which uses a grammar of the language to restrict, e.g., crossover [Whigham, 1995a]. This way it is also possible to guarantee correct operation of GP operators on individuals. This method is also well suited for evolution of machine code. There are also other methods for *structure preserving crossover* which might be used [Koza, 1992d].

These tree-based GP methods for evolution of machine language might seem slightly forced. Most researchers have used a crossover method, instead, which is similar to GA crossover due to the linear structure of machine code programs. This crossover operator exchanges pieces of machine code in a linear representation.

The JB Language

One of the earliest approaches to evolution of computer programs similar to machine code is the JB language [Cramer, 1985]. Cramer formulated his method as a general approach to evolve programs, but his register machine language is in many ways similar to a simple machine code language and will thus serve as a good illustration for register machine GP. He uses a string of digits as the genome. Three consecutive digits represent an instruction. The first integer in the triple determines the type of instruction. This is similar to the syntax of a machine code instruction which has specific bit fields to determine the type of instruction. There are five different instructions or statements in JB. `INCREMENT` adds one to a specified register. The register number is given by the second digit of the triple. `ZERO` clears a register while `SET` assigns a value to a register. There are also `BLOCK` and `LOOP` instructions that group instructions together and enable `while-do` loops. Figure 9.16 shows a very short JB example. Neither `INCREMENT` nor `CLEAR` uses the last digit of its triple. JB

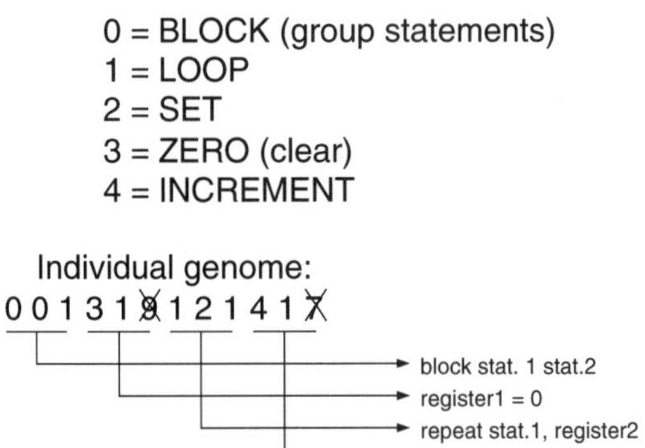

Figure 9.16
Program representation in JB

employs a variable-length string crossover. In part to avoid infinite loops, Cramer later abandoned this method in favor of an approach using a tree-based genome.

The GEMS System

One of the most extensive systems for evolution of machine code is the GEMS system [Crepeau, 1995]. The system includes an almost complete interpreter for the Z-80 8-bit microprocessor. The Z-80 has 691 different instructions, and GEMS implements 660 instructions, excluding only special instructions for interrupt handling and so on. It has so far been used to evolve a "hello world" program consisting of 58 instructions. Each instruction is viewed as atomic and indivisible, hence crossover points always fall between the instructions in a linear string representation. Figure 9.17 illustrates the crossover method used in GEMS, where new offspring are created by exchanging a block of instructions between two fit parents.

9.2.5 An Example: Evolution in Machine Language

A typical application of GP is symbolic regression. Symbolic regression is the procedure of inducing a symbolic equation, function, or program that fits given numerical data. A GP system performing symbolic regression takes a number of numerical input/output relations, called fitness cases, and produces a function or program that is consistent with these fitness cases. Consider, for example, the following fitness cases:

```
f(2) = 2
f(4) = 36
f(5) = 80
f(7) = 252
```

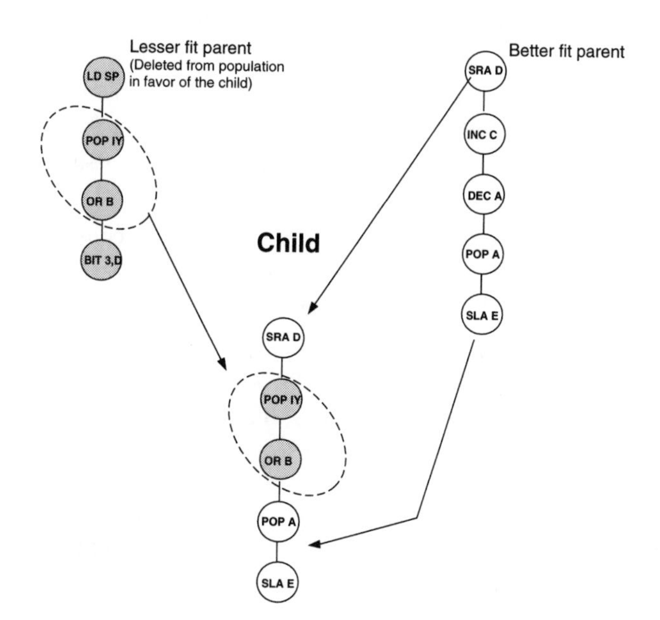

Figure 9.17
The crossover of GEMS

One of the infinite number of perfect solutions would be $f(x) = (x-1)^2 + (x-1)^3$. The fitness function would, for instance, be the sum of the difference between an individual's (function's) actual output and the target output specified by a fitness case. If a matching formula is found by the GP system it can be used to predict the values between or outside the given range of points.

The first issue we need to decide is which instructions from machine language we would like to include in the function set. When evolving a program in machine language, we should always spend some time assuring that the representation of the problem and the fitness cases are well suited for the available machine code instructions. Sometimes it is advantageous to map the fitness cases to another representation which better corresponds to primitives of the language. By introducing a `CALL` instruction, one can use any sort of user-defined function even when working with machine code evolution. However, the power of machine code GP is exploited to the highest degree only when native machine code instructions are used for calculation. In our symbolic regression problem, we choose three arithmetic instructions for the function set $+, -, \times$.

We must also decide on the number of registers an individual has access to. Hardware usually provides an upper limit on the number of registers accessible. This limit is between 2 and 64 registers, depending on the processor type and the programming environment. If there are not enough registers available for the application, one has to include instructions for stack manipulation. In our case, we can assume that two registers (x, y) will be enough.

After the number of registers, the type of the registers has to be fixed. Three types exist: input registers, calculation registers, and output registers.[5] The input value to the function is assigned to the x register while the y register is cleared.

If the machine code instruction format allows for constants inside the instructions, then we also need to give an initialization range for these instructions. The initialization range can be smaller than the maximal number expected in the formula because the system can build arbitrary constants by combining them.

The usual parameters for GP runs, like population size and crossover or mutation probabilities, have to be fixed as well. The population is initialized as random programs with random length. The maximum length of programs during initialization is another parameter of the system. In our example, two average individuals are the parents in Figure 9.18. The figure shows a possible crossover event between the parents. In this instance, a perfect individual is found that satisfies all the fitness cases and has a fitness 0.

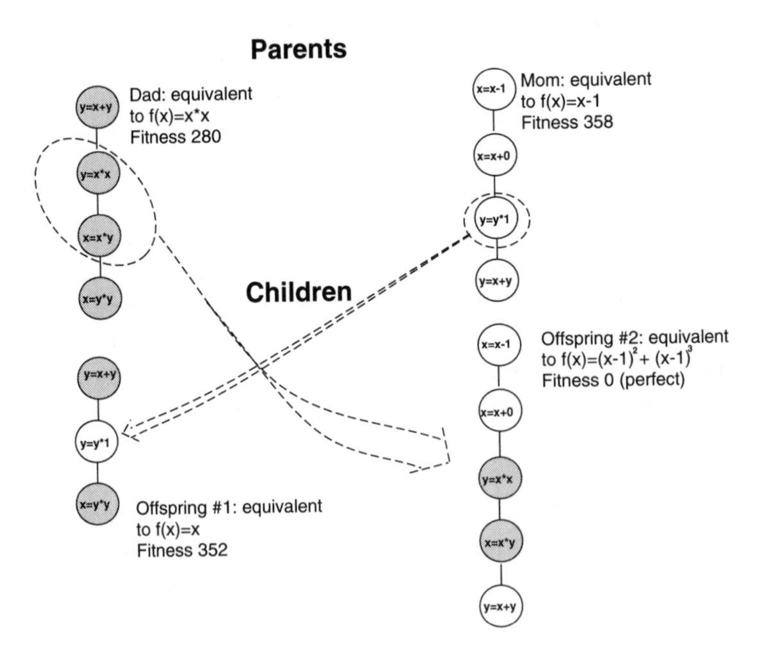

Figure 9.18
Example crossover with fitness of parents and offspring

[5] Sometimes, specific state registers need to be used. Here this is not the case and we defer a discussion of those to Chapter 11.

9.3 GP with Graph Genomes

Graphs are another important structure on which GP has been applied. In this section we shall discuss two examples, the PADO system of Teller and Veloso [Teller and Veloso, 1995b], and the developmental system of Gruau [Gruau, 1992b].

9.3.1 PADO

The graph-based GP system PADO (Parallel Algorithm Discovery and Orchestration) employs several different innovative ideas. The topic we discuss in this section is that PADO works not with S-expressions but with programs. Teller and Veloso say they prefer this way to evolve a superset of those elements of traditional hierarchical GP functions.

The programs of PADO are regarded as N nodes in a directed graph, with as many as N arcs going out from each node. Each node consists of an action part and a branch-decision part. Figure 9.19 shows a program, a subprogram, and the memory organization for each node.

Each program has a stack and an indexed memory for its own use of intermediate values and for communication. There are also the following special nodes in a program:

- Start node
- Stop node
- Subprogram calling nodes
- Library subprogram calling nodes

There are also parameters stating, for example, the minimum and maximum time for a program to run. If a particular program stops earlier, it is simply restarted from the start node, with values accumulated in stack or memory reused. Library subprograms are available to all nodes, not just the one which is calling, whereas subprograms without that provision are for a program's "private" use only.

A population of these programs competes for performance on signal classification tasks. In fact, no one program becomes responsible for a decision; instead, a number of programs are "orchestrated" to perform the task. Teller claims that the system is able to classify signals better thus than with LISP-like constructs such as S-expressions and ADFs [Teller and Veloso, 1995a]. In addition, due to the special representation of programs in directed graphs, loops are easily incorporated.

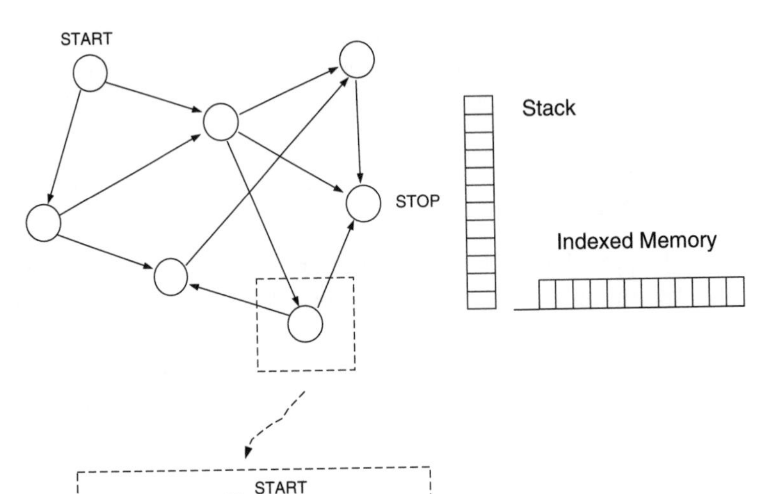

Figure 9.19
The representation of a program and a subprogram in the PADO system. Each node is comprised of: Number, Action, Branch, Arc1, Arc2, ..., Branch Constant.

Special care has to be taken to provide good mutation and crossover operators for such a system. Teller and Veloso use a co-evolutionary approach they call *smart operators* [Teller, 1996] (see Chapter 6). There are more sophisticated breeding details, which we shall omit here, as well as additional evaluation procedures specifically curtailed to facilitate pattern recognition, but the above is the gist of PADO's graph representation.

9.3.2 Cellular Encoding

An interesting scheme for a developmental approach toward graph evolution has been proposed by Gruau [Gruau, 1992b] [Gruau, 1993] [Gruau, 1994a]. Gruau designed a rewriting system that can work on graph structures as exemplified by neural networks, electrical circuits, or finite state automata. Because the notion of graphs is ubiquitous in computer science, the application area of this method is potentially large.

The general idea is that what ultimately should result as a phenotype is separated from the genotype of the tree-based individual. The phenotype is a (possibly) cyclic graph, something strictly forbidden in tree structures by their very definition, which forces the separation between genotype and phenotype.

Grammar Trees as Genotype

The genotype is a so-called grammar tree, which contains rules for rewriting, like the models of Whigham (see Section 9.4.2) or Jacob (see Section 9.4.3). What is rewritten, however, is not a string but a graph, whose nodes (cells) or edges (connections) are undergoing rewriting. The former is called division by Gruau, the latter change of weights/links/connections. These operations are applied in a parallel manner so that grammar trees develop over time like L-systems [Lindenmayer, 1968]. Figure 9.20 shows a selection of operations in this rewriting system.

Another aspect of Gruau's work in cellular encoding is the ability to define modules once that subsequently can be applied in various locations in the grammar tree. We shall discuss an example of the development of a neural network for a walking insect-like robot in Chapter 12.

9.4 Other Genomes

A wealth of other structures has been used for evolution. This section will discuss a few of them.

9.4.1 STROGANOFF

Iba, Sato, and deGaris [Iba et al., 1995b] have introduced a more complicated structure into the nodes of a tree that could represent a program. They base their approach on the well-known *Group Method of Data Handling* (*GMDH*) algorithm for system identification [Ivakhnenko, 1971]. In order to understand *STructured Representation On Genetic Algorithms for NOnlinear Function Fitting* (*STROGANOFF*) we first have to understand GMDH.

Suppose we have a black box with m inputs $x_1, x_2, \ldots, x_m$ and one output y; see Table 9.6. The black box receives signals on its input lines and responds with a signal on its output line. We can model the process by saying that the system computes a function f from its inputs:

$$y = f(x_1, x_2, \ldots, x_m) \tag{9.3}$$

(a) Initial graph (b) Sequential Division 'S' (c) Parallel Division 'P'
(d) Division 'T' (e) Division 'A' (f) Division 'A'
(g) Division 'H' (h) Division 'G' (i) Recurrent Link 'R'
(j) Cut input link 'C 3' (k) Change Input Weight 'D 3' (l) Change Output Weights 'K 2'

Figure 9.20
Side effects of nodes of cellular encoded grammar trees ([Gruau, 1995], copyright MIT Press, reproduced with permission)

Unfortunately, we cannot know anything about the system's internal working. We can only observe the system's reactions to a number N of input stimuli.

Table 9.6
Input/output behavior of the black box to be modeled

# Obs.	Input	Output
1	$x_{11}x_{12}\ldots x_{1m}$	y_1
2	$x_{21}x_{22}\ldots x_{2m}$	y_2
k	$\ldots$	y_k
N	$x_{N1}x_{N2}\ldots x_{Nm}$	y_N

To tackle the problem of estimating the true function f with an approximation $\hat{f}$, GMDH now assumes that the output can be constructed by using polynomials of second order in all input variables. Thus the assumption is that the system has a binary tree structure of the kind shown in Figure 9.21. Intermediate variables z_i compute (recursively) input combinations.

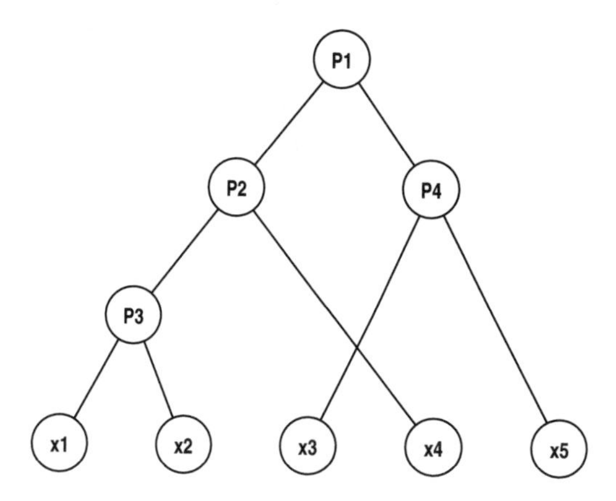

Figure 9.21
Group Method of Data Handling (GMDH) using 5 input variables $x_1 \dots x_5$

The polynomials are of second order in their input variables, which by themselves might be the result of other second-order polynomials:

$$P_j(x_1, x_2) = z_j = a_0 + a_1 x_1 + a_2 x_2 + a_3 x_1^2 + a_4 x_2^2 + a_5 x_1 x_2 \quad (9.4)$$

assuming for the moment a polynomial in inputs x_1, x_2. Given the observations of $x_1, x_2, \dots, x_m$, y, we can now adjust the parameters a_1, a_2, a_3, a_4, a_5 by conventional least mean square algorithms. The parameter fitting aims to minimize the difference between observed output y_i and output $\hat{y}_i, i = 1 \dots m$ generated by the polynomial tree structure. Multiple regression analysis may be used, for instance, to achieve this goal. Now that the best parameter set of all P_j has been found, it is considered an individual of a GP population.

The STROGANOFF method applies GP crossover and mutation to a population of the above polynomial nodes. A sketch of how crossover works is shown in Figure 9.22. Note that a node returns a complicated function of its inputs that is subject to a local search process before being finally fixed. Thus, the features of the function set are not determined from the very beginning, except for their overall form. Instead, a local search process that Iba et al. call a "relabeling procedure" is used to find the best combination of parameters to fit the output function, provided a structure is given. The overall form, on the other hand, is the same for all non-terminal nodes.

The fitness of the GP algorithm, now operating on polynomial trees as shown in Figure 9.23, is not necessarily the degree of fitting

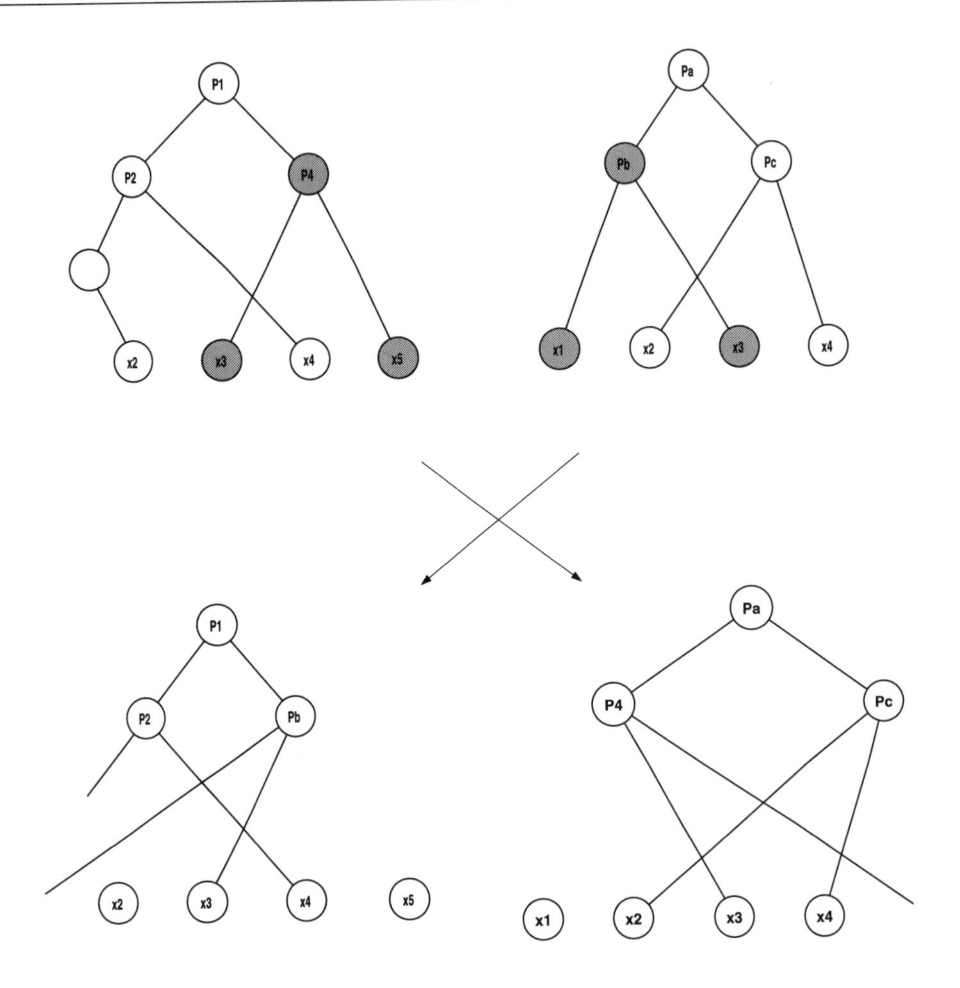

Figure 9.22
Crossover of trees of GMDH nodes, 5 variables $x_1 \ldots x_5$*. In offspring only part of the functions has to be recomputed for evaluation:* y'_1, y'_4 *(left side),* $\bar{y}'_1, \bar{y}'_2$.

that is used in the multiple regression part. Instead, it can use other measures, such as complexity of the individual – Iba et al. use the minimal description length (*MDL*) criterion – or a combination of fitness and complexity, to compare the quality of solutions.

In Section 12.2 we will see an application example of this method.

9.4.2 GP Using Context-Free Grammars

Whigham [Whigham, 1995a] [Whigham, 1995b] has introduced a very general form of grammar-based genetic programming. He uses context-free grammars as the structure of evolution in order to overcome the closure requirements for GP. By the use of a context-free grammar, typing and syntax are automatically assured throughout the evolutionary process, provided the genetic operators follow a simple rule.

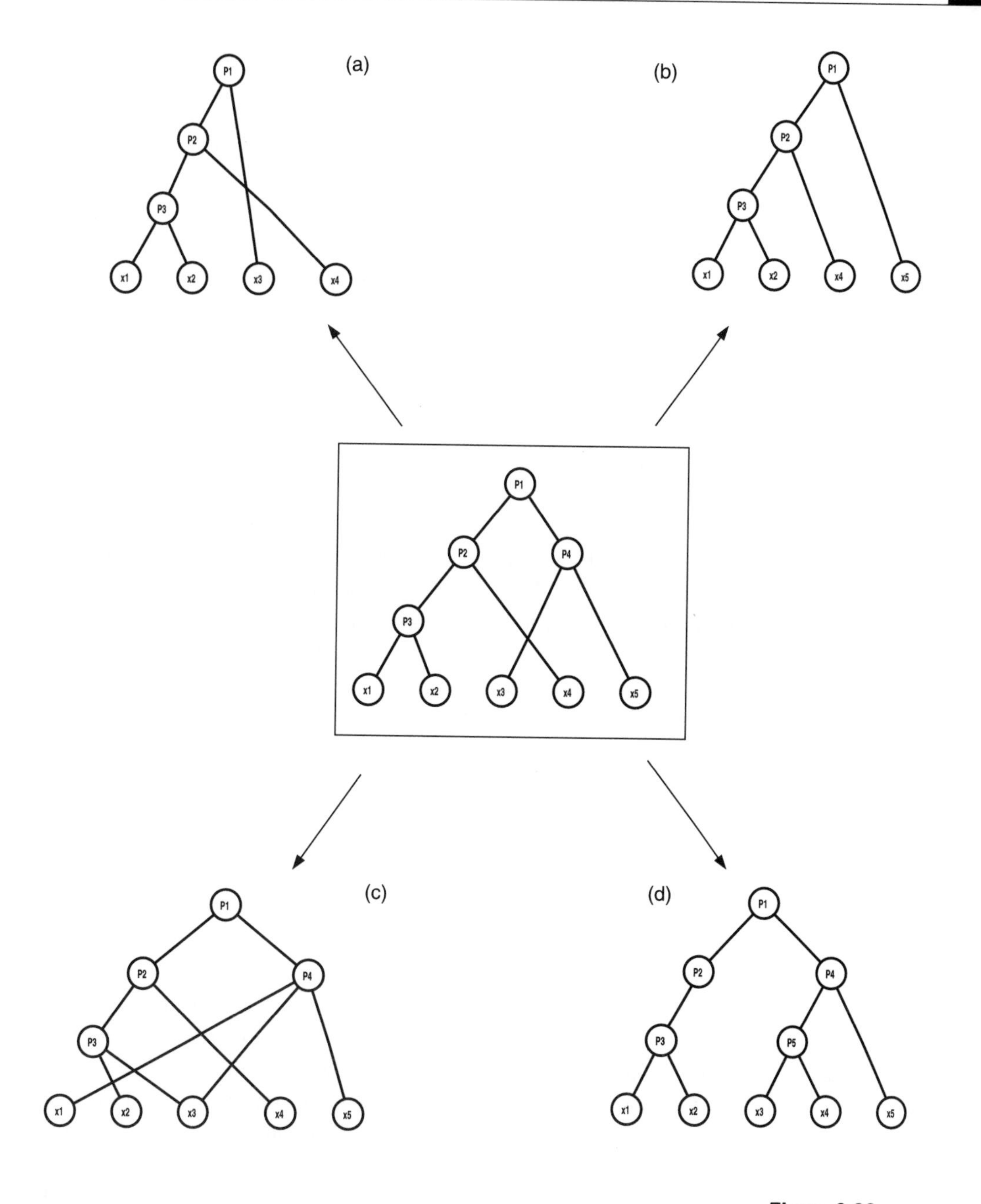

Figure 9.23
Different mutations of a tree of GMDH nodes, 5 variables $x_1 \ldots x_5$

In context-free grammar GP, the central role of a function possessing arguments is taken over by a production rule generating new symbols. A context-free grammar can be considered a four-tuple (S, Σ, N, P) [Gustafson et al., 1986], where $S \cup N$ is the set of non-terminal symbols, with S the starting symbol, Σ the set of terminal symbols, and P the set of production rules.

Terminal of a Context-Free Grammar

Definition 9.2 *A* **terminal of a context-free grammar** *is a symbol for which no production rule exists in the grammar.*

Production Rule

Definition 9.3 *A* **production rule** *is a substitution of the kind* $X \to Y$ *where* $X \in N$ *and* $Y \in N \cup \Sigma$.

There might be more production rules applicable to a symbol $X \in N$ that can be expressed using the disjunction $|$; for example, $X \to Y|z$ means that the non-terminal symbol X can be substituted by either the non-terminal Y or the terminal z (we adopt the convention that non-terminals are written with capital letters).

What has been a terminal in the conventional tree-based approach to GP has become a terminal in context-free grammar, too, but, in addition, all functions of conventional GP have now become terminals as well. Thus, a sort of developmental process has been introduced into evolution, with production rules applied until all symbols have reached (from left to right) terminal character. A functional expression of traditional GP can then be read off from left to right following all leaves of the structure.

As an example, we discuss the following grammar:

$$S \to B \tag{9.5}$$

$$B \to +BB| - BB| * BB|\%BB|T \tag{9.6}$$

$$T \to x|1 \tag{9.7}$$

where S is the start symbol, B a binary expression, T a terminal, and x and 1 are variables and a constant. The arithmetic expression discussed earlier can be considered a possible result of repeated application of these production rules as shown in Figure 9.24.

Crossover in context-free grammar ensures syntactic closure by selecting, in one of the parents, a non-terminal node of the tree, and then searching, in the second parent, for the same non-terminal. If the same non-terminal is not found, then no crossover operation can take place; otherwise, crossover is allowed at precisely this node. It is easy to see that diversity comes in when a non-terminal allows more than one production rule to be applied. In this way, differences between two individuals can develop in the first place.

Mutation involves selecting one non-terminal and applying randomly selected productions until (at the latest) the maximally allowed depth of a tree is reached.

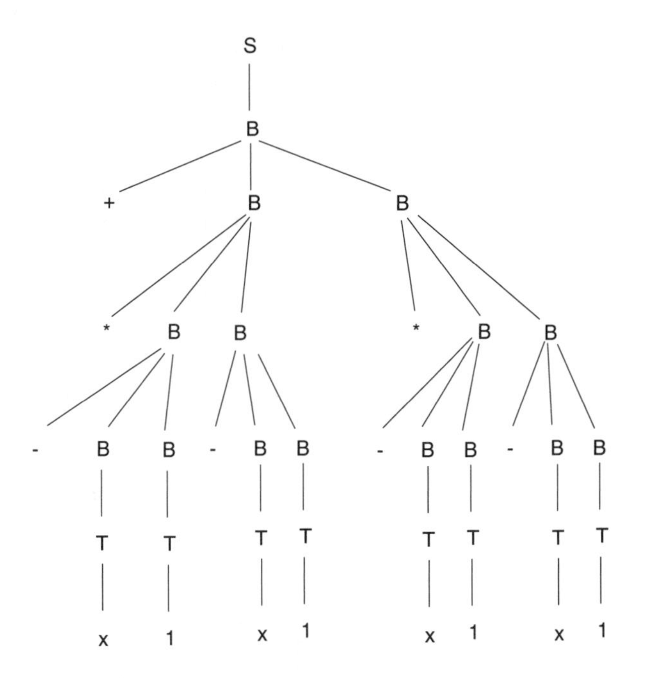

Figure 9.24
Arithmetic expression of Figure 9.14 expressed as a grammatical structure. S: start symbol, B: binary expression, T: terminal, x,1: variables, a constant

Whigham [Whigham, 1995a] has also considered a modification of the context-free grammar by incorporating new beneficial productions into the grammar at certain stages of the evolutionary process (*epochs*). Wong and Leung introduced an even more general system by using context-sensitive grammars as the basis of his GP system [Wong and Leung, 1996].

9.4.3 Genetic Programming of L-Systems

Lindenmayer systems (also known as *L-system* [Lindenmayer, 1968] [Prusinkiewicz and Lindenmayer, 1990] have been introduced independently into the area of genetic programming by different researchers [Koza, 1993a] [Jacob, 1994] [Hemmi et al., 1994b]. L-systems were invented for the purpose of modeling biological structure formation and, more specifically, developmental processes. The feature of rewriting all non-terminals in parallel is important in this respect. It allows various branches to develop independently of each other, much as a true developmental process does.

Different L-Systems

L-systems in their simplest form (0L-systems) are context-free grammars whose production rules are applied not sequentially but simultaneously to the growing tree of non-terminals. As we have seen in the previous section, such a tree will grow until all branches have produced terminal nodes where no production rule can be applied any more. The situation is more complicated in context-sensitive

L-systems (IL-systems), where the left and right contexts of a non-terminal influence the selection of the production rule to be applied. Naturally, if more than one production rule is applicable to a non-terminal, a non-deterministic choice has to be taken.

The goal of evolution is to find the L-system whose production rules generate an expression that is most suitable for the purposes envisioned. Thus, L-systems are individuals themselves. They are evaluated after all non-terminals have been rewritten as terminals according to the grammar. Figure 9.25 shows a typical individual where each *LRule* consists of a non-terminal predecessor and a terminal or non-terminal successor (in the simplest case of 0L-systems).

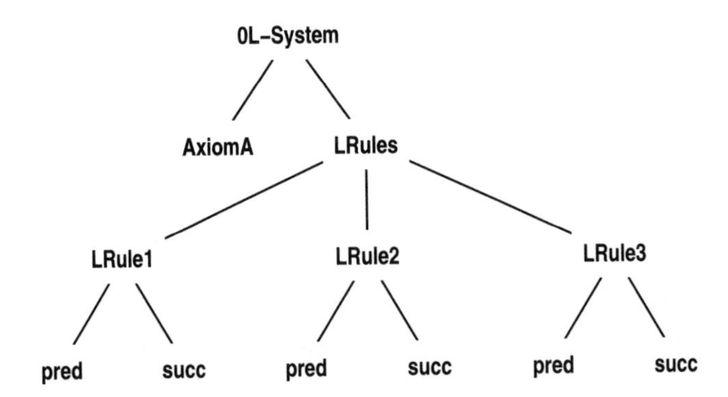

Figure 9.25
Context-free L-system individual encoding a production rule system of Lindenmayer type

Note that individuals of this kind cannot be varied arbitrarily, but only following the "meta-grammar" of how L-systems are encoded. For instance, the Axiom A branch is *not* allowed to change into an LRule. Rather, it can change only into a different axiom, say, Axiom B. The situation is very similar to the variation of ADF trees in Koza's treatment [Koza, 1994a]. There, for example, a result-producing branch in one tree can be recombined only with the result-producing branch of another tree, not with its function-defining branch.

Jacob [Jacob, 1994] gives a general treatment of how allowed variations may be filtered out of the total of all variations.

Exercises

1. Give four examples of different mutation operators acting on trees.
2. Give four examples of different crossover operators acting on trees.
3. Describe a linear GP system.
4. Describe the STROGANOFF GP system.
5. Give an example of a GP system that has a graph genome.
6. Describe the genotype–phenotype mapping in the BGP method.
7. Give two reasons why it is beneficial to evolve machine code.
8. Describe cellular encoding.
9. Why is it not possible to allow unrestricted crossover in GP using a context-free grammar as genome?

Further Reading

K.E. Kinnear, Jr. (ed.),
ADVANCES IN GENETIC PROGRAMMING.
MIT Press, Cambridge, MA, 1994.

J.R. Koza,
GENETIC PROGRAMMING.
MIT Press, Cambridge, MA, 1992.

J.R. Koza et al.,
GENETIC PROGRAMMING 1996. PROCEEDINGS OF THE FIRST ANNUAL CONFERENCE.
MIT Press, Cambridge, MA, 1996.

J.P. Rosca (ed.),
PROCEEDINGS OF THE WORKSHOP ON GENETIC PROGRAMMING: FROM THEORY TO REAL-WORLD APPLICATIONS.
Tahoe City, CA, 1995.

E.S. Siegel and J.R. Koza (eds.),
WORKING NOTES FOR THE AAAI SYMPOSIUM ON GENETIC PROGRAMMING.
AAAI Press, 1995.

10 Advanced Genetic Programming

Contents

The basic approach to genetic programming has many limitations. Over time, GP researchers have become aware of those limitations and there is widespread consensus among them that despite its successes, there is vast room for improvement in the GP algorithm. In this introduction we shall quickly highlight three limitations of the basic approach to genetic programming, before we delve into a more detailed and systematic discussion.

Convergence and Diversity

One of the most serious problems of standard GP algorithms is the convergence of a population. An often used rule states that what has not been achieved in a GP run in 50 generations will never be achieved. Although this is an informal statement, it points to a serious weakness of the paradigm: in a population that is recombined repeatedly, sooner or later uniformity will develop. Diversity will be the keyword for finding remedy to this situation, which can be found by parallelizing genetic programming using demes.

The Granularity of Programs

Another problem of GP is its dependence on the terminal and function set initially chosen for a certain task. It may well be that the abstractions residing in this terminal and function set are too high or too low to allow for a sensible evolution of individuals. If they are too low, the system will find the task too complicated to simultaneously move in the various directions necessary to improve a solution. This is the complexity issue, which would actually call for a correlated change in individuals as could be achieved by using modules that might be modified as wholes. If the abstractions are at too high a level, the task cannot be broken down appropriately and the solutions developed will always be more abstract, i.e., more general, than the fitness cases would dictate. The system will thus be forced to ignore part of the information in the fitness cases shown to it, with all the negative consequences following from that. We shall discuss this topic in Section 10.2.1.

Closure and Typing

A more practical consideration is related to the requirement of closure of genetically evolved programs. If the set of terminals and functions were of such a kind that not all functions could bear all terminals as arguments, the system would become brittle. Typing is one of the ways out of this dilemma (see Section 10.2.10), where each node carries its type as well as the types it can call, thus forcing functions calling it to cast the argument into the appropriate type.

We have mentioned just a few examples of the potential for improvement of the basic algorithm of genetic programming. In Chapter 6 we already discussed potential improvements for the crossover operator. This chapter will focus on three principal areas for improvement:

1. Improving the speed of GP;
2. Improving the evolvability of programs;
3. Improving the power of GP search.

10.1 Improving the Speed of GP

GP is computationally intensive. That arises from the need to evaluate each individual's fitness over many different fitness cases. There have been a number of different approaches to accelerating GP runs.

10.1.1 Run Termination

Often there is a need for exploring parameter choices to be set for an entire GP run. This can be done serially by choosing different sets of parameters for subsequent runs. In such a situation it is very helpful to have various criteria for signaling the end of a run. For example, one such signal could be the intron explosion we introduced in Chapter 7. Fitness improvement is effectively over once exponential intron growth sets in. In this way, criteria for early run termination can accelerate the exploration of the parameter space of GP considerably. In one set of experiments, where the maximum number of generations was set to 200, the early termination criterion reduced the run time by 50% [Francone et al., 1996].

10.1.2 Parallelization

Another way to accelerate GP and, at the same time to keep diversity high, is to use parallelization. Parallel populations, also known as *demes*, might possess different parameter settings that can be explored simultaneously, or they might cooperate with the same set of parameters, but each work on different individuals. Parallelization is possible for all EAs, of course, due to their population-based approach.

Parallel Populations

Koza has developed a system that provides fast performance on 64 Power PC processors arranged in a toroidal mesh. The host in charge of the entire run is a Pentium PC. This system is modeled as an island GA [Andre and Koza, 1996a]. That is, each processor has its own population, which is separate from the populations of the other processors. Each processor sends a "boatload" of "migrants" to the four adjacent processors on the toroid in each generation. So the isolation of the individual demes, or populations, is tempered by a weak migration of individuals from processor to processor.

A typical implementation of this system would involve 64 demes of population 800 each. The total population over all demes would be 51 200 [Koza et al., 1996b]. Of course, the parallelization of the system results in a great speed-up in processing such a large population. Koza and Andre have reported intriguing results of a more than linear speed-up due to this arrangement. It is possible that a larger population allows the system to find solutions with fewer evaluations [Keith and Martin, 1994].

10.1.3 Parallel Fitness Evaluations

Another way to parallelize a GP system is to evaluate programs in parallel. In [Juille and Pollack, 1996] a system is reported that uses the 4096 processors of a MasPar MP-2 parallel computer (SIMD) to implement parallel evaluation of trees. The machine in question has a peak performance of 17 GIPS. The idea of Juille and Pollack was to implement a virtual processor with its own specific memory organization in order to simultaneously execute a population of programs on this SIMD architecture. Depending on the specific problem they used (trigonometric identities and symbolic integration), approximately 1000 S-expressions could be evaluated per second. For two other problems (tic-tac-toe and the spiral problem) the authors used more sophisticated fitness evaluations.

Oussaidene et al. [Oussaidene et al., 1996] report in another study where they used farming of fitness cases to slave processors from a master processor that, for problems of 1000 and more fitness cases on an IBM SP-2 machine, a nearly linear speed-up can be obtained.

10.1.4 Machine Code Evolution

Evolution of programs in machine code is treated in more detail elsewhere (see Chapters 9 and 11). Here we note only that, due to the fast execution of machine code on processors, a considerable speed-up can be reached by directly evolving machine code. Though figures differ somewhat depending on which system is compared to machine code evolution, an acceleration factor of between 60 and 200 compared to a traditional GP system can be safely assumed.

10.1.5 Stochastic Sampling

In many GP problems, the evaluation of a program individual is a time-consuming process. Consider, for example, the problem of teaching a robot the correct behavior in an unknown environment. An entire population of programs has to be evaluated in order to find out which behavior is appropriate (has a high fitness) in this environment, and which is not. A common procedure to compute this fitness measure is to let each program control the robot for a certain time, allowing the robot to encounter sufficiently many situations to be able to judge the quality of this program. The same process is repeated for all programs in the population, with the robot suitably prepared for generating equal conditions for all programs.

A Simple Calculation

A simple calculation should help to illustrate the difficulties. Suppose the robot can perform an elementary behavior within 500 ms, i.e., every 500 ms there will be a command issued from the robot that depends on its state and the inputs present at the outset of this period. Because a task such as navigating or searching in the real world is complex, at least 100 different situations (the fitness cases) should be examined in order to get a measure of fitness for the individual program. Thus, 50 s will be consumed for the evaluation of one individual. Given a relatively small population of 50 programs used for the GP runs, we end up needing at least 40 minutes to evaluate one generation of behavioral programs. In order to do so, however, we had to prepare an initial starting state for the robot every 50 s. Without counting the time taken for that, a moderate test run with, say, 20 generations would require more than 13 hours to be invested, a rather full workday for an experimenter. As many researchers have found, evaluation is a tedious process, even without the preparation for identical initial conditions.

There is one radical step to shorten the time of an individual's evaluation: allow each individual program to control the robot for just one period of 500 ms, i.e., evaluate only one fitness case. This way evaluation is accelerated by a factor of 100. In addition, we can skip – under these non-deterministic conditions – the preparation into a normalized initial state. Each program will have to steer the robot under different conditions anyway!

This example can be generalized to any type of application. A way to accelerate fitness evaluation is to evaluate each individual with a different (possibly small set of) fitness case(s), a method we call "stochastic sampling" [Nordin and Banzhaf, 1995c]. The assumption is that over a number of sweeps through the entire set of fitness cases, accidental advantages or disadvantages caused by the stochastic assignment of fitness cases to individuals will cancel out and individuals will develop that are fit on the basis of the entire set of fitness cases.

A side effect of this method is the tendency toward generalization. We can expect that the drive toward more general solutions is much stronger with this method than with keeping the fitness cases constant. A similar effect has been observed with the on-line versus batch learning methods for supervised neural networks [Hecht-Nielsen, 1988].

Stochastic sampling is also much closer to fitness evaluation in nature and natural selection, where evaluation can take place only locally in space and time. The non-determinism implicit in such an approach can even be used to good effect, e.g., for escaping local minima.

Dynamic Subset Selection

In [Gathercole and Ross, 1994] [Gathercole and Ross, 1997a] another approach toward selection of fitness cases is proposed. Through observation of the performance of a GP run Gathercole and Ross's system obtains data about which fitness cases are difficult and which are easy (there is also an age factor weighed in). Subsequently, the system preferentially selects those fitness cases which have a higher rank based on these criteria. So in this approach, a "sorting" of fitness cases takes place which leads to a two-fold acceleration in learning, first based on the fact that only subsets of the entire training set are selected, and second based on the fact that the system chooses the harder ones.

Limited Error Fitnes

Another interesting suggestion [Gathercole and Ross, 1997a] [Gathercole and Ross, 1997b] by the same authors is to restrict fitness evaluation by keeping track of the cumulative error an individual has collected in the course of evaluation. If, during evaluation, the cumulative error should exceed an adaptive threshold, evaluation is stopped, and all fitness cases that have not been evaluated are registered as failures. The error threshold adapts to the performance of the best individual of a generation.

10.2 Improving the Evolvability of Programs

GP has established itself as a powerful learning tool – often with very basic function sets. For example, using only `Plus`, `Minus`, `Times`, `Divide` and bitwise Boolean operators, a linear GP system has outperformed neural networks on pattern recognition problems. But the GP system was not allowed `If-then-else`, `For`, `Do-until`, `Do-while`, or other looping constructs, whereas a multilayer feedforward neural network has addition, subtraction, multiplication, division, and conditional branching capabilities built into its neuron squashing and threshold functions and its transfer functions. GP can accomplish a lot with a very restricted set of operators. That said, there are many ways that GP can improve the expressive power of evolvable programs.

10.2.1 Modularization

A natural tool for humans when defining an algorithm or computer program is to use a modularization technique and divide the solution into smaller blocks of code. Human programmers find it useful to subdivide a program hierarchically into functions. This is an example of a *divide and conquer* strategy – one of the most basic problem solving strategies known. GP systems may have some tendency toward self-modularization.

Modularization in general is a method by which functional units of a program are identified and packaged for reuse. Since human programmers have to struggle constantly with the increasing complexity of software – as humans generally have to when dealing with the complexity of their environment – the natural tool of thought, namely, to bundle entities into abstractions to be used later on as units, is also useful in programming. Modular approaches are mainstream in software engineering and result, generally speaking, in simpler and more generic solutions.

Let us first look at a general definition of modules as given by Yourdon and Constantine [Yourdon and Constantine, 1979].

Module

Definition 10.1 *A* **module** *is a lexically contiguous sequence of program statements, bounded by boundary elements, having an aggregate identifier.*

A module has at least the following properties:

- it is logically closed;
- it is a black box (referred to as the information hiding principle in computer science); and
- it offers an interface to other modules.

An *interface* between modules is the set of operations and data types offered to other modules by a module.

Encapsulation of Code against Inefficiency

Modularization approaches are central when trying to tackle two of the main problems of simple GP: scaling and inefficiency. Modularization methods weigh in the benefits of encapsulating code and ultimately generalizing from one to more applications of the same (encapsulated) piece of code. If GP is able to work with building blocks, as some researchers have suggested (and others have doubted), then modules will be synonymous with building blocks.

Building Blocks

A key to understanding why modularization and encapsulation are useful in GP has to do with the effects of crossover on building blocks. All modularization techniques are ways of encapsulating blocks of code. Such encapsulated blocks become subroutines and can be called repeatedly from the main program or from other subroutines. This means that a program can reduce its length by putting frequently used identical blocks of code into a subroutine. As we have argued earlier in Chapter 7, programs of shorter effective length have better chances of survival than programs with larger effective length. Thus it might be expected that some modularization techniques work spontaneously: as soon as working code is encapsulated, the odds of it being destroyed by crossover are zero. Hence, encapsulation is a crossover protection, and more generally, a variation protection mechanism. In addition to that, shorter programs are better at generalizing when applied to unseen data. Thus modularization in GP can be seen to affect generalization in a positive way. Re-use of modules is further facilitated by placing modules into a library from where they can be called by different individuals.

Different modularization techniques have been suggested for genetic programming. Automatically defined functions (ADF) are the most thoroughly evaluated method [Koza, 1994a]. Other examples of modularization techniques include encapsulation [Koza, 1992d] and module acquisition [Angeline and Pollack, 1993].

10.2.2 Automatically Defined Functions

Automatically defined functions (ADFs) have been proposed by Koza and represent a large proportion of his more recent work [Koza, 1994a]. ADFs are inspired by how functions are defined in LISP during normal manual programming. The program individual containing ADFs is a tree just like any program in regular tree-based GP. However, when using ADFs, a tree is divided into two parts or branches:

1. The *result-producing branch*, which is evaluated during fitness calculation; and

2. the *function-defining branch*, which contains the definition of one or more ADFs.

These two branches are similar to program structures in, e.g., C or Pascal. There, a *main* part is accompanied by a part for the definition of functions. The main part corresponds to the result-producing branch while the definition part naturally corresponds to the function-defining branch. Both of these program components participate in evolution, and the final outcome of GP with ADFs is a modular program with functions.

Figure 10.1 gives an example of a program individual with an ADF. At the top one can see the root node `Program` which is just a place holder to keep the different parts of the tree together. The node `defun` to the left is the root node of an ADF definition. If there were more than one ADF then we would have more than one `defun` node under the `Program` node. In other words, we may need more than one ADF definition in the function-defining part of the program. Likewise, it is also possible to have more than one result-producing branch. The `Values` function determines the overall output from this individual. If the result-producing branch returns a single value only, then it is called the *result-producing branch.* The ADF definition branch has a similar `Values` node which wraps the result from the ADF body. In most examples `Values` is a dummy node returning what it receives from below in the tree.

Result-Producing Branch

The node labeled ADF0 in Figure 10.1 is the name of the single ADF we use in the individual. This is a name of a function and it will be a part of the *function set* of the result-producing branch to allow this branch to call the ADF. Just to the right of the ADF0 node is the argument list. This list defines the names of the input variables to the ADF0 function. These variable names will then be a part of the terminal set of the ADF0 function body. The principle is similar to a function definition in C or Pascal where we need to give the function a name and define its input variables together with a function body defining what the function does. All evolution takes place in the bodies of the ADFs and the result-producing branch (see Figure 10.1).

Depending on how the system is set up it is also possible to have hierarchies of ADFs, where an ADF at a higher level is allowed to call an ADF at a lower level. Normally, one must take care to avoid recursion though an ADF that calls itself either directly or via a chain of other ADFs. In principle, it would be possible to allow recursive ADF calls, but this is an area of GP that needs more exploration.

It is evident that the ADF approach will work only with a special, syntactically constrained crossover. We cannot use the simple GP crossover crossing over any subtree with any other subtree. For instance, we must keep all of the nodes that do not evolve in Figure 10.1 intact. There are separate function sets for the result-producing branch and the function-defining branch, and the terminal sets are also different. Crossover must not move subtrees between the two different branches of the ADF. Hence, crossover is done by first selecting a subtree in either the function-defining or the result-producing part of one parent. The crossover point in the second parent then has to be chosen in the same *type* of branch, ensuring that we switch subtrees only between ADF branch and ADF branch or result-producing branch and result-producing branch.

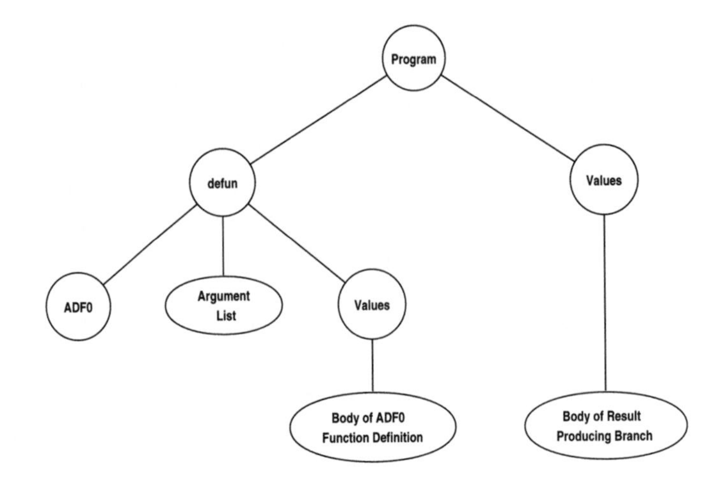

Figure 10.1
A typical automatically defined function definition

Before starting a run with a GP system that uses ADFs we must determine the number of ADFs and the number of arguments each is allowed to use. In other words, we have to specify the shape of the nodes that do not evolve in Figure 10.1. During initialization the system will keep track of this definition and will produce only individuals with this shape. The function-defining bodies and the result-producing bodies are then generated randomly.

Figure 10.2 shows an example individual where the cube function has evolved as an ADF. This ADF is then used in the result-producing branch to realize the x^6 function.

ADFs have been shown to outperform a basic GP algorithm in numerous applications and domains. Also, Kinnear has compared the performance of ADFs to the performance of another modularization technique called "module acquisition" defined below (see Section 10.2.4). In his study [Kinnear, Jr., 1994], ADFs outperformed both a basic GP algorithm and a GP algorithm with module acquisition. But in accordance with what we said earlier, ADFs seemed to give a performance advantage only when the introduction of functions reduced the length of possible solutions sufficiently [Koza, 1994a].

In summary, these are the steps needed when applying GP with ADFs:

1. Choose the number of function-defining branches.
2. Fix the number of arguments for each ADF.
3. Determine the allowable referencing between ADFs if more than one ADF is used.

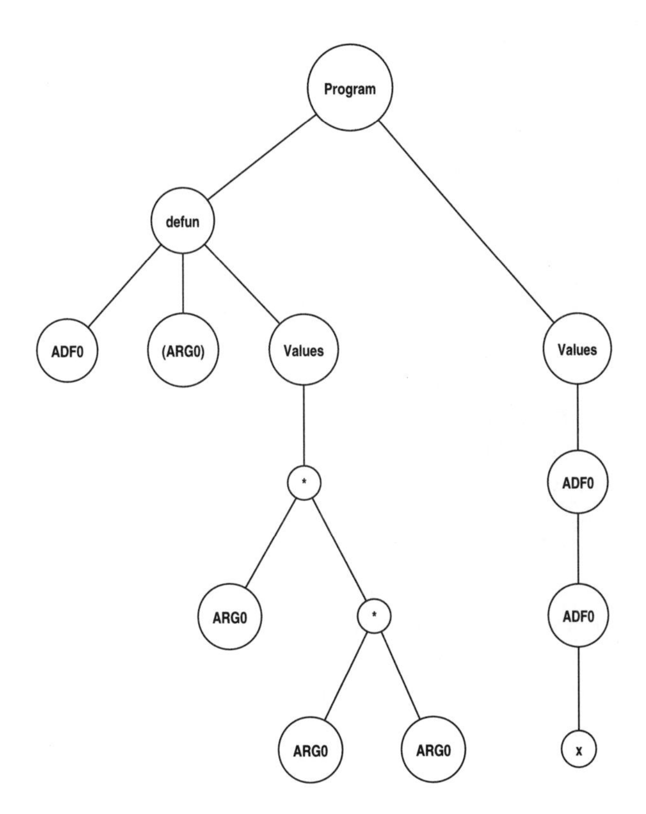

Figure 10.2
Example of an ADF program tree

4. Determine the function and terminal sets for both of the branches (remember they are different).

5. Define the fitness measure and fix parameters and the termination criterion, as in any other GP run.

Architecture Altering Operations

A weakness of the ADF approach is that the architecture of the overall program has to be defined by the user beforehand. The architecture consists of the number of function-defining branches in the overall program, the number of arguments (if any) possessed by each function-defining branch, and, if there is more than one function-defining branch, the nature of the hierarchical references (if any) allowed by them. This requirement adds another set of parameters to the initial parameters of a GP run. It would be much better if the complete structure of an individual could evolve, including all ADF specifications. Koza [Koza, 1995a] has proposed *architecture altering operations* as a method to achieve this goal. Architecture refers to the complete structure of the program individual. He proposed six architecture altering genetic operations that can add initial ADF structures, clone them, and change the number of parameters. Again, nature was used as a source of inspiration by referring to gene duplication [Ohno, 1970].

Gene Duplication

Evolution by gene duplication works by occasional duplication of a gene that does something useful. The risk of such an event is small – an additional copy of the same gene only marginally affects an individual. However, the copy can be mutated away from its original function into a different function because the original gene is still functional. Using architecture altering operations and gene duplication, no additional prespecified information is necessary for an ADF run compared to a basic GP run. Even iteration-performing branches could be added with this method [Koza and Andre, 1996b].

A system with architecture altering operations can be benchmarked against an ADF system with predefined size and shape, but the outcome will strongly depend on which predefined architecture it was compared to. One recent study [Koza, 1995b] compared, for one problem, the architecture altering method to the best known predefined architecture. The result was that the new method performed slower. It is too early to draw any conclusion, though, and one should always keep in mind that the benefits of specifying fewer parameters are very important.

10.2.3 Encapsulation

The original idea of encapsulation in GP is due to Koza [Koza, 1992d] who introduced it as an elementary operation into hierarchical GP. All of the following, however, is also applicable to linear GP.

Encapsulation of Subtrees into Terminals

The operation of encapsulation consists of selecting an individual from the population, selecting a non-terminal node within that individual and replacing the subtree below that node with a newly defined terminal node that contains the subtree removed. The new terminal is applicable in other individuals from this moment on. Removing the entire subtree has the effect of admitting only functions with arity 0 as encapsulated code. Nevertheless, if the newly defined terminal should turn out to contain useful code, reuse has been made possible.

An example of encapsulation is shown in Figure 10.3 for the arithmetic expression of Figure 9.14. By replacing a subtree with a new node E_0, crossover can no longer recombine parts of the subtree. Analogously, other operations are prohibited from changing this part of the code. Provided the subtree contains useful operations, this is beneficial. At present it is unclear whether encapsulation really confers a significant advantage on a GP system.

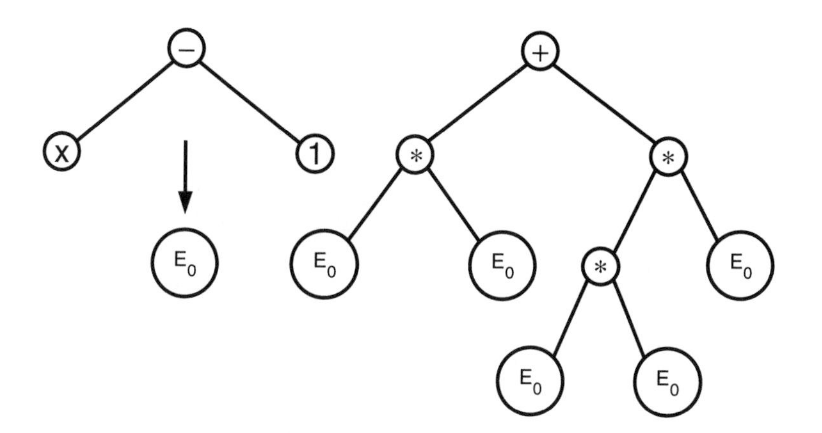

Figure 10.3
E_0 is used as a terminal to symbolize an encapsulated subtree of the arithmetic expression of Figure 9.14. The full tree can be shortened through the substitution.

10.2.4 Module Acquisition

Encapsulation into Library Functions

Another form of preparing code for reuse is module acquisition ciscussed in [Angeline and Pollack, 1992] [Angeline and Pollack, 1993]. Module acquisition can be used as an elementary operator to act on an individual. From a chosen individual a subtree is selected and a part of this subtree (up to a certain depth) is defined as a module. This operation has also been called compression. The parts of the subtree below the module are considered arguments to the module. In this way, module acquisition offers a way to create new functions, much as Koza's encapsulation operation generates new terminals.

The Library Is Available to the Entire Population

Angeline and Pollack go another way, however. In their approach, the new module is placed in a library of modules from where it can be referenced by individuals in the population. Figure 10.4 gives a sketch of the procedure. If a module provides some fitness advantage in the population, it will spread to more individuals, thus increasing the number of references to it in the library. As long as there is any one individual in the population to refer to this module, it is kept in the library.

The Counter-Operator: Expansion

Much in the spirit of encapsulation, once a module is defined it ceases to evolve any further. Module acquisition thus gives operator protection to the modules, although parameters are allowed to change. The compression operation carries a parameter itself, the depth at which the module is cut off. There might even be two parameters used, one for minimal depth, one for maximal depth of modules. We note in passing that the authors also introduce an operator countering the effect of compression: expansion. Expansion selects a single module to expand in an individual and expands it one level. So other modules are not affected nor are any modules that might be inside the expanded module.

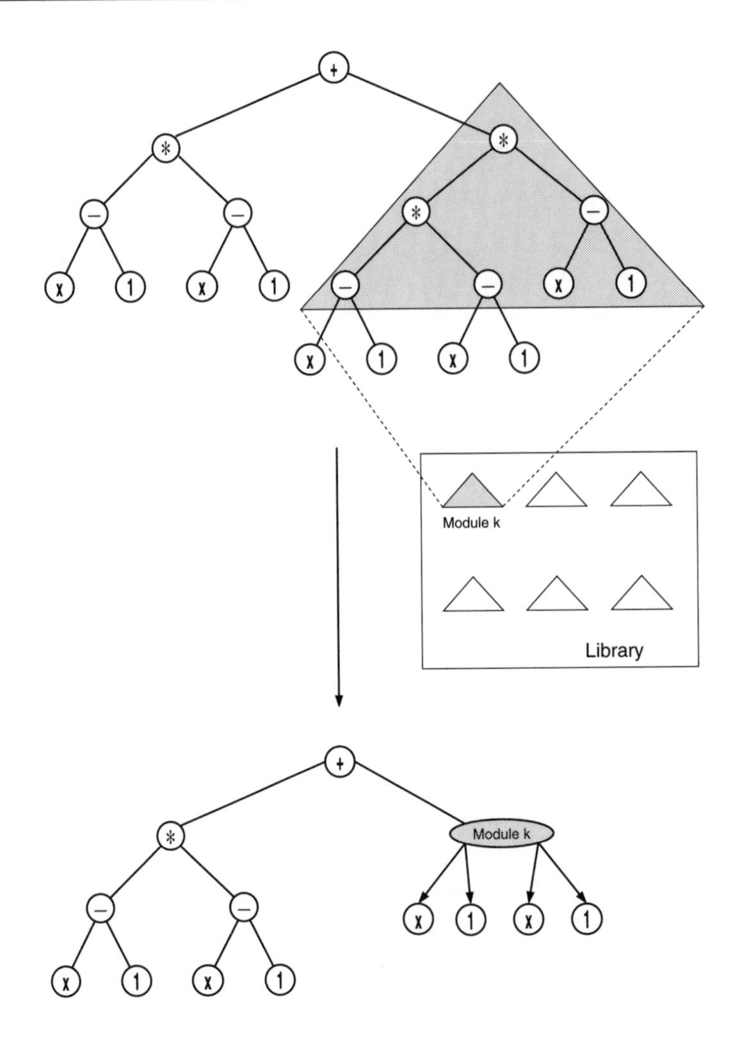

Figure 10.4
How a module is acquired. A node is selected, and its subtree, down to a certain depth, is considered a module and moved into the library. The module might contain terminal branches, as well as functional branches that become arguments of the module.

One very important similarity to the encapsulation method mentioned in the foregoing section is that newly defined modules are available globally, i.e., to the entire population. Kinnear has done a comparative study on the use of module acquisition but concludes that there is no obvious advantage [Kinnear, Jr., 1994]. However, it is too early to close the books on module acquisition, because the one study done was based on a limited sample of runs and problems.

10.2.5 Adaptive Representation

Adaptive representation is another technique for modularization in GP [Rosca and Ballard, 1994b]. Based either on heuristic information about the problem domain or, preferably, on certain statistical information available in the population, small subtrees are named to be candidates for modules. These subtrees are extracted from the population in order to serve as new functions or subroutines for further evolution.

Epochs

Every step of this kind can be done by use of a partial new seeding of the population with individuals containing the newly defined functions. Thus, the EA has a new time scale called an *epoch* lasting a number of generations until a new set of candidates for modules is extracted from the population. Rosca et al. [Rosca and Ballard, 1994b] claim that their method is the only bottom-up method for modularization developed so far that really catches what should be contained in modules.

Viability of Modules

To determine which parts of programs might serve as viable modules, Rosca et al. introduce the learning of adaptive representation based on differential offspring–parent fitness and on the notion of activity of certain parts of the programs. In a nutshell, the former criterion identifies individuals in the population that presumably possess highly fit subtrees. The notion is that, once an individual has made a jump in fitness compared to its parents, this should be due to a useful combination of genetic material in this individual.

Salient Blocks of Code and Block Activation

Now that the genetic material has been narrowed down, the search for salient blocks of code is the second step. The criterion applied here is block activation, that is, the execution frequency of a particular part of the code within the individual. The measurement of block activation of new blocks of code is, according to Rosca [Rosca, 1995a], done in $\mathcal{O}(N)$ time. Nodes with the highest value of block activation are considered candidates, provided all nodes in the subtree have been activated at least once.

Once blocks of code have been selected, they are generalized by dropping variables, which is an important heuristic. Rosca argues that this way he is able to create modules of a variable number of arguments.

He claims to find extremely interesting modules using the adaptive representation method. In experimental results presented on an agent behavior problem in a dynamic world [Rosca, 1995a] – a Pac-man-like problem earlier described by Koza [Koza, 1992d] – Rosca shows that adaptive representation runs maintain higher diversity in the population and discover appropriate modules much faster than conventional GP.

10.2.6 Automatically Defined Macros

Spector has suggested and evaluated a variant of ADFs called Automatically Defined Macros (ADMs) [Spector, 1996]. *Macros* are modularization structures which are part of most computer languages and define transformations to be performed on the source code before it is compiled or interpreted. A common macro transformation is substitution, where the programmer substitutes frequent code fragments by macros. The macros are *expanded* by a preprocessor before compilation or interpretation. Many LISP dialects contain powerful macro capabilities that can be used to implement new control structures. If a program is purely functional, then there is no difference in result between a macro and a function. However, if the macro uses side effects or is sensitive to its calling context, then it can produce unique effects.

Spector shows how substitution macros can be evolved simultaneously with the main program in a method analogous to the ADF method. The evolved macros can be seen as special control structures, producing, for example, specialized forms of iteration or conditional execution.

One disadvantage of macros is that the macro expansion process and the execution of expanded macros can sometimes take much more time than calls to functions, depending on the implementation. In addition, the caching of values provided by arguments to functions, but not by arguments to macros, can sometimes be valuable. So macros should only be used where side effects or context sensitivity are an important part of the application. Spector shows how the ADM method has advantages over ADFs in certain simulated control tasks while ADFs produce better results in other control tasks. He concludes that ADMs are likely to produce better results in non-functional domains while ADFs may be better suited for functional domains with few side effects.

10.2.7 Modules and Crossover

The discussion above has shown how we can protect code from being destroyed by variation operations like crossover or mutation. The main answer of authors was encapsulation. By freezing code and compressing it into terminals and functions, this goal certainly can be reached, since, by definition, functions and terminals are the "atomic units" of GP. Also, modularization using ADFs has been shown to increase protection of code [Teller, 1996].

But we might turn the question around, and ask: if a GP system has evolved (without the help of encapsulation) a solution to a problem, a solution that contains chunks of code separated by crossover-protecting code segments like introns,[1] are these chunks of code actually modules? If we go back to the definition of modules given

[1] Crossover protection is by far the most important protection, given the large amount of crossover used in GP.

earlier in this chapter, a definition which is now adopted widely in computer science, they do fit this definition exactly.

In fact, crossover can be expected to be the only operator to find these chunks of code, since contiguity is one of its hallmarks. In addition, it can be argued that intron segments separating effective pieces of code will evolve preferentially at those places where interactions between subsequent instructions are minimal and disruption of the sequence is not harmful. Those pieces of code that have minimal interaction will therefore be automatically singled out. But what is missing is the aggregate identifier, though this could be added when making the module explicit. The boundary elements need to be added at the same time.

We are therefore led to the conclusion that chunks of code which evolved together and have an actual function, i.e., are not introns, can be considered modules or candidate modules.

10.2.8 Loops and Recursion

Loops and recursion are useful to any computer system. Without them, a computer would simply consume all of its program memory in a few seconds and then stop. This does *not* mean that we could not have any useful program parts without iterative structures or that GP would be useless without loops, but for a computer system to be complete it needs iteration. GP, too, must be able to address these issues in order to scale up to more complex problem domains. One of the challenges of GP is to evolve complete computer applications, and if this is ever going to be possible, then iteration is necessary. But loops and recursion are also beneficial to smaller programs of the type currently addressed with GP. Iteration can result in more compact programs that among other things allow better generalization.

Infinite Loops

The problem with iteration is that it is very easy to form infinite loops. Worse, it is theoretically impossible to detect infinite loops in programs. The halting theorem described in Section 3.3.2 states that we cannot decide whether a computer program will halt or not. In GP this could mean that the entire system might wait for the fitness evaluation of an individual that has gone into an infinite loop. Even if we were able to detect infinite loops, there would still be enough finite loops executing far too long to be acceptable in a learning situation. Hence, we must find a way to control the execution of programs that contain loops and recursion.

There are in principle three ways to address the issue of infinite and "near-infinite" loops:

- The simplest solution is to define a limit on the number of iterations for each program and fitness case. A global variable is incremented for each iteration and execution is aborted when it exceeds a predefined maximum value.

- The iteration limit can be defined more flexibly by allowing the program to distribute execution time among different fitness cases and even to make its own judgments of the tradeoff between quality of solutions and execution time. This method is referred to as aggregate computation time ceiling.

- In some cases, it is possible to ensure that the problem domain and the function and terminal sets are constructed such that there could be no infinite or very long loops. It might be necessary to change the original representation and to switch primitives to achieve this goal.

An example from the last category is Brave's system for evolution of tree search programs [Brave, 1994] [Brave, 1995]. A tree search program navigates in a tree data structure in order to find a specific node or to obtain a specific state [Nilsson, 1971]. Each program is recursive in its structure, but due to the limited depth of the input trees each program is destined to terminate.

Kinnear uses a specialized loop construct (`dobl`) to evolve sorting algorithms [Kinnear, Jr., 1993b]. The loop construct operates on indices of lists with limited length, and hence also limits the number of iterations to a finite number.

Koza has also performed experiments where he evolved recursive sequences, and he has been able to produce the Fibonacci sequence [Koza, 1992d]. This experiment is not really an instance of a recursive program but it touches recursive issues. His programs belong to the third category where termination is guaranteed. In recent work, Koza demonstrated the evolution of iteration [Koza and Andre, 1996b].

Time-Bounded Execution

Categories 1 and 2 above are examples of time-bounded execution, a technique that can be used both with iteration and recursion. The next section examines these methods in more depth. Time-bounded execution in GP is when a program is limited in how many execution steps or time units it is allowed to perform, or as a milder restriction, if the program is punished for long execution times.

10.2.9 Memory Manipulation

Programming in most common computer languages such as Pascal, C, FORTRAN, or Ada relies on the assignment of memory or structures in memory. All commercial computers are register machines that operate by assigning values to memory or to internal registers of the CPU. This means that even purely functional languages will end up being implemented by memory assignment. Consequently, manipulation of memory and variables is an important concept in computer science.

The earliest method for evolution of computer programs that was evaluated on a large scale – tree-based GP – used programs with LISP syntax. The reason is that LISP has one of the simplest syntaxes of any computer language. The simpler the syntax the easier is it to ensure that crossover and mutation result in syntactically correct programs. The crossover operator working with LISP must in principle ensure only that the parentheses are balanced. This will result in syntactically correct programs. A crossover operator exchanging subtrees is such an operator: it always leaves the number of parentheses balanced in offspring.

LISP is a functional language in its purest form and hence does not use assignment. In the tree representing a LISP program the terminal node variables always represent the same value. In an *imperative* language such as C, a variable can be assigned at different moments and will hence correspond to different values. Complete LISP systems give an interface to assignment through special functions such as the *set function* for assignment or storage of values.

Assignment of Variables

Koza uses a `Set-Value` function to set the content of a single variable [Koza, 1992d]. The variable is then included in the terminal set where it can be read like any other variable. In other GP paradigms, variables and side effects are a more integrated part of the representation. In machine code GP, for instance, each instruction results in the assignment of memory or registers. It is a natural step from the assignment of a variable to the manipulation of indexed memory.

Complex Data Structures and Abstract Data Types

GP differs from other evolutionary techniques and other "soft computing" techniques in that it produces symbolic information (i.e., computer programs) as output. It can also process symbolic information as input very efficiently. Despite this unique strength, genetic programming has so far been applied mostly in numerical or Boolean problem domains.

A GP system can process fitness cases consisting of any data type for which we can supply the necessary function set. These could be strings, bitmap images, trees, graphs, elements of natural language, or any other data type that we can process with computer programs. For instance, GP has been used to evolve mathematical proofs where the fitness cases are trees representing statements about arithmetic [Nordin and Banzhaf, 1997a] and to evolve sorting algorithms operating on lists [Kinnear, Jr., 1993b].

GP can also be used to evolve its own abstract data types. A human programmer working in a high-level programming language can define his or her own abstract data structures fitting the problem domain he or she is working in. Data structures often help to structure a problem and result in higher quality code. Langdon has succeeded in evolving abstract data types with a GP system. He argues that these structures may help to increase the scalability of GP [Langdon, 1995c]. Abstract data structures have also been shown in several cases to be superior to an unstructured, flat indexed memory [Langdon, 1995a] [Langdon, 1996b].

In Langdon's work, abstract data structures are evolved from a flat index memory similar to the one used by Teller. The system has a handful of primitives at its disposal in order to address and move around in the indexed memory. The genome consists of multiple trees. The number of trees is predefined and corresponds to the functions expected to be necessary to manipulate the data structure. The stack data structure, for example, has five trees in its genome, corresponding to the five stack manipulation operations that one would like to evolve (push, pop, top, makenull, empty). All functions are evolved simultaneously. Crossover can exchange subtrees freely between the different trees in the genome. Figure 10.5 illustrates the multiple tree structure of the genome with abstract data types.

Cultural GP

Spector and Luke present an approach that may be called "cultural GP" [Spector and Luke, 1996a] [Spector and Luke, 1996b]. The principle here is to increase GP performance by methods that are an analog of cultural learning, i.e., of learning by *non-genetic* information transfer between individuals. To that end, Teller's indexed-memory technique gets used in a modified form [Teller, 1994a] [Teller, 1993]. For some problems, Spector and Luke show that the computational effort for problem solving gets reduced by employing cultural GP.

In the context of cultural GP, culture is viewed as the sum of all non-genetically transferred information. For instance, when a program transmits a numerical value to another program, this is an instance of culture. Referring to Dawkins [Dawkins, 1989], Spector and Luke use the term *meme* to designate the cultural analog of a gene: a meme is a cultural information unit.

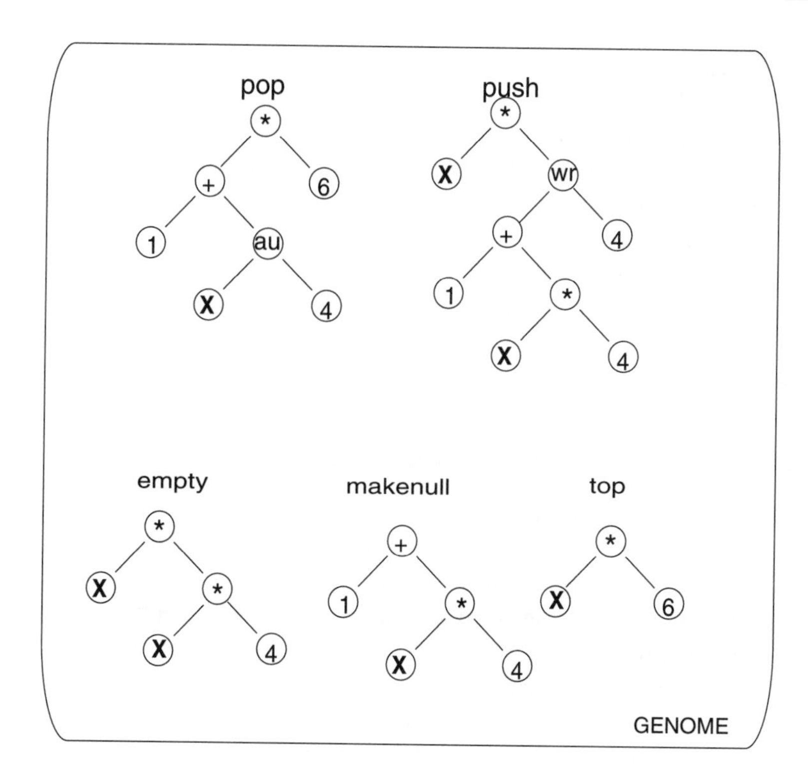

Figure 10.5
The multiple tree structure of the genome with abstract data types

A key issue in the context of culture is the interaction between evolutionary and cultural learning. It can be argued that:

- the dissemination speed of genetic information is slow compared to that of cultural information;
- hence, the gene pool is more stable compared to the meme pool.

Considering these properties, the synthesis of both learning methods may be superior to an approach using either one of them. One may hope, for instance, that beneficial memes would spread fast and could get exploited by many individuals, while, due to selection, deleterious memes – also spreading fast – would not "poison" the complete population.

Indexed Memory

The basic idea behind implementing cultural GP is to let all individuals have access to the same indexed memory. This memory gets initialized once only, that is, when a run starts. That way it serves as an information storage within and *across* generations. In particular, a certain individual may transfer information to itself. It would store information during its evaluation on a certain fitness case, and it would read this information during a subsequent evaluation on another fitness case. It could also transfer the information to another individual of the same generation or to an individual of a subsequent generation. Thus, the meme pool (indexed memory) would evolve over time like the gene pool.

Spector and Luke empirically investigated the efficiency of cultural GP on sample problems from symbolic regression, control, and artificial intelligence. They found that in some cases the computational effort was significantly lower with cultural GP than without.

The method of cultural GP surely is an interesting approach that deserves further attention, since it represents a significant extension of GP toward a software analog of organic evolution that incorporates more features than just genetic learning.

10.2.10 Strongly Typed Genetic Programming

Most modern programming languages, such as C++, have mechanisms preventing the programmer from mixing different data types unintentionally. For instance, trying to pass an integer argument to a function that is declared a string function will result in an error. If the programmer wants to violate the types he or she has declared, he or she must say so explicitly. Besides helping the programmer to avoid mistakes, these strongly typed programming languages make the source code more readable. Strongly typed genetic programming (STGP) introduces typed functions into the GP genome [Montana, 1994]. Using types might make even more sense in GP than with a human programmer because the GP system is completely random in its recombination. The human programmer has a mental model of what he is doing, whereas the GP system has no such guidance. Type checking also reduces the search space, which is likely to improve the search.

The use of strongly typed GP is motivated by the closure property of the function set. All functions in the individuals must be able to gracefully accept all input that can be given from other functions in the function set. This could be rather difficult in real-life applications with many different types of information represented as many different data structures.

A common example is a mix of Boolean and numeric functions. Let us say that we want to evolve a customer profile from a customer database. For instance, we would like to know to which customers we should mail our latest product offer. The function set consists of Boolean functions (`AND`, `OR`, ...), arithmetic functions (`+, -, *, /`), comparison functions (`>, <, =`) and conditionals (`IF THEN ELSE`). We might want to evolve criteria such as *IF the latest customer order is bigger than the average customer order times 10 OR IF the customer is a regular costumer, THEN send him the offer.* This kind of rule mixes numerical and Boolean information. It probably makes little sense to recombine the two different sorts of data with crossover. Interpreting a numerical as a Boolean value (0=`TRUE`; 1=`FALSE`) does not make much sense here and is not likely to improve the search. If we instead use a strongly typed GP system, we can guarantee that crossover will switch only two subtrees that return the same type.

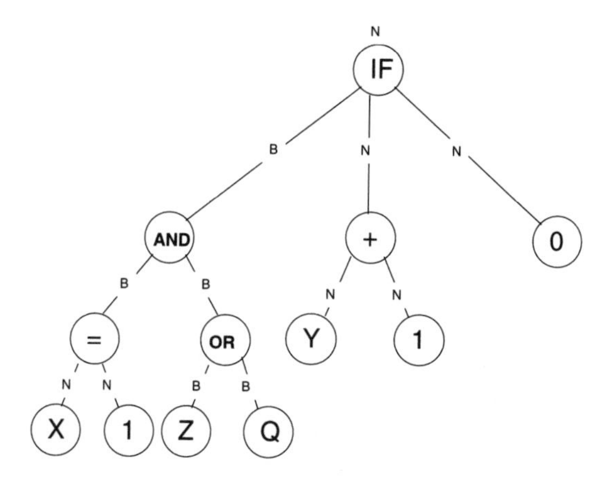

Figure 10.6
A strongly typed GP individual

In Figure 10.6 we see how each link between the nodes in the program tree is typed as either a Boolean or a numerical value. The comparison functions take a numerical and produce a Boolean, the arithmetical functions both take and produce numerical values, while the Boolean functions use only Boolean inputs and outputs. Conditionals may be defined in different ways. The leftmost input should be a Boolean, while the other inputs and the output are either completely Boolean or completely numerical. Functions that can manipulate several different data types are called *generic* types.

10.3 Improving the Power of GP Search

One of the central issues in machine learning is: how does an algorithm move from point n in the search space to point $n + 1$? In GP, the basic GP operators crossover and mutation are used to that end. Genetic programming search can be improved by employing additional mechanisms, some of which we present below. Among them are software analogs of ontogeny and co-evolution as well as hybrids built from different search paradigms.

10.3.1 Ontogeny

While evolutionary biology is concerned with processes that result in genotypic adaptation to an environment, developmental biology investigates a process that modifies the structure and, thus, the behavior of an organism during its entire lifetime. This process is called *ontogeny*.

Many GP systems ignore ontogeny altogether, because it certainly adds enormous complexity to the evolutionary process. Here we shall discuss those GP approaches that make explicit use of ontogenetic ideas. It has been shown empirically for certain problems that these systems improve the search in GP. The interested reader will find additional material in [Gruau, 1994a] [Teller, 1994a] [Iba et al., 1995a] [Zomorodian, 1995] [Keller and Banzhaf, 1996] [Spector and Stoffel, 1996b] [Spector and Stoffel, 1996a].

Development

From an abstract point of view, a structure is the carrier of a function. An individual in an evolutionary algorithm is such a structure. The interpretation of the structure in the context of a problem domain determines the function that will be carried by that structure. For instance, this structure can be a real-valued vector, a cellular automaton rule, a 3D-wire-frame model of a real object, a tree, a network, etc. In GP, the function of an individual is defined by the semantics of the program language.

In our opinion, developmental approaches to GP try to address at least one of the following problems:

- How can the search be left unconstrained, whereas the solutions need to be highly constrained through syntax?
- How can one use a compact genetic representation that unfolds into a complex individual program?
- Is there a way to include the "environment" into the process of generating complexity? Can the environment even help in organizing an individual program?

Developmental biology, as one very important branch of biology, will provide more insights into the ontogeny of an individual, and GP stands to learn a lot from those models. Not only GP but also other EAs will benefit as well. We anticipate theories of development to have a considerable impact on genetic programming and evolutionary algorithms in general in the coming years.

As one example of what has already been achieved, we shall discuss in the rest of this section a model due to Spector et al.

Ontogenetic Programming

Spector and Stoffel [Spector and Stoffel, 1996b] present an approach they call ontogenetic programming. By including program *self-modification* operators into the function set, a program may evolve that can change its structure during run time. Such a program represents an analog of an organism that changes its structure during its lifetime, for instance, when it grows. A change in the program structure may modify the program's future behavior. Since the self-modification operators are elements of the function set, the self-modification of a certain program may evolve itself.

Ontogeny vs. Phylogeny

Ontogenetic modification of an individual program – versus phylogenetic modification of the population, which is what every GP system does – may enhance the performance of a program under certain circumstances. For instance, if different expressions of a certain program feature are needed over the program's run time, ontogenetic program modification may yield it.

For an example, the authors describe a program performing as an adventure-game agent. Depending on the properties of the game environment, different tactical agent behavior in different game stages may be helpful in reaching the strategic goal of the game. Thus, a development of the individual may be needed.

Ontogeny vs. Conditionals

Different behaviors of the same program might be realized, however, in a much simpler way than by ontogenetic modification, by using conditionals. On the other hand, ontogenetic modification allows for a potentially unlimited number of different behaviors over the run time and is therefore significantly more flexible.

Stoffel and Spector use their own system called *HiGP* [Stoffel and Spector, 1996] to validate the ontogenetic approach. A HiGP program has a linear structure. The self-modification operators are tailored to this structure. Stoffel and Spector describe three such operators, although many others are possible: `segment-copy` replaces a program part by the copy of another part, `shift-left` and `shift-right` rotate the program instruction-wise.[2] On a problem of binary sequence prediction, ontogenetic programming has enhanced program performance considerably.

[2] In fact, Spector has recently been focusing on rather different ontogenetic operators, including an `instruction` instruction that replaces itself – or another instruction – with an instruction obtained from a lookup table. He has also been adding functions that allow a program to know how "old" it is so that it can incorporate timed developmental strategies [Spector, personal communication, July 1997].

The concept of ontogenetic programming can also be used with S-expression-based programs, although the implementation is easier for linear program structures.

10.3.2 Adaptive Parsimony Pressure

Zhang and Mühlenbein [Zhang and Mühlenbein, 1996] have suggested a simple method for adaptively controlling parsimony pressure during evolution. The method is simple, well grounded in theory and empirically validated [Zhang and Muehlenbein, 1995] [Blickle, 1996]. The theoretical analysis of the concepts draws from the notion of the minimal description length (MDL) principle [Rissanen, 1984] and it considers the programs as Gaussian models of the data. For details the reader may consult the original sources.

Let $E_i(g)$ denote the error produced by a given individual i in generation g. Similarly, let $C_i(g)$ denote the complexity of an individual i in generation g. For $E_i(g)$ and $C_i(g)$ the following assumptions should hold:

$$0 \leq E_i(g) \leq 1 \tag{10.1}$$

$$C_i(g) > 0 \tag{10.2}$$

Then fitness is defined as the error plus an adaptive parsimony term $\alpha(g)$ times complexity:

$$F_i(g) = E_i(g) + \alpha(g) \cdot C_i(g) \tag{10.3}$$

where $\alpha(g)$ is determined as follows. Let N be the size of the training set and ϵ be the maximal error allowed in an acceptable solution. The ϵ-factor is given before training and is an important ingredient of the method: it could be said to represent the expected noise in the training data. Given this, the parsimony factor should be chosen according to the following equation:

$$P(g) = \begin{cases} \frac{1}{N^2} \frac{E_{best}(g-1)}{C'_{best}(g)} & \text{if } E_{best}(g-1) > \epsilon \\ \frac{1}{N^2} \frac{1}{E_{best}(g-1) \cdot C'_{best}(g)} & \text{otherwise.} \end{cases}$$

$E_{best}(g-1)$ is the error of the best performing individual in the preceding generation while $C'_{best}(g)$ is an estimation of the size of the best program, estimated at generation $(g-1)$. $C'_{best}(g)$ is used to normalize the influence of the parsimony pressure.

This method has some similarities to the way a human may solve a problem or program an algorithm. First, we are more directed toward finding a solution that works, and when this is achieved we might try to simplify the solution and to make it more elegant and generic. Similarly, the parsimony pressure decreases while the error is falling. In contrast, when the fitness approaches zero the parsimony pressure decreases to encourage small, elegant, generic solutions.

10.3.3 Chunking

An almost obvious method to improve the efficiency of a GP system when dealing with very large sets of fitness cases is to divide the fitness case set into smaller chunks.[3] A solution can then be evolved for each of the smaller chunks, and the overall program will be some kind of concatenation of all the solution generated for the chunks. For instance, consider the programmatic compression of images [Nordin and Banzhaf, 1996], where we try to evolve a GP program that produces a bitmap image – close to a target image – when executed. This application has fitness case sets of hundreds of thousands of input/output pairs, one for each image pixel. The image can be chunked into subimages and a program for each subimage can be evolved. The overall solution to the compression problem is then a loop activating the evolved results for each of the chunks in the correct order.

10.3.4 Co-Evolution

The idea behind co-evolution of fitness functions is to get rid of the static fitness landscape usually provided by a fixed set of fitness cases. In fitness co-evolution, all fitness cases are considered to be the result of an algorithm that can be subjected to the same principles of variation, evaluation, and selection as the population to be evolved.

Thus, a separate population encoding a variety of fitness tests for the original population could be co-evolved by allowing the performance of tests to influence their survival probabilities. In other words, the environment for the second population is given by the first, and vice versa. Because both populations are allowed to evolve, and weaknesses of the first population are exploited by the second, as are weaknesses of the second population by the first, an "arms race" will develop [Dawkins, 1987].

Figure 10.7 shows a sketch of different evaluation patterns for both populations. In (a), all members of the first population are systematically tested against all the members of the second. In (b), all are tested against the best performing member of the other population. In (c), randomly selected members are tested with randomly selected members from the other population.

[3] This usage of the word *chunking* should not be confused with that in other fields of artificial intelligence, for instance, in connection with the SOAR system.

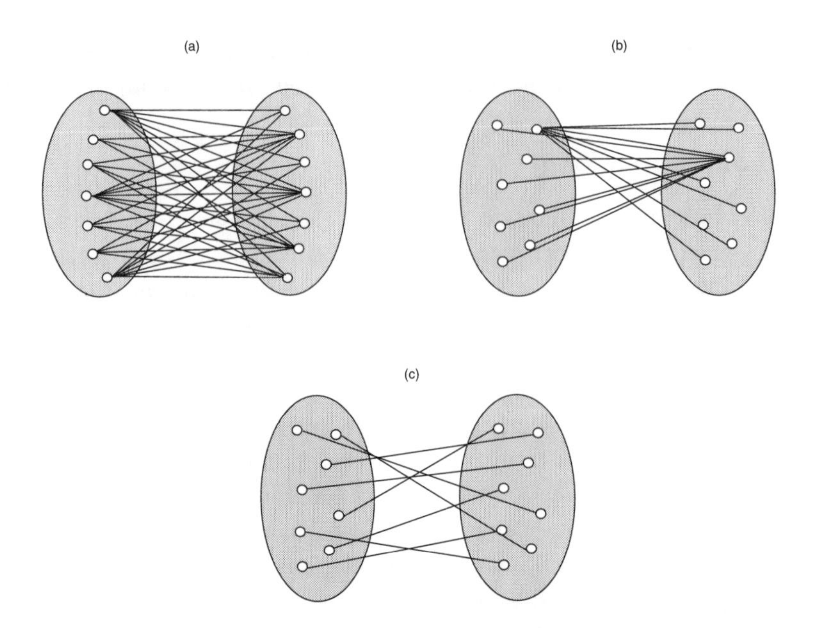

Figure 10.7
Three scenarios for competitive co-evolution with two populations: (a) all vs. all; (b) random; (c) all vs. best. Similar patterns can be used within one population. (from [Sims, 1994])

Both populations will improve their fitness in response to the criteria set forth in their respective evaluation function, which is allowed to shift dynamically to more sophisticated levels. Hillis [Hillis, 1989] [Hillis, 1992], Angeline and Pollack [Angeline and Pollack, 1993] and Sims [Sims, 1994] report a considerable improvement in the degree of optimization reached when performance is judged dynamically. Ronge [Ronge, 1996] has examined the artificial ant problem under the provision that trails providing food pellets to a population of searching ants themselves develop to escape the searching ants.

In Jannink [Jannink, 1994] pseudo-random number generators are co-evolved with testers. Testing programs try to guess the sequence of numbers generated by an algorithm. The problem has a long history with roots going back to von Neumann and Morgenstern's seminal work [von Neumann and Morgenstern, 1944] on game theory. It can be shown that a random strategy is the best a player can apply in order to test another player's random strategy. Thus, both populations, pseudo-random number generators and pseudo-random number testers will be driven in the direction of better pseudo-random number series.

Competitive co-evolution can be set up even within one population. Juille and Pollack [Juille and Pollack, 1996] have reported good results by putting individuals from one population into opposite camps for a game. What they aimed at was to reward differential fitness instead of absolute fitness. Therefore, fitness was computed using tournaments between individuals, but only those fitness cases were counted on each side that were unique hits. A number of tournaments were executed for each individual and the scores were summed up to yield a fitness for the individual.

So far, we have considered competitive co-evolution, which has a long-standing tradition in population biology under the heading of predator–prey models. However, cooperative co-evolution is another legitimate strategy. Here, it is not the weakness of the counterpart that is exploited, but the strength of the other population that helps improve one's fitness. So each population should benefit the other one, and itself profit from the thriving of the other one. A typical application is the co-evolution of solutions to a search problem with genetic search operators that themselves are allowed to evolve [Teller, 1996].

10.3.5 Hybrid Approaches

We noted earlier that there are different kinds of learning[4] in nature – genetic learning, learning from experience, and learning from culture. Many types of learning are going on simultaneously. Some researchers have developed hybrid systems combining genetic learning with other types of learning.

We distinguish two classes of hybrid approaches. In the first class, GP is blended with other general-purpose search algorithms, such as simulated annealing (SA) or stochastic iterated hill climbing (SIHC). We will discuss this issue first.

Later, we discuss a second class of hybrid approaches where GP is merged with special-purpose search algorithms. The specialty lies in the fact that, given a certain choice of terminals and functions (including their side effects), various algorithms can be employed to add explorative power to GP.

GP and General-Purpose Search Algorithms

This subsection covers combinations of GP with other algorithms for general-purpose search. Like GP, other general-purpose search algorithms for program induction have to deal with the following problems:

- Local minima
- Size of search space

Thus, the class of algorithms combinable with GP is restricted. Basically, all the algorithms that have been blended with GAs are also

[4]We take learning in a natural system to be synonymous with search in an artificial system.

candidates to be blended with GP. Table 10.1 shows which techniques have been combined with GP so far.

Search technique	Problem domain	Source
stochastic iterated hill climbing	6-,11-multiplexer	[O'Reilly and Oppacher, 1996]
	sort	[O'Reilly and Oppacher, 1996]
	block stacking	[O'Reilly and Oppacher, 1996]
cultural transmission	regression	[Spector and Luke, 1996a]
	lawnmower	[Spector and Luke, 1996a]
	Wumpus agent world	[Spector and Luke, 1996a]

Table 10.1
Combination or comparison of GP with other general-purpose search techniques

GP and Special-Purpose Search Algorithms

Special-purpose algorithms are designed in the realm of a particular problem domain. They are intended to provide problem-specific operations working only on a specific representation of solutions. Their advantage is that they work efficiently and fast; on the other hand, they usually suffer from being local search tools that can be easily trapped by local minima. It is here that GP offers a remedy.

GP can benefit heavily from an insertion of efficient local search operations; see Table 10.2. Iba et al. [Iba et al., 1994] have done a study on the problem class of Boolean problems. The ordinary GP approach to Boolean problems is augmented by a specific local search procedure that does not change the tree structure of the individuals but periodically employs what they call a relabeling of nodes. Every now and then, the entire population is subjected to this relabeling, where the nodes of all trees are exchanged with those nodes that guarantee optimal performance, given the structure of the trees. This local algorithm is the adaptive logic network algorithm first published in 1979 by Armstrong et al. [Armstrong and Gecsei, 1979].

The key idea of their GP system is to use a local search procedure periodically to improve all trees of the population simultaneously. Thus, structured advantages are not lost due to overspecification. Iba et al. [Iba et al., 1994] report favorable results for low periods of 2–3 generations before an adaptive logic network algorithm is applied in the population: with much smaller populations or tree depths, a perfect solution can be found. Up to 50 times fewer evaluations have to take place for this hybrid GP as compared to conventional GP.

In a second comparison, Iba et al. used the GMDH algorithm [Iba et al., 1994] for relabeling nodes in a system identification problem. We explained this approach in more detail in Chapter 9. The

Search technique	Problem domain	Source
ALN	6-,11-mult	[Iba et al., 1994]
	6-,11-mult, non-stationary	[Iba et al., 1994]
	even 3,4,5 parity	[Iba et al., 1993]
	Emerald's robot world	[Iba et al., 1993]
GMDH	regression: Heron formula	[Iba et al., 1994]
	Glass-Mackey time series prediction	[Iba et al., 1995a]
	pattern recognition	[Iba et al., 1995b]

Table 10.2
Combination of GP with special-purpose search techniques. ALN = adaptive logic network algorithm

problem of symbolic regression of the Heron formula

$$S = \sqrt{\frac{(a+b+c)(a+b-c)(a-b+c)(-a+b+c)}{16}} \tag{10.4}$$

was successfully solved with 10 times fewer evaluations than in conventional GP [Iba et al., 1994].

As we have seen in the GA community, we can expect that other local search methods will be employed together with GP to yield better results that could not have been achieved by either method alone.

Exercises

1. Give two limitations to the basic GP approach.
2. Describe two methods for parallelizing GP.
3. What is stochastic sampling?
4. Name three modularization techniques.
5. Describe the basic ADF technique. What is an architecture altering operation?
6. What is similar and what is different between the encapsulation method and the module acquisition method?
7. Give two examples of how to do time-bounded execution.
8. How does a strongly typed GP system work?
9. Give two examples of competitive co-evolution in populations. What is cooperative co-evolution?
10. Which concepts of developmental biology could be included in GP?
11. What is the idea behind hybridizing GP with other algorithms? Give two examples.

Further Reading

P.J. Angeline and K.E. Kinnear, Jr. (eds.),
ADVANCES IN GENETIC PROGRAMMING 2.
MIT Press, Cambridge, MA, 1996.

J.R. Koza,
GENETIC PROGRAMMING II.
MIT Press, Cambridge, MA, 1994.

J.R. Koza et al.,
GENETIC PROGRAMMING 1997. PROCEEDINGS OF THE SECOND ANNUAL CONFERENCE.
Morgan Kaufmann Publishers, San Francisco, CA, 1997.

11 Implementation — Making Genetic Programming Work

Contents

Tradeoffs to Consider

This chapter addresses fundamental "how to" issues of GP. A GP system has to store representations of the individuals it is evolving, perform genetic operations upon them, interpret the stored individuals as programs, and perform fitness calculations using the interpreted individuals. Coding a GP system that is fast, memory efficient, easily maintainable, portable, and flexible is a difficult task. As a matter of fact there are many tradeoffs, and this chapter will describe how major GP systems have addressed those tradeoffs.

The chapter will discuss the following subjects:

1. We will present examples to show why GP uses CPU cycles and memory so profusely. This is the central problem faced by GP programmers. In the examples of CPU and memory use, the numbers we present are approximate and could vary substantially from system to system and from machine to machine. The purpose of these numeric examples is to explain and to underline the magnitude of the difficulties that must be faced in GP.

2. We will describe systematically various low-level representations of evolving programs. The three principal models that have been used for storing GP individuals during evolution are (LISP) lists, compiled language (such as C++) data structures, and native machine code. We will discuss how they have been implemented and what the relative advantages and disadvantages are.

3. We will give an overview of the parameters that must be set during GP runs and suggest rules of thumb for setting those parameters.

The thread that will run through the entire chapter is that GP programs need speed, flexibility, portability, and efficient use of memory, but that there are tradeoffs among these goals. Inevitably, in discussing tradeoffs, the authors will often point out what may seem to be a drawback of a particular approach. In doing so, they do not mean to criticize the decision that led to these approaches or to discourage their use.

11.1 Why Is GP so Computationally Intensive?

A brief examination of the nature of GP will make it clear why GP is so very intensive in its use of CPU time and memory. Three factors are at work.

1. **Large Populations**
   GP evolves a population or populations of individuals. Typical population sizes range between 500 and 5000. But they can be much larger to good effect. In Koza's parallel implementation on 64 Power PCs, populations in excess of 600 000 individuals have been evolved successfully [Andre and Koza, 1996b]. One of our own experiments went up to population sizes of 1 000 000 individuals in AIMGP.

2. **Large Individual Programs**
   GP works with variable length individuals (programs). This fact frequently results in large GP program sizes for two reasons. First, the problem to be solved often appears to require reasonably long solutions. Second, invariably the problem of bloat occurs during GP runs. Much of that problem is connected to the accumulation of introns in the GP individuals. The growth of introns is ultimately exponential during a GP run unless controlled.

3. **Many Fitness Evaluations**
   Almost all of the time consumed by a GP run is spent performing fitness evaluations. Fitness evaluations are necessary because they provide the metric by which GP performs selection on the population. To understand why fitness evaluations are so CPU intensive we have to add two more factors. First, GP fitness evaluations typically take place using a training set of fitness cases. The size of the training set varies enormously, depending on the problem – from 50 to 5000 or more fitness cases is not unusual. Second, every GP individual is normally evaluated on each fitness case in each generation.

11.1.1 Why Does GP Use so Much CPU Time?

Let us look at what these three factors mean for CPU use during a sample GP run. Our sample run will last $G = 50$ generations and have a training set of $F = 200$ fitness cases, a population size of $P = 2000$ and an average individual size that grows from $\langle N_{init} \rangle = 50$ nodes to $\langle N_{max} \rangle = 800$ nodes during the run due to bloat. We assume each of the nodes will be evaluated during evaluation.

Even in this average-size run, the demand on the processor is very substantial. For example, in the initial generation of this run, each individual must be evaluated for each training instance. In other words, for each individual in the population in the first generation, $\langle N_{init} \rangle \times F = 10^4$ nodes must be executed. To perform fitness evaluation during the first generation, the GP system must evaluate all individuals. So, the GP system must execute $\langle N_{init} \rangle \times F \times P = 2 \cdot 10^7$ nodes just to evaluate the individuals in the first generation. Therefore, 10^8 node evaluations would be required for the first 5 generations of the example run. Table 11.1 shows how this node number grows as bloat sets in.

Generations	Ave. no. of nodes per individual	Nodes evaluated (millions)	Cum. nodes evaluated (millions)
0 – 4	50	100	100
5 – 9	50	100	200
10 – 14	100	200	400
15 – 19	100	200	600
20 – 24	200	400	1,000
25 – 29	200	400	1,400
30 – 34	400	800	2,200
35 – 39	400	800	3,000
40 – 44	800	1,600	4,600
45 – 49	800	1,600	6,200

Table 11.1
Example of node calculations with a population of 2000 GP individuals

By the 50th generation, the GP system will have performed over $6 \cdot 10^9$ node evaluations! Had we used the very large population size (600 000) employed by Andre and Koza, the number would have been $1.86 \cdot 10^{12}$ node evaluations just through generation 50. It should, therefore, come as no surprise that GP programmers have labored hard to make GP node evaluation as efficient and fast as possible.

11.1.2 Why Does GP Use so Much Main Memory?

The high CPU time consumption is not the only problem. GP also consumes a lot of memory. Ideally, every GP individual should be stored in RAM in as compact a manner as possible. Further, that compact representation should make application of genetic operations simple and interpretation of the individual as a program easy and fast. Needless to say, it is not quite that easy. We will first take a look at the practical limit of efficiently representing an individual in memory. Then we will suggest some of the problems that have prevented GP programmers from reaching that ideal.

Limits of Efficient Memory Use

A compiled computer program is no more than a sequence of instructions in native machine code. Processor designers have spent years compressing a large amount of information into the native machine code format designed for their processor. Thus, we may safely regard the amount of memory that would be used by a GP individual, expressed as native machine code, as the practical minimum amount of memory that must be devoted to an individual. A 100-node individual would occupy 400 bytes of memory, expressed as machine code, assuming that each node is equivalent to and may be expressed as one 32-bit machine code instruction.

Most GP systems do not begin to approach the efficiency of memory usage that is implicit in a machine code representation of a particular program. This might seem unusual. We are, after all, talking about genetic *programs*. But with one exception, genetic programs are *not* stored in memory directly as machine code. Rather, most GP systems represent the individuals/programs symbolically in some sort of high-level data structure. That data structure is interpreted by the GP system as a program when it is time to conduct a fitness evaluation. The decision to represent the GP individual in data structures that are less efficient memory-wise is often a deliberate tradeoff made by the software designer to effect, for instance, easier application of the genetic operators or portability.

A Memory Usage Example

An example will be useful. One structure often used in GP systems is an expression tree. Essentially, each node in an individual is stored in an array or a linked list that identifies what function the node represents and where the node gets its inputs from. Both the function and the inputs are identified with pointers. In modern programming languages, pointers consume, say, four bytes of memory. If the average arity of the nodes in an individual is two, then it takes, on average, three pointers to represent each node – two pointers to point to the input nodes and one pointer for the function represented by the node. For a 100-node GP individual, therefore, it would take at least 1200 bytes of memory to represent the individual symbolically rather than in machine code. As a practical matter, more is necessary – 1600 bytes would be a conservative estimate.

We may now look at the approximate memory requirements of the same example run that we discussed in the above table. In the first generation, a population of $P = 2000$ individuals would occupy $\langle N_{init} \rangle \times P \times 16$ bytes = 1.6 megabytes. But, by the 50th generation, the population averages $\langle N_{max} \rangle = 800$ nodes for each individual and would occupy around $\langle N_{max} \rangle \times P \times 16$ bytes = 25.6 megabytes of memory. This exceeds the memory available on many modern computers for an application. Thus, the run would degenerate into disk swapping.

Effects of Extensive Memory Use

There are two prominent effects of extensive memory use:

1. **Memory access is slow**
   Every byte of data used to describe an individual during evolution causes the GP system to slow down. Every time the

processor must look to RAM for data, the processor usually has to stop and wait several cycles – say it averages three cycles waiting for a response from RAM. If each node is defined by 16 bytes of data held in RAM, it takes four RAM accesses to get that data on a 32-bit machine. This means it takes approximately twelve cycles of the processor to get those 16 bytes of data from RAM. During that time, the processor could have executed at least twelve instructions. Given these assumptions, the run in the above table would have required 74.4 billion processor cycles just to access the RAM for loading nodes.[1]

2. **Garbage collection**
   There is a more subtle memory problem caused by many GP representation schemes like LISP list systems and C tree systems. That problem is garbage collection. One of the common features shared by systems interpreting C and by LISP lists is that memory is being constantly allocated and deallocated for the nodes and the elements of the lists. When small chunks of memory are being allocated and deallocated constantly, it leads to RAM that resembles Swiss cheese, where the holes in the cheese represent available RAM, and the cheese itself represents allocated RAM. Over the course of evaluating several billion nodes during a GP run, one would expect a lot of small holes to develop in the cheese: the memory gets *fragmented.* Thus, it gets increasingly harder to find contiguous memory chunks large enough for housing certain data structures.

   Dealing with fragmented memory is a task known as *garbage collection*, a practice of moving the "cheese" around in RAM such that one gets large "holes" again. This process may be very time consuming. This problem has caused researchers to abandon LISP lists for storage of GP individuals. Instead, they use high-level symbolic data structures (arrays) to contain symbols that are then interpreted as a program.

11.2 Computer Representation of Individuals

Ultimately, all programs are just information about what operations on which data to perform with the processor. A GP individual may, therefore, be viewed as only a collection of information that should be interpreted as a program. Perhaps the most important decision a

[1] This consideration does not take into account cache effects.

GP programmer must make is how to store that information. This decision affects how simple it is to create and maintain genetic operators, how fast the system runs, how efficient the use of memory is, and many other important aspects of performance. Programmers have represented the GP individual during evolution in three different ways. Here is a brief description of the three approaches and their various advantages and disadvantages:

1. **LISP lists**
   LISP is a high-level programming language that is very popular with artificial intelligence programmers. The GP individual in this approach is represented as a LISP list. LISP lists are very convenient for the representation of tree-based programs – they make crossover and mutation simple to implement. LISP is not so convenient for genome structures other than tree-based structures. As long as the programmer sticks to tree structures, however, a LISP-based system is easy to maintain and makes execution of the individual simple since LISP has a built-in interpreter.

2. **Data structures in compiled languages such as** C, PASCAL, **or** FORTRAN
   In this approach, information about the individual is stored symbolically in a data structure such as a tree, an array, or a linked list. When it comes time to evaluate the individual for fitness, compiled programs are much faster than LISP. On the other hand, C data structures, for instance, are much more labor intensive to provide than LISP lists. This approach requires the programmer to write genetic operators that operate on variable length structures and that usually engage in large-scale pointer manipulation – always a tricky task. The C programmer must also write and maintain his or her own interpreter to convert the symbolic program information stored in the data structure into a usable program.

3. **Native machine code**
   In this approach, the GP individuals are stored as arrays in memory. The arrays actually contain machine code instructions in the form of binary code, which operate directly on the CPU registers. There is no high-level interpretation or compilation step involved since the instructions get interpreted – that is, executed – directly by the processor. This approach is very fast (about sixty times faster than compiled C code [Nordin and Banzhaf, 1995b]) and is very compact in its use of memory. However, programming this type of system is more

difficult than either LISP list or C data structure systems, and much greater effort is needed to assure portability, flexibility, and maintainability.

The next several sections of this chapter will discuss four different actual implementations of a GP system using these three different representations of the individual. The emphasis here will be on the details of how the GP individuals are stored, executed, and crossed over.

11.3 Implementations Using LISP

Because of the prominence of Koza's work, references to LISP and LISP type concepts dominate large parts of the GP literature. In fact, many researchers who do not use LISP systems nevertheless report their results using LISP S-expressions. LISP is, in many ways, a natural language for AI and tree-based machine learning applications because of its simplicity and its strong support of dynamic data structures.[2] Indeed, our discussion of the LISP approach will be much simpler than our discussion of the C and machine code approaches for one reason – most of the support for storing and manipulating tree structures is handled by the LISP language itself.

Before we look at how LISP implementations of GP work, it will be useful to define a few terms.

11.3.1 Lists and Symbolic Expressions

A LISP list is an ordered set of items inside parentheses. For the purpose of this work, a symbolic expression (S-expression) may be regarded as a list. In LISP, individuals are written and stored as S-expressions. Here is an example of a very simple S-expression, which could be a simple GP program:

$$(\times a b) \tag{11.1}$$

This expression is equivalent to the algebraic expression:

$$(a \times b) \tag{11.2}$$

Therefore, if the value of a is 2 and the value of b is 13, the above S-expression evaluates to 26.

[2]Other researchers have devised GP systems for other high-level languages such as PROLOG and MATHEMATICA. We will address the LISP systems here because of their prominence.

S-expressions can be more complex than this simple example. In particular, parentheses may be nested. Note how the above S-expression is nested into the S-expression below:

$$(-(\times ab)c) \tag{11.3}$$

This S-expression is equivalent to the algebraic expression:

$$(a \times b) - c \tag{11.4}$$

These three equivalent expressions are shown in Figure 11.1.

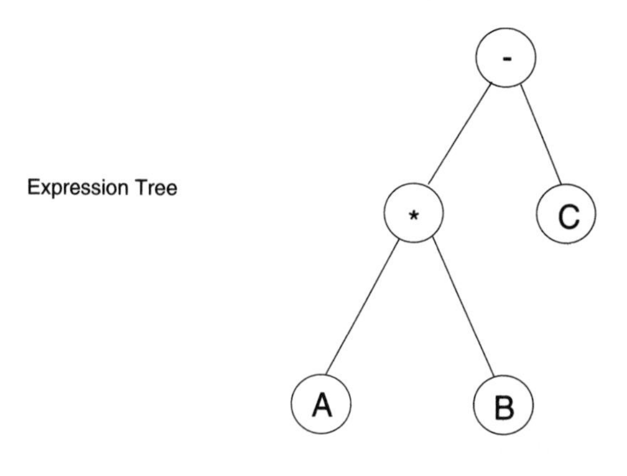

S-Expression (- (* A B) C)

Algebraic Notation (A * B) - C

Figure 11.1
A program statement as an expression tree, an S-expression, and an algebraic expression

11.3.2 The "How to" of LISP Implementations

S-expressions have several important properties that make them useful for GP programmers. To begin with, the GP programmer does not need to write any code interpreting the symbolic information in the S-expression into machine code that may be executed. The LISP interpreter does that for the programmer.

Furthermore, LISP S-expressions make it very easy to perform genetic operations like tree-based crossover. An S-expression is itself an expression tree. As such it is possible to cut any subtree out of an

S-expression by removing everything between any pair of matching parentheses. Note how this works for the S-expression in Figure 11.1. Crossing over the subtree under the $\times$ symbol is accomplished in an S-expression by clipping out the string ($\times$ a b). The process of performing this operation is shown in Figure 11.2.

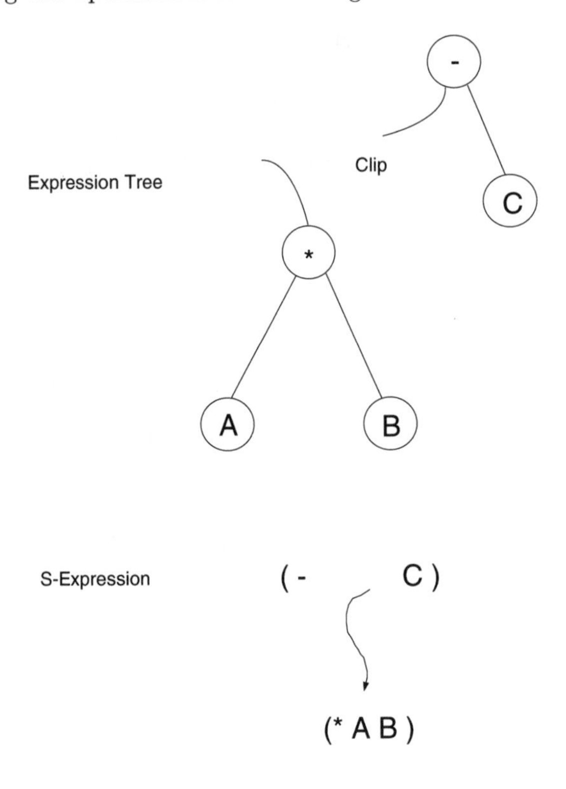

Figure 11.2
Clipping out a subtree from an expression tree and an S-expression

Clipping out a subtree in LISP is thus very simple compared to performing crossover on a GP individual that is held in a C array. Exchanging subtrees during crossover is also simple. It is only necessary to make sure that the portion to be clipped out of each S-expression is between matching parentheses in each individual. Then the two segments are clipped and exchanged as shown in Figure 11.3.

Likewise, mutation is simple. The subtree $(* \ a \ b)$ could easily be mutated by changing the * to a + in the S-expression. In that case, the resulting subtree would be $(+ \ a \ b)$. The reason for this simplicity is that manipulating tree structures is something that LISP was designed to do.

11.3.3 The Disadvantages of LISP S-Expressions

We have already indicated above that LISP S-expressions can cause memory problems. This is because lists are constantly being created

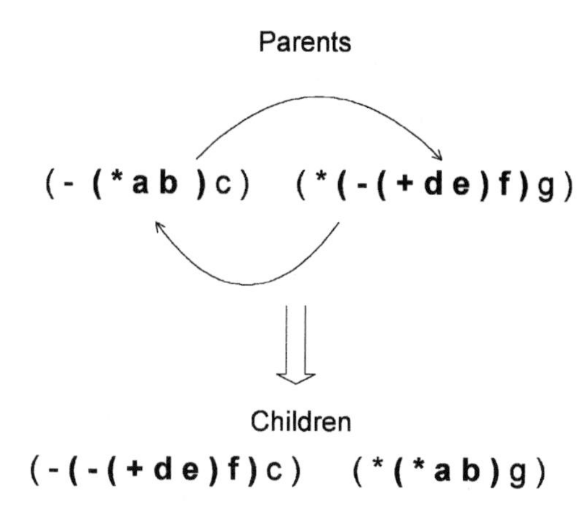

Figure 11.3
Exchanging two subtrees in LISP crossover

and destroyed during evolution. Although many flavors of LISP have built-in garbage collection, GP may create garbage faster than it can be collected.

Another problem with LISP is speed. C-based GP is more than ten times as fast as LISP-based GP. Finally, although LISP lists provide the advantage of simple tree manipulation, LISP does not have the same advantage in manipulating other GP type genomes such as a linear graph genome.

11.4 Some Necessary Data Structures

Many GP systems today are written in a compiled language, which often is C. We have already described C systems that use pointers to link together the nodes of a tree. Other researchers have used a quite different approach in compiled language GP. They combine a more linear C data structure such as a linked list or an array with a stack. This chapter will focus on implementation issues in two of these systems:

1. The systems described in [Keith and Martin, 1994], which implemented tree-based GP in an array/stack arrangement, and

2. Teller's PADO system which implements GP as a directed graph of program execution nodes that are held in linked-list structures. PADO also relies on a stack to implement the system [Teller and Veloso, 1996].

Many readers will already be familiar with arrays, linked lists, and stacks. Understanding these basic data structures and how the memory for them is allocated and de-allocated is essential to understanding the remainder of this chapter. If the terms *push*, *pop*, *pointer to next*, and *pointer to prior* are familiar to you, then you may skip to the next section.

11.4.1 Arrays

An array is an ordered list of items of the same type with each of the items having an index. Such a type may be integer, real, or character, for instance. Figure 11.4 shows a generic array and an array filled with integers.

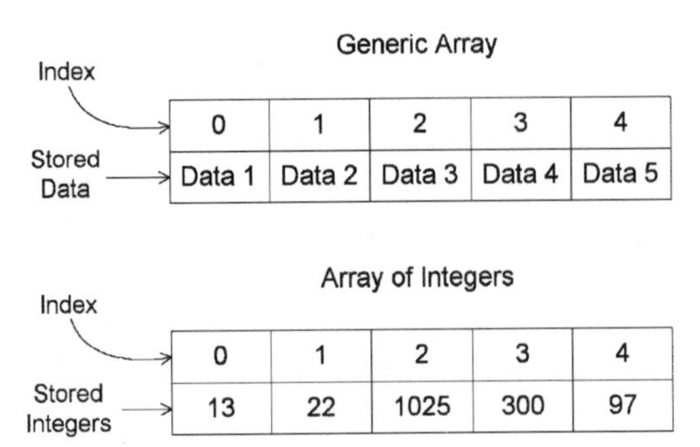

Figure 11.4
Two ways of looking at arrays

To access an item, the programmer simply uses its index. For example, to access the fourth integer in the array of integers shown in Figure 11.4, the programmer could write **array[3]**. Of course, the value of **array[3]** in Figure 11.4 is 300. It should be obvious that a GP individual can be stored in an array. In fact, some GP systems do just that.

There are several problems that arise in using arrays for storing GP individuals:

- **Variable size individuals**
  Although it is possible in C++ to create arrays of undetermined size, resizing arrays at run time requires a lot of programming and memory allocation and de-allocation. This is an important factor in GP where the evolving individuals are variable in length. If a program holds the individuals in the GP run in arrays, crossover is constantly changing the size of the individual, allocating and de-allocating memory. This can cause garbage collection problems.

- **Complex crossover operators**
  Figure 11.3 has shown how crossover is performed in LISP S-expressions. Note that the sizes of the subtrees that are ex-

changed in that figure are different. There is no obviously easy way, as there is in LISP, to clip out part of an array containing, say, four elements, and exchange those four elements with five elements taken from another array. Both arrays have to be resized, which implies programming work and computation overhead. We will see two different solutions to this problem below.

- **Variable amounts of data per node**
  In GP, different nodes often have different numbers of inputs. For example, a `Plus` node has two inputs. On the other hand, an `If/Then/Else` node has three inputs. This makes an array representation difficult to keep track of, because the `Plus` node will need at least three array elements to store it (one element to identify the type of function and the other two to identify the inputs to the operator). Using the same logic, the `If/Then/Else` node will use up at least four array elements. This would require the GP programmer to add logic to keep track of what operators are where in the array, both for crossover and for execution.

None of the above problems is insoluble, and we will look at how programmers have addressed them. Consistent with the theme of this chapter, we conclude that although arrays are easy to manipulate and to access, there are tradeoffs in using arrays to hold GP individuals.

11.4.2 Linked Lists

The details of creating and traversing linked lists are beyond the scope of this section. However, it is important to know what such a list looks like, because linked lists can provide important flexibility in creating a GP system, as we will see when we examine the PADO system.

A linked list is a sequence of elements. Each element contains some data. In addition, it contains a pointer to the next element in the list and, sometimes, a pointer to the previous element in the list. Technically, a pointer is simply an integer, stored at one place in memory, that identifies a second location in memory. Using pointers, a program can hop around in memory in order to access data in a very flexible manner. Figure 11.5 shows a doubly linked list.

It is obvious that, from any element in Figure 11.5, the programmer can move forward or backward to any other element in the list just by following the pointers.

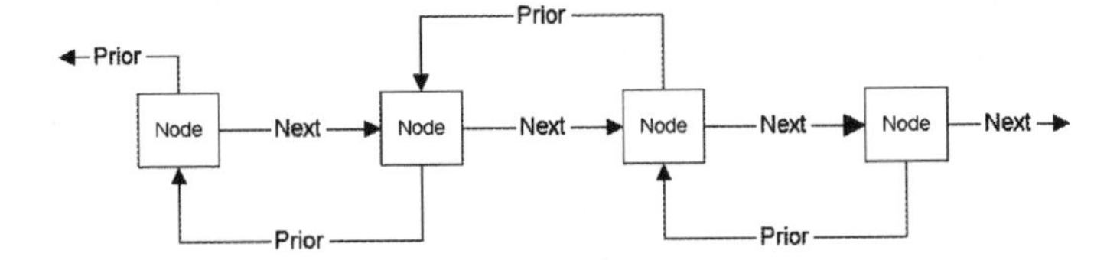

Figure 11.5
A doubly linked list, a versatile storage structure for GP individuals

A linked list is a very flexible type of data structure because any information may be attached to an element. It has several advantages and disadvantages over arrays for storing GP individuals:

- **Ease in resizing**
  Resizing a linked list requires only that memory be allocated or de-allocated for the nodes to be added or deleted. For example, Figure 11.6 shows how a node may be easily inserted into the middle of a linked list.

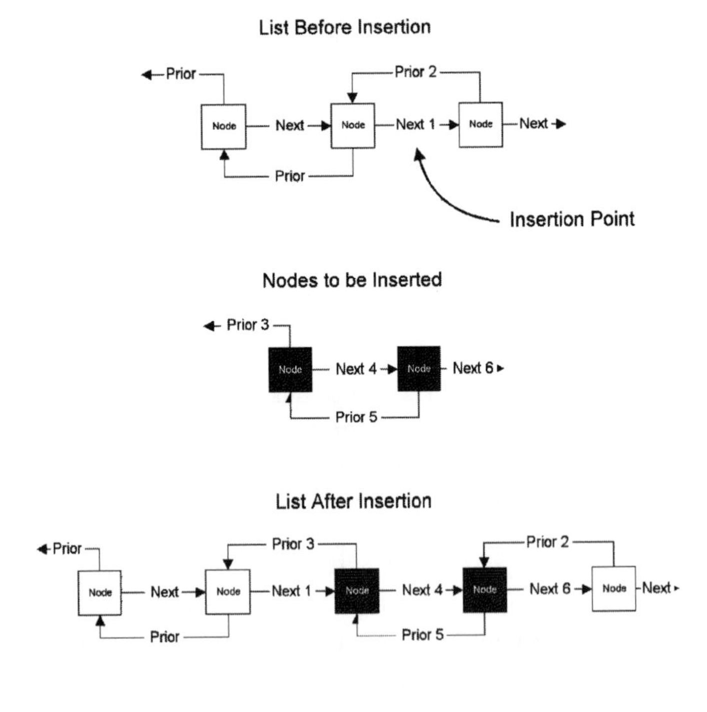

Figure 11.6
Inserting GP nodes into a linked list

The node insertion is a simple matter of reassigning four pointers (the relevant pointers are numbered so that the insertion may be clearly followed). The only memory allocation that occurs is the one for the inserted node. It is equally easy to clip a node out of a linked list.

- **Simple to crossover**
  Crossover between two linked lists is simpler than for an array. Consider Figure 11.7. All that is required to crossover these two linked lists is to rearrange the pointers between the nodes. Not only is this simpler than array crossover, it is also much less likely to cause garbage collection problems than the use of an array. Note that both of the linked lists in Figure 11.7 change size after crossover. A linked list would not, as would an array, have to create a new (and different sized) array for each individual, allocate memory for them, transfer the data from the old arrays to the new arrays, and then de-allocate the old arrays.

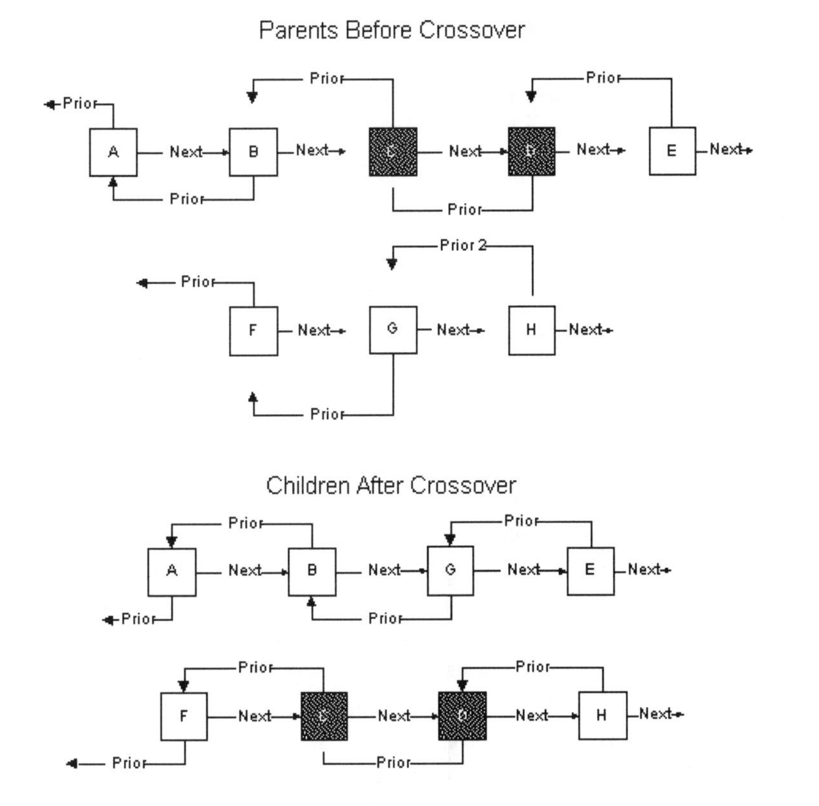

Figure 11.7
Crossover between two doubly linked lists

- **Flexible in size of elements**
  We noted above that an element in a linked list can contain whatever data may be desired. Thus, for an `If/Then/Else` operator, one linked list element can have four items of data, while, for a `Plus` operator, the linked list element needs to carry only three items of data. The corollary of this is that the pro-

grammer may establish a one-to-one correspondence between the elements of a linked list and the program operators.

- **Fast insertion/deletion**
  A linked list is faster to insert into or delete from than an array.

- **Demanding in memory requirements**
  A linked list consumes more memory than an array representation because the latter does not use pointers.

- **Slow access**
  A linked list is slower to access than an array. This can be very expensive during fitness evaluations where every element is accessed repeatedly.

Of course, it is not necessary to use either a linked list or an array only. The PADO system uses a linked list to store individuals and to perform genetic operations. But it uses an array structure to execute the individual programs.

11.4.3 Stacks

Two of the systems that we will consider use a *stack* for temporary data storage. A stack can be thought of as an ever-changing pile of data. Both the data items and their number change in the pile. A stack is a little like using a pile of books to store your books instead of using a bookcase. You would place a book onto the pile by putting it on top of the top book. On the other hand, you could get to any book in the pile by successively taking the top book off the pile and reading it until you reached the book.

What seems odd about this arrangement is that you can only get to the top book and you can only store a book by putting it on top. While a top-of-the-pile-only arrangement would be an odd way to store books, it is a powerful way to store short-term data during execution of a program. What we just described is also known as LIFO: Last In – First Out. This acronym comes from the property of a stack that the last item you put into – that is, on top of – the stack will be the first item you can remove again.

In the computer, a stack is an area of memory that holds data much like the pile of books stores books. When you *push* an item onto the stack, the top of the stack now holds that item for later use. That is like when you put the book on top of the pile of books. Figure 11.8 shows graphically how a piece of data may be stored by being pushed onto the stack for later use.

When you *pop* an item off the stack, you get the top item to use in program execution. Popping data off a stack is like taking the top

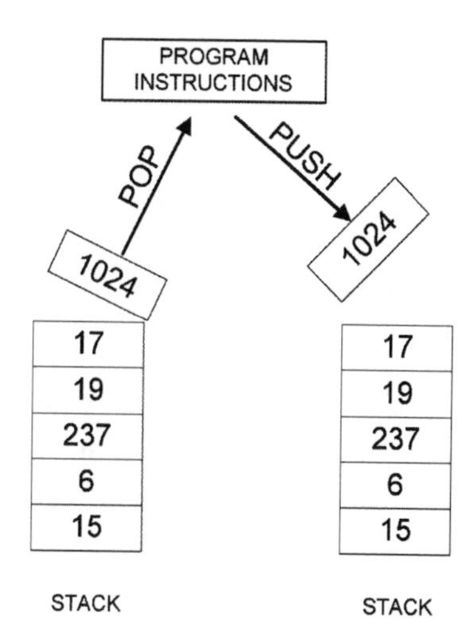

Figure 11.8
Push and pop operations on a stack

book off the pile. Figure 11.8 shows an item being popped off the stack.

Technically, a stack is realized by using a stack pointer pointing to the address in memory address space where the presently newest value resides. With these basics in mind, we are now prepared to describe various C-based GP systems.

11.5 Implementations With Arrays or Stacks

In 1994, Keith and Martin looked at five different approaches to programming GP in C++ [Keith and Martin, 1994]. We have already looked briefly at the tree approach, where every node contains a pointer to each of its input nodes. Keith and Martin also considered two systems where the GP individuals were stored in an array and the tree structure was implicit in the ordering of the elements of the array. One of these approaches used a stack and postfix ordering to effect the implicit tree structure. The other used prefix ordering and recursive evaluations of the next node to accomplish the same thing.

11.5.1 Postfix Expressions with a Stack

Postfix ordering occurs when the operator follows its operands. For example, the following postfix ordering of nodes for evaluation:

$$wx \times yz + \times SQRT \qquad (11.5)$$

is equivalent to the following algebraic expression:

$$\sqrt{(w \times x) \times (y + z)} \qquad (11.6)$$

One advantage of postfix ordering is that, if one evaluates the postfix expression from left to right, one will always have evaluated the operands to each operator before it is necessary to process the operator. Thus, one need not spend time finding the operands so that they can be evaluated prior to the operator processing.

Keith and Martin proposed that one way to represent GP individuals would be to hold a symbolic representation of each operator and operand – the *elements* of an expression – in an array in postfix order. The top array in Figure 11.9 shows how the above postfix expression would appear in an array.[3]

Figure 11.9
Postfix and prefix expression representation

0	1	2	3	4	5	6	7
W	X	MUL	Y	Z	+	MUL	√

0	1	2	3	4	5	6	7
√	MUL	+	Z	Y	MUL	X	W

Postfix Evaluation

One question remains: where do intermediate and final results occuring during execution get stored? For example, element 0 evaluates the value of the input w. But w is not needed until after the system has also gotten the value of x from element 1 of the array. Only then is the system ready to evaluate the product of w and x as required by the multiplication operator in element 2 of the array.

Stack Evaluation

A stack is a simple way to effect this kind of storage, and Keith and Martin do exactly that. The logic of evaluation is set up so that values are pushed onto and popped off the stack so as to maintain the logic of the tree structure. For example, element 0 in the array represents the value of the input w which is pushed onto the stack. The same is true for element 1 with respect to input x. Element 2 represents popping two values from the stack (w and x), multiplying them and pushing the result onto the stack. Now the product of w and x is safely stored in the stack for later use as an input. Execution proceeds in that manner right down the array.

Postfix Crossover

There is no intuitive LISP-like way to perform crossover in this postfix arrangement. Take a moment and try to figure out a general

[3]The programmer would probably not use symbols like letters and strings – MUL, for instance – that we have used. Rather, each element

rule that would always crossover valid subexpressions and is as simple as the LISP "match the parentheses" rule. Actually, the answer is easier than it seems, but it is still far from being as simple as the LISP rule. The answer is: start at any node, and move to the right to another node. If the number of items on the stack is never less than zero, and the final number of items on the stack is one, then the visited nodes represent a subtree, that is, a subexpression.

The principal advantage of this arrangement is the compact individual representation in memory. It is a little slower than the tree arrangements [Keith and Martin, 1994], and the compactness of the representation could make the system difficult to extend. The main reason it is slower is that each element must be evaluated for what type it is – the element type needs to be stored by a symbolic representation. That symbol must be interpreted, and Keith and Martin do so with a switch statement, which takes a substantial amount of time to evaluate.

The other principal drawback to this method is that it does not allow for skipping evaluation of parts of the program that do not need to be evaluated. For example, assume that $x = 10$. In the following expression, it is unnecessary to evaluate y, `If x < 10 Then y Else z`. However, in a postfix representation, y would be evaluated nevertheless.

11.5.2 Prefix Expressions with Recursive Evaluation

Keith and Martin also proposed a prefix expression representation with recursive evaluations. Prefix ordering is the opposite of postfix ordering. The operator precedes the operands. The second array of Figure 11.9 shows the same expression we used above but expressed in prefix ordering. A problem here is that the operands must be known in order to process the operator, while the operands *follow* the operator.

Keith and Martin solve that problem with their `EvalNextArg` function. Every time `EvalNextArg` is called, it increments a counter that identifies the current element in the array. Then it calls itself in a manner appropriate for the type of the identified element. When it calls itself, it automatically jumps to the next element in the array.

Evaluating a Prefix Array

Let us look at how this works more specifically. To begin evaluating the array in Figure 11.9, `EvalNextArg` gets called for element 0 of the array. The function reads element zero and interprets it to mean "take the square root of the next element in the array." To do

---

would be represented by, say, an integer. However, for clarity, we will use symbols.

that, `EvalNextArg` might execute the following pseudocode: `return SQRT(EvalNextArg)`.

The effect of this pseudocode is to attempt to take the square root of element 1 in the array.[4] However, element 1 in the array interprets to `MUL`. It has no inherent return value of which there is a square root until the multiplication has been performed. This means that the square-root operator must wait for its operands until further evaluations have been performed. The further evaluations happen in the following manner. `EvalNextArg` might execute the following pseudocode at this point: `return EvalNextArg * EvalNextArg`.

Note what happens here. The system tries to multiply the values in elements 2 and 3 of the array. A review of the array shows that further evaluations will still be necessary because element 2 is a `+` operator, and it has no immediate value until it, too, has been evaluated. So, the system calls `EvalNextArg` over and over again until it has completely evaluated the individual.

Prefix Crossover

Crossover is performed in a way similar to that of the postfix approach. Starting at any element, take the arity of each element minus one and sum these numbers from left to right. Wherever the sum equals minus one, a complete subexpression is covered by the visited nodes. For instance, for elements two, three, and four, this gives $1 + (-1) + (-1) = -1$, and, indeed, $+ \ z \ y$ is a complete prefix expression.

This arrangement is superior to the postfix arrangement in that it allows for skipping over code that does not need to be evaluated. It shares the compactness of representation of the postfix approach. The interested reader may review the original article, which includes an innovative opcode approach to storing the values in the array.

11.5.3 A Graph Implementation of GP

Another GP system employing arrays and stacks is PADO (Parallel Algorithm Discovery and Orchestration) [Teller, 1996]. Using graphs, it looks unlike the tree-based systems whose implementation we have considered so far in this chapter. Graphs are capable of representing complex program structures compactly. In addition, PADO does not just permit loops and recursion, it positively embraces them. This is not a trivial point, since other GP systems have experimented with loops and recursion only gingerly because of the great difficulties they cause.

A graph structure is no more than nodes connected by *edges*. The edges, sometimes also called arcs, may be thought of as pointers

[4] Remember `EvalNextArg` automatically goes to the next element in the array every time it is called.

between two nodes indicating the direction of a movement from one node to the other. Each edge represents a part of the flow of program control. The reader may note that tree genomes and linear genomes can also be represented as graphs. In a tree structure, for instance, there may be several incoming but only one outgoing edge at each node. In a graph system, however, there may be several incoming *and* several outgoing edges for each node. Figure 5.3 showed a diagram of a small PADO program.

A few points about the system are important before we can look at implementation issues.

- **Start and end nodes**
  There are two special nodes, `Start` and `End`. Execution begins at `Start`, and when the system hits `End`, the execution of the program is over.[5] The flow of execution is determined by the edges in the graph. More about that later.

- **PADO use of the stack**
  Data is transferred among nodes by means of a stack. Each of the nodes executes a function that reads from or writes to the stack. For example, the node `A` reads the value of the operand a from RAM and pushes it onto the stack. The node `6` pushes the value 6 onto the stack. The node `MUL` pops two values from the stack, multiplies them, and pushes the result onto the stack.

- **PADO use of indexed memory**
  Data may also be saved by the system in the indexed memory. The node labeled `Write` pops two arguments from the stack. It writes the value of the first argument into the indexed memory location indicated by the second argument. `Read` fetches data from the memory location.

Note that there are really two things that each node must do: it must perform some function on the stack and/or the indexed memory, and it must decide which node will be the next to execute. This latter function is really a matter of choosing between the outgoing edges from the node. Consider Figure 5.3 again. The `MUL` node may transfer control to `Write`, `Read`, or `4`. For this purpose each node has a branch-decision function which determines, depending on the stack, memory cells, constants, or the foregoing node, which of the edges to take.

[5] In some implementations, PADO repeats execution a fixed number of times or until a certain condition is met. So, it is not quite accurate to say that the program always ends when the `End` node is reached.

From an implementation viewpoint, PADO is much more difficult than tree or linear genomes. For example, there are a lot more pointers to keep track of. Imagine crossing over the program representation in Figure 5.3 with another program representation. What to do with all of the nodes and, worse, all of the pointers that have been cut? How do you keep track of the pointers, and where do they get reattached? The designer has come up with good solutions to these problems, but the point is that the problem was a tough one relative to other GP systems.

Dual Representation of Individuals

Because of the difficulty of coding this system, an individual is represented in two different structures. For the purpose of storage, crossover, and mutation, individuals are stored in a linked list. Each element in the list contains all of the information necessary to identify a PADO node. On the other hand, for the purpose of execution, individuals are stored as arrays. The purpose of this dual representation is that the linked list provides simple crossover and mutation but would slow down the system if it were the basis of program execution. On the other hand, while an array provides for fast execution, it would be complicated to keep track of all of the pointers during crossover in the array.

The minute details of implementation of this interesting and powerful system are beyond the scope of this section. The reader may consult [Teller and Veloso, 1995a] [Teller and Veloso, 1995b] [Teller, 1996] [Teller and Veloso, 1996] for more information.

11.6 Implementations Using Machine Code

While the systems presented above represent individuals as high-level data structures, machine code-oriented genetic programming systems use low-level representations. Let us take a closer look at how such systems are implemented.

11.6.1 Evolving Machine Code with AIMGP

No matter how an individual is initially represented, it is always represented finally as a piece of machine code, because a processor has to execute the individual for fitness evaluation. Depending on the initial representation there are at least the following three GP approaches, the third of which is the topic of this section.

1. The common approach to GP uses a technique where an individual representation in a problem-specific language is executed by a virtual machine, as shown in Figure 11.10 (top). This solution gives high ability to customize the language depending

on the properties of the problem at hand. The disadvantage of this paradigm is that the need for the virtual machine involves a large programming and run time overhead.

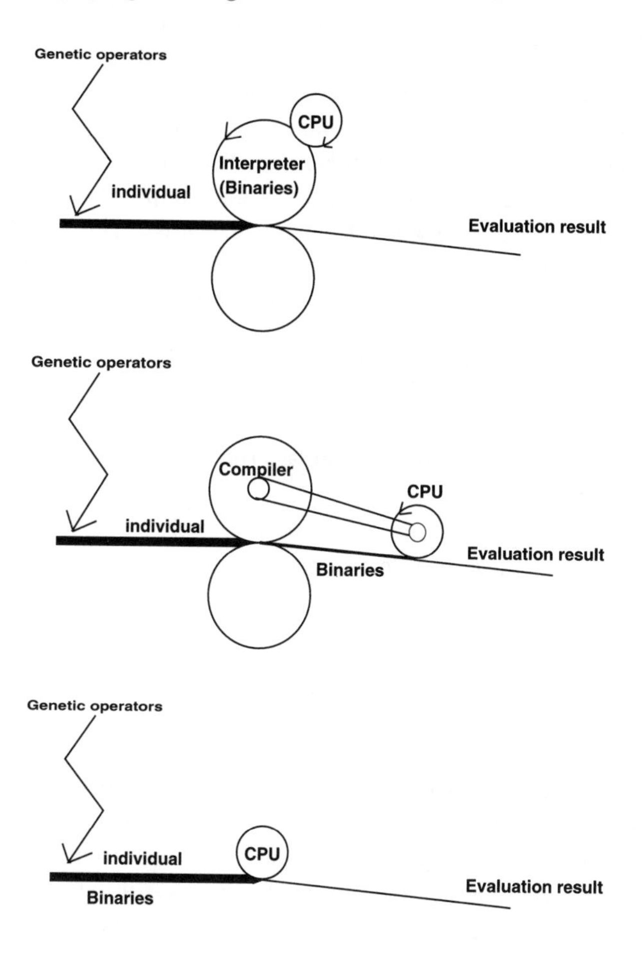

Figure 11.10
Three approaches to make GP work

2. Another approach is to compile each individual from a higher-level representation into machine code before evaluation, as shown in Figure 11.10 (middle). This approach can provide genetic programming with problem-specific and powerful operators and also results in high-speed execution of the individual. The compilation itself is an overhead, of course, and compilers do not produce perfectly optimized machine code. Nevertheless, the speed-up can be considerable if an individual runs, on average, for a long time during fitness evaluation. Long execution times may be due to, say, a long-running loop or may be due to a large number of fitness cases in the training set. Problem-specific operators are frequently required.

This approach has been used by [Keller and Banzhaf, 1996] and [Friedrich and Banzhaf, 1997]. While the first approach evolves code in a potentially arbitrary language which then gets interpreted or compiled by a standard compiler like ANSI-C, the second uses its own *genetic compiler*. Thus, there is no separate compilation step for the system and execution times approach those of the system that manipulates machine code directly. Figure 11.11 shows how tokens representing machine code instructions are compiled – by a fast and simple genetic compiler – to machine code which then gets executed. This method has advantages when working with CISC processors, where machine language features variable length instructions.

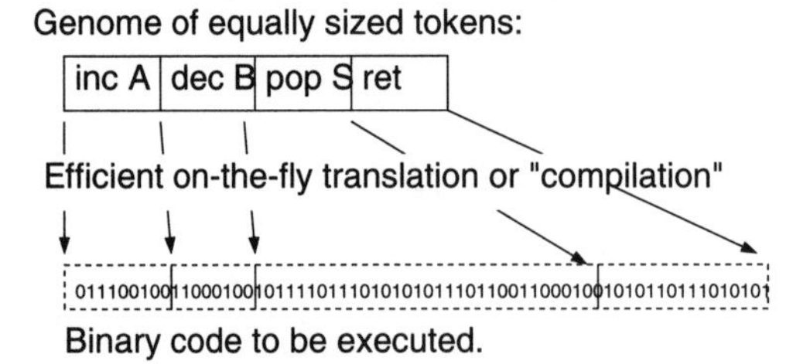

Figure 11.11
Machine code GP with genetic compilation

3. The central approach of this section – automatic induction of machine code with GP (AIMGP) – represents individuals as machine code programs which are directly executable. Thus, each individual is a piece of machine code. In particular, there are no virtual machines, intermediate languages, interpreters, or compilers involved. This approach was earlier introduced as the compiling genetic programming system [Nordin, 1994].[6]

   AIMGP has been implemented in C. Thus, individuals are invoked with a standard C function call. The system performs repeated type casts between pointers to arrays for the manipulation of individuals and between pointers to functions for the execution of the individuals as programs.

AIMGP has accelerated individual execution speed by a factor of 2000 compared to LISP implementations [Nordin and Banzhaf, 1995b].

[6]The approach was earlier called "compiling" because it composes machine code programs which are executed directly. Because the approach should not be confused with other approaches using a compiler for the mapping into machine code, we have renamed it.

It is the fastest approach available in GP for applications that allow for its use [Nordin, 1997]. (See `www.aimlearning.com`.)

11.6.2 The Structure of Machine Code Functions

As said earlier, the individuals of AIMGP consist of machine code sequences resembling a standard C function. Thus AIMGP implements linear genomes, and its crossover operator works on a linear structure. Figure 11.12 illustrates the structure of a function in machine code. The function code consists of the following major parts:

1. The **header** deals with administration necessary when a function gets entered during execution. This normally means manipulation of the stack – for instance, getting the arguments for the function from the stack. There may also be some processing to ensure consistency of processor registers. The header is often constant and can be added at the initialization of each individual's machine code sequence. The genetic operators must be prevented from changing the header during evolution.

2. The **footer** "cleans up" after a function call. It must also be protected from change by the genetic operators.

3. The **return instruction** follows the footer and forces the system to leave the function and to return program control to the calling procedure. If variable length programs are desired, then the return operator could be allowed to move within a range between the minimum and maximum program size. The footer and the return instruction must be protected against the effects of genetic operators.

4. The **function body** consists of the actual program representing an individual.

5. A **buffer** is reserved at the end of each individual to allow for length variations.

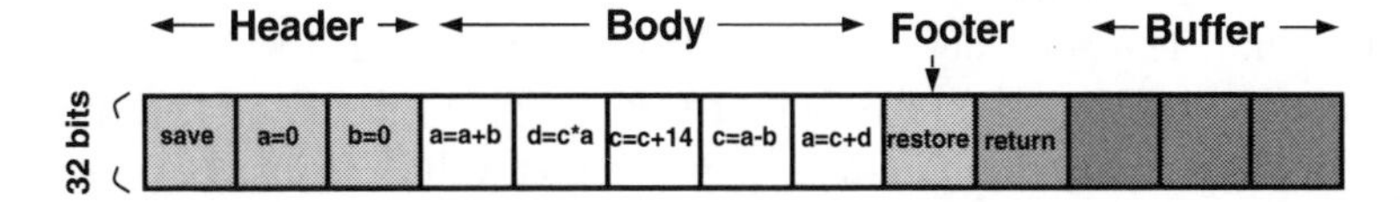

Figure 11.12
Structure of a program individual in AIMGP

11.6.3 Genetic Operators

AIMGP has the following two genetic operators:

1. A **mutation** operator changes the content of an instruction by mutating op-codes, constants or register references. It randomly changes one bit of the instruction provided certain criteria are fulfilled. The operator can change only the instruction to a member of the set of approved instructions to assure that there will be no illegal instructions, bus errors, unwanted loops or jumps, etc. Furthermore, the operator ensures arithmetic consistency, such as protection against division by zero.

2. The **crossover** operator works on variable length individuals. Two crossover methods have been used in AIMGP. The first method (protected crossover) uses a regular GA binary string crossover where certain parts of the machine code instruction are protected from crossover to prevent creation of illegal or unwanted instructions, as shown in Figure 11.13. The second method (instruction crossover) allows crossover only *between* instructions, in 32-bit intervals of the binary string. Hence, the genome is snipped and exchanged so as to respect the 32-bit machine code instruction boundaries. Figure 11.14 illustrates the latter crossover method [Nordin and Banzhaf, 1995b].

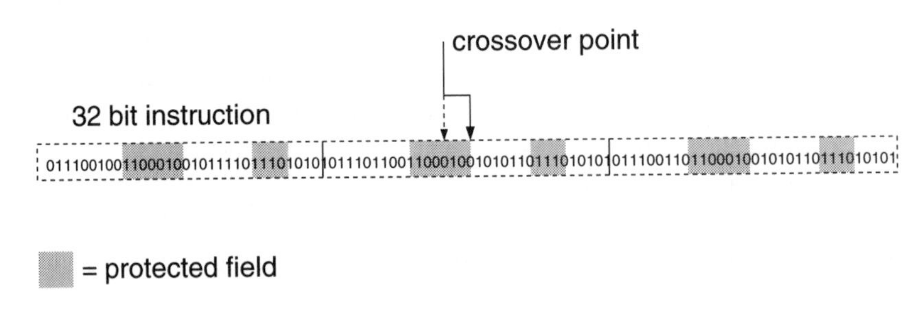

Figure 11.13
Protected crossover. Hitting an illegal location for crossover is prohibited by deflecting crossover to a neighboring legal crossover point.

11.7 A Guide to Parameter Choices

Once a system has been coded, one must choose parameters for a GP run. The good news is that GP works well over a wide range of parameters. The bad news is that GP is a young field and the effect of using various combinations of parameter values is just beginning to be explored. We end this chapter by describing the typical parameters, how they are used, and what is known about their effects.

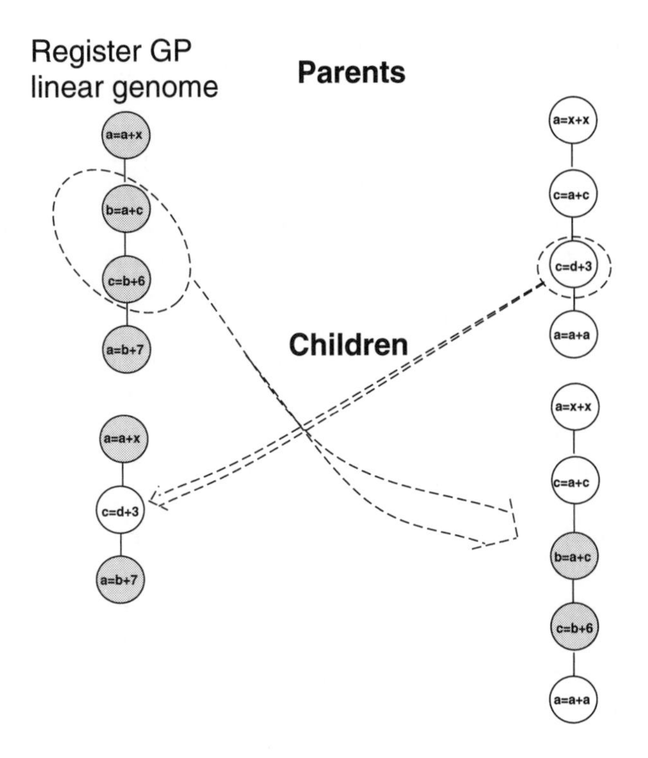

Figure 11.14
Instruction crossover. Crossover is allowed only between instructions.

Much of what GP researchers know about parameters is anecdotal and based on experience. We will suggest rules of thumb for the various parameters based on our experience. Our aim here is not merely to list parameters that are well described elsewhere, but to look at those parameters where real experience makes it possible to give some practical advice about real implementation issues.

Population Size

Population size is an important parameter setting in a GP run for several reasons. To begin with, bigger populations take more time when evolving a generation. Also, bigger populations have more genetic diversity, explore more areas of the search space, and may even reduce the number of evaluations required for finding a solution.

Positive results have been achieved with population sizes ranging from $P = 10$ to $P = 1\,000\,000$ individuals. Between 10 and 100 000 individuals, the authors have experienced a near linear improvement in performance of the system. A starting point of $P = 1000$ is usually acceptable for smaller problems. But as the problem grows more difficult, the population size should grow. A rule of thumb in dealing with more difficult problems is that, if a problem is sufficiently difficult, then the population size should start at around $P = 10\,000$. This number should be increased if the other parameters tend to exert heavy selection pressure. On the other hand, if there is a lot of noise in selection (very small tournaments and the like), then smaller populations will suffice.

A larger number of training cases requires an increase in the population size. In a smaller problem, say, less than 10 fitness cases, a population size of 10 to 1000 usually suffices. Between 10 and 200 fitness cases, it is better to use $1000 \leq P \leq 10\,000$. Above 200 training cases, we recommend using $P > 10\,000$. Koza summarizes his experiences with population size in Chapter 26 of [Koza, 1992d]. He uses $50 \leq P \leq 10\,000$ in his book, but reports 500 individuals as the commonest setting.

Maximum Number of Generations

Early in the history of GP, it was argued that the limit on generations should be quite low because nothing happens after generation $G_{max} = 50$. That has not been our experience. In some runs, interesting evolution has been delayed to as late as generation 1000. Other times, evolution will seem to stagnate and then suddenly pick up and begin improving the population again. We have seen examples where interesting things happen after generation 10 000.

That said, it is impractical to run most GP systems for that many generations – there is not enough CPU time available. There are two possible practical solutions:

1. Start testing with a relatively low setting for G_{max}, such as $50 \leq G_{max} \leq 100$ generations. If you are not getting the results you want, first raise the population size and then raise the number of generations.

2. Monitor the run to determine whether it may safely be terminated. Explosive growth of introns almost always marks the end of effective evolution. This growth may be measured indirectly by measuring the percentage of total crossover events that are destructive. As a rule of thumb, when destructive crossover falls to below 10% of all crossover events, no further effective evolution will occur. So it is possible to set G_{max} quite high but to catch the runs in which evolution is finished and terminate them early.

Terminal and Functions Set

A few rules of thumb for the terminal set and function set have served the authors well:

- Make the terminal and function set as small as possible. Larger sets usually mean longer search time. The same is true for the number of registers used in AIMGP.

- It is not that important to have (all) customized functions in the function set: the system often evolves its own approximations.
- It is very important, however, that the function set contains functions permitting non-linear behavior, such as if-then functions, Boolean operators on numbers, and sigmoid squashing functions.
- The function set should also be adapted to the problem in the following way: problems that are expected to be solved by smooth curves should use function sets that generate smooth curves, and functions that are expected to be solved by other types of functions should have at least one representative of these functions in the function set.
- Sometimes transformations on data are very valuable, for instance, fast Fourier transforms [Oakley, 1996].

Mutation and Crossover Balance

The typical settings of mutation and crossover probabilities in GP involve very high rates of crossover and very low rates of mutation. Experiments suggest that a different balance ($p_c = 0.5, p_m = 0.5$) between the two operators may lead to better results on harder problems [Banzhaf et al., 1996] and that the worst results are obtained when either operator is left out.

The proper balance between these operators is, therefore, a wide open question and may be very problem dependent. A rough rule of thumb would be to start with 90% crossover and 10% mutation. If the results are not pleasing, increase the mutation rate.

Selection pressure is another parameter to be put under some conditions. If tournament selection is applied, the size of the tournament will determine the selection pressure. The authors have very good experiences with low selection pressure. Tournaments of 4 individuals regularly perform very well.

Parsimony Pressure

Research in the area of parsimony pressure is not fully conclusive at this time. Some researchers have reported good results with parsimony pressure. Our own experience is that constant parsimony pressure usually gives worse results and makes local optima more likely. However, variable parsimony pressure produces very nice, short, and elegant solutions. Some researchers have reported good results with adaptive parsimony, which is applied only when a solution that performs well is found.

Maximum Program Size

As a general rule, the maximum depth of trees or the maximum program size should be set such that the programs can contain about ten times the number of nodes as the expected solution size. This allows both for estimation error in predicting the solution size and for intron growth.

Initial Program Size

Typically, the initial program size should be very small compared to the maximum size. This allows the system to build up good individuals piece by piece. But for complex problems, when no success results from this approach, we suggest trying longer programs at the start to allow the system to start with some complexity already and to avoid local minima early on.

Exercises

1. Why is GP computationally expensive?
2. What is garbage collection, and why is it necessary in some GP implementations?
3. Give three examples of how to represent an individual in GP.
4. Why is it easier to do crossover when using linked lists than when using arrays to represent individuals in a tree?
5. Define postfix and prefix representations.
6. Describe two methods of implementing tree structures suitable for GP.
7. Describe three important parameters of GP runs.

12 Applications of Genetic Programming

Contents

12.1 General Overview

In this chapter we discuss a selected number of applications of genetic programming. The selection is arbitrary but is intended to give a taste of what is already discussed in GP. Regretfully, due to the large number of practically relevant applications of genetic programming, we had to select among very many important contributions. This section and the next, however, are devoted to an overview of the diversity of applications researchers have tackled with GP.

To start with, it is clear that the number of applications must be correlated with the number of papers published in GP. This will give a lower bound on applications, because it can be safely assumed that a considerable percentage of applications never get published at all.

So let us first look at the development of GP publications in general. Table 12.1 summarizes the history of of GP publications since 1989 in part using data given in [Alander, 1995].

Table 12.1
Development of GP literature since 1989

Year	Number of publications
1989	1
1990	12
1991	18
1992	30
1993	40
1994	95
1995	140 (est.)
1996	220 (est.)
1997	150 (1st half, est.)

12.2 Applications from A to Z

Table 12.2 gives an overview of applications according to our classification. As one can see, genetic programming has spawned numerous interesting applications in the short time of its existence.

It can be safely assumed that at least the same growth factor applies to applications as it does to GP in general.

Table 12.3–12.6 on the following pages present the different applications of genetic programming, including the sources where more information can be found. Some entries are repeated as they fit under more than one heading.

Application domain	First publication	Cum. number
algorithms	1992	8
art	1993	5
biotechnology	1993	9
computer graphics	1991	7
computing	1992	17
control (general)	1992	4
control (process)	1990	5
control (robots and agents)	1992	27
control (spacecraft)	1996	2
data mining	1996	6
electrical engineering	1994	9
financial	1994	9
hybrid systems	1993	9
image processing	1993	14
interactive evolution	1991	4
modeling	1994	7
natural languages	1994	4
optimization	1994	7
pattern recognition	1994	20
signal processing	1992	5

Table 12.2
Summary overview of applications of GP in different areas

The following sections discuss some applications in more detail in order to give an impression of the diversity of problems GP has been applied to. The authors have made an arbitrary selection and do not claim to cover all topics exhaustively.

We group the selected applications roughly in the following domains:

1. Science-oriented applications
2. Computer science-oriented applications
3. Engineering-oriented applications

12.3 Science-Oriented Applications of GP

12.3.1 Biochemistry Data Mining

Problem Domain

In many areas of science and technology, so much knowledge has accumulated that the methods of data mining are needed in order to discover interesting and valuable aspects of the data that would have gone undiscovered otherwise.

Application domain	Year	Application	Source
algorithms	1996	acyclic graph evaluation	[Ehrenburg, 1996]
	1997	caching algorithms	[Paterson and Livesey, 1997]
	1996	chaos exploration	[Oakley, 1996]
	1994	crossing over between subpopulations	[Ryan, 1994]
	1994	randomizers (R)	[Jannink, 1994]
	1996	recursion	[Wong and Leung, 1996]
	1993	sorting algorithms	[Kinnear, Jr., 1993a, Kinnear, Jr., 1994]
	1992	sorting networks	[Hillis, 1992]
art	1994	artworks	[Spector and Alpern, 1994]
	1993	images	[Sims, 1993a]
	1995	jazz melodies	[Spector and Alpern, 1995]
	1995	musical structure	[Spector and Alpern, 1995]
	1994	virtual reality	[Das et al., 1994]
biotechnology	1996	biochemistry	[Raymer et al., 1996]
	1995	control of biotechnological processes	[Bettenhausen et al., 1995a]
	1995	DNA sequence classification	[Handley, 1995]
	1993	detector discovering and use	[Koza, 1993b]
	1994	protein core detection (R2)	[Handley, 1994a]
	1994	protein segment classification	[Koza and Andre, 1996a]
	1994	protein sequence recognition	[Koza, 1994b]
	1993	sequencing	[Handley, 1993a]
	1996	solvent exposure prediction (R2)	[Handley, 1996b]
computer graphics	1994	3D object evolution	[Nguyen and Huang, 1994]
	1994	3D modeling (R)	[Nguyen et al., 1993]
	1991	artificial evolution	[Sims, 1991a]
	1991	artificial evolution	[Sims, 1991b]
	1993	computer animation	[Ngo and Marks, 1993]
	1995	computer animation	[Gritz and Hahn, 1995]
	1997	computer animation	[Gritz and Hahn, 1997]
computing	1995	computer security	[Crosbie and Spafford, 1995]
	1994	damage-immune programs	[Dickinson, 1994]
	1996	data compression	[Nordin and Banzhaf, 1996]
	1992	data encoding	[Koza, 1992d]
	1996	data processing structure identification	[Gray et al., 1996a]
	1991	decision trees	[Koza, 1991]
	1996	decision trees	[Masand and Piatesky-Shapiro, 1996]
	1996	inferential estimation	[McKay et al., 1996]
	1994	machine language	[Nordin, 1994]
	1994	monitoring	[Atkin and Cohen, 1993]
	1995	machine language	[Crepeau, 1995]
	1996	object orientation	[Bruce, 1996]
	1996	parallelization	[Walsh and Ryan, 1996]
	1997	parallelization	[Ryan and Walsh, 1997]
	1996	specification refinement	[Haynes et al., 1996]
	1995	software fault number prediction	[Robinson and McIlroy, 1995a]
	1994	virtual reality	[Das et al., 1994]

Table 12.3
GP applications overview, part I (R means "repeated")

Various data mining methods exist. One of them is (automatic) clustering of data into groups such that from the structure of those clusters one can draw appropriate conclusions. Clustering methods are a general tool in pattern recognition, and it can be argued that data in a database are patterns organized according to a homogeneous set of principles, called features.

The problem of data mining thus becomes a problem of feature extraction, and it is this point of view that is discussed in this appli-

Application domain	Year	Application	Source
control	1995	boardgame	[Ferrer and Martin, 1995]
	1995	cooperating strategies	[Haynes et al., 1995]
	1994	steady states of dynamical systems	[Lay, 1994]
	1992	vehicle systems	[Hampo and Marko, 1992]
control (process)	1990	control strategy programs	[Koza and Keane, 1990]
	1996	modeling chemical process systems	[Hinchliffe et al., 1996]
	1995	process engineering	[McKay et al., 1995]
	1996	process engineering	[McKay et al., 1996]
	1994	stirred tank	[Lay, 1994]
control (robotics)	1993	autonomous agents	[Atkin and Cohen, 1993]
	1994	autonomous agents	[Atkin and Cohen, 1994]
	1994	autonomous agents	[Fraser and Rush, 1994]
	1994	autonomous agents	[Ghanea-Hercock and Fraser, 1994]
	1994	autonomous agent	[Rush et al., 1994]
	1994	battle tank	[D'haeseleer and Bluming, 1994]
	1994	corridor following	[Reynolds, 1994a]
	1995	juggling	[Taylor, 1995]
	1997	manipulator motion	[Howley, 1997]
	1995	motion	[Nordin and Banzhaf, 1995c]
	1992	motion in critter population	[Reynolds, 1992]
	1996	motion planning	[Faglia and Vetturi, 1996]
	1997	motion and planning	[Banzhaf et al., 1997a]
	1997	navigation	[Bennett III, 1997]
	1997	navigation	[Iba, 1997]
	1994	obstacle avoiding	[Reynolds, 1994b]
	1993	planning	[Handley, 1993b]
	1994	planning	[Handley, 1994b]
	1994	planning	[Spector, 1994]
	1996	sensor evolution	[Balakrishnan and Honavar, 1996]
	1992	subsumption	[Koza, 1992a]
	1994	terrain flattening	[Lott, 1994]
	1992	trailer back-up	[Koza, 1992b]
	1996	wall-following	[Ross et al., 1996]
	1997	wall-following	[Dain, 1997]
	1993	walking and crawling	[Spencer, 1993]
	1994	walking and crawling	[Spencer, 1994]
control (spacecraft)	1996	maneuvering	[Howley, 1996]
	1997	maneuvering	[Dracopoulos, 1997]
data mining	1996	databases	[Raymer et al., 1996]
	1996	internet agents	[Zhang et al., 1996]
	1995	predicting DNA	[Handley, 1995]
	1997	rule induction	[Freitas, 1997]
	1995	signal identification	[Teller and Veloso, 1995c]
	1995	time series	[Lee, 1995]

Table 12.4
GP applications overview, part II

cation. As a particular example where it has been applied successfully, we shall discuss the biochemistry database CONSOLV containing data on water molecules bound to a number of proteins.

Using the classification technique of K-nearest neighbors (Knn) and a GP system to feed this classification scheme, a very successful feature analysis can be done resulting in an identification of important features as well as a good classification of untrained data entries.

The authors of this study did earlier work using a genetic algorithm for the same task [Punch et al., 1993] but concluded that GP would be better suited for the goals to be achieved. One important

Application domain	Year	Application	Source
electrical engineering circuit design	1997	analog source identification circuit	[Koza et al., 1997b]
	1994	circuit design	[Ehrenburg and van Maanen, 1994]
	1996	circuit design	[Koza et al., 1996a]
	1996	circuit design	[Koza et al., 1996b]
	1994	circuit simplification	[Coon, 1994]
	1997	controller circuit	[Koza et al., 1997a]
	1994	evolvable hardware	[Hemmi et al., 1994a]
	1996	facility layout	[Garces-Perez et al., 1996]
	1996	decision diagrams	[Drechsler et al., 1996]
financial market	1996	bargaining	[Dworman et al., 1996]
	1994	horse race prediction	[Perry, 1994]
	1996	hypothesis	[Chen and Yeh, 1996]
	1997	investment behavior	[Lensberg, 1997]
	1995	share prediction	[Robinson and McIlroy, 1995a]
	1994	strategies	[Andrews and Prager, 1994]
	1994	trade strategies	[Lent, 1994]
	1996	trade models	[Oussaidene et al., 1996]
	1997	volatility models	[Chen and Yeh, 1997]
hybrids	1996	cellular automata rules	[Andre et al., 1996b]
	1996	fuzzy logic controllers	[Alba et al., 1996]
	1996	L-systems	[Jacob, 1996b]
	1997	learning rules	[Segovia and Isasi, 1997]
	1994	neural network training	[Bengio et al., 1994]
	1994	neural network training	[Gruau, 1994b]
	1994	neural network training	[Zhang and Muehlenbein, 1994]
	1997	neural network training	[Esparcia-Alcazar and Sharman, 1997]
	1994	regular languages	[Dunay et al., 1994]
image processing	1995	analysis	[Robinson and McIlroy, 1995b]
	1996	analysis	[Bersano-Begey et al., 1996]
	1996	classification	[Zhao et al., 1996]
	1996	compression	[Jiang and Butler, 1996]
	1996	compression	[Nordin and Banzhaf, 1996]
	1996	edge detection	[Harris and Buxton, 1996]
	1993	feature extraction	[Tackett, 1993]
	1995	feature extraction	[Daida et al., 1995]
	1996	feature extraction	[Daida et al., 1996b]
	1997	image enhancement	[Poli and Cagnoni, 1997]
	1994	magnetic resonance image processing	[Thedens, 1994]
	1995	recognition	[Teller and Veloso, 1995a]
	1994	structure of natural images	[Gordon, 1994]
	1994	visual routines	[Johnson et al., 1994]
interactive evolution	1992	dynamical systems	[Sims, 1992a]
	1991	interactive image evolution	[Sims, 1991a, Sims, 1991b]
	1992	procedural models	[Sims, 1992b]
	1993	procedural models	[Sims, 1993b]

Table 12.5
GP applications overview, part III

strength of their method, the authors claim, is that it is useful in noisy environments [Pei et al., 1995].

Genetic algorithms can do an optimization based on a linear weighting of features, whereas genetic programming can do non-linear weighting and an adjustment of the function [Raymer et al., 1996].

Task

The task Raymer et al. considered was to generate a good scaling of features for a Knn classifier of data entries in the biochemical database CONSOLV. This database contained data on the environment of a set of 1700 randomly selected water molecules bound to

Application domain	Year	Application	Source
modelling	1995	biotechnological fed-batch fermentation	[Bettenhausen et al., 1995b]
	1995	macro-mechanical model	[Schoenauer et al., 1995]
	1997	metallurgic process model	[Greeff and Aldrich, 1997]
	1995	model identification	[Schoenauer et al., 1996]
	1995	model induction	[Babovic, 1995]
	1994	spatial interaction models	[Openshaw and Turton, 1994]
	1995	system identification	[Iba et al., 1995b]
natural languages	1994	confidence of text classification (R)	[Masand, 1994]
	1994	language decision trees	[Siegel, 1994]
	1996	language processing	[Dunning and Davis, 1996]
	1997	sense clustering	[Park and Song, 1997]
optimization	1994	database query optimization	[Kraft et al., 1994]
	1996	database query optimization	[Stillger and Spiliopoulou, 1996]
	1994	job shop problem	[Atlan et al., 1994]
	1996	maintenance scheduling	[Langdon, 1996a]
	1996	network (LAN)	[Choi, 1996]
	1995	railroad track maintenance	[Grimes, 1995]
	1994	training subset selection	[Gathercole and Ross, 1994]
pattern recognition	1994	classification	[Tackett and Carmi, 1994]
	1996	classification	[Abramson and Hunter, 1996]
	1997	classification	[Gray and Maxwell, 1997]
	1995	dynamics extraction	[Dzeroski et al., 1995]
	1994	feature extraction	[Andre, 1994a]
	1994	filtering	[Oakley, 1994b]
	1994	combustion engine misfire detection	[Hampo et al., 1994]
	1996	magnetic resonance data classification	[Gray et al., 1996b]
	1996	myoelectric signal recognition	[Fernandez et al., 1996]
	1993	noise filtering	[Oakley, 1993]
	1994	optical character recognition	[Andre, 1994b]
	1996	object classification	[Ryu and Eick, 1996]
	1997	object detection	[Winkeler and Manjunath, 1997]
	1997	preprocessing	[Sherrah et al., 1997]
	1994	signal filtering	[Oakley, 1994a]
	1994	text classification	[Masand, 1994]
	1996	text classification	[Clack et al., 1996]
	1996	time series	[Masand and Piatesky-Shapiro, 1996]
	1996	time series prediction	[Mulloy et al., 1996]
	1996	visibility graphs	[Veach, 1996]
signal processing	1992	control vehicle systems	[Hampo and Marko, 1992]
	1996	digital	[Esparcia-Alcazar and Sharman, 1996]
	1993	signal filtering	[Oakley, 1993]
	1993	signal modeling	[Sharman and Esparcia-Alcazar, 1993]
	1996	waveform recognition	[Fernandez et al., 1996]

Table 12.6
GP applications overview, part IV (R means "repeated")

20 different proteins. Four features were used to characterize each of the water molecules in their ligand-free structure:

1. the crystalographic temperature factor, called the B-value
2. the number of hydrogen bonds between the water molecule and the protein
3. the number of protein atoms packed around the water molecule, called the atomic density
4. the tendency of the protein atoms to attract or repel the water molecule, called the hydrophilicity

Based on those features of the ligand-free configuration, the water molecules binding to active sites were classified into either *conserved* or *displaced*, predicting whether they participate in ligand-active site binding (*conserved*) or not (*displaced*). The authors claimed that, if the active-site water molecules could be classified correctly and thus predicted, this would have broad applications in biotechnology [Raymer et al., 1996]. From the 1700 water molecules in the database, only 157 were binding to the active site, and those were used to participate in training of the system.

Figure 12.1 shows the idea behind scaling of features for Knn classification. The scaling is done in order to maximize classification correctness in the training set. Validation is done using a separate part of the database that has not been involved in the training.

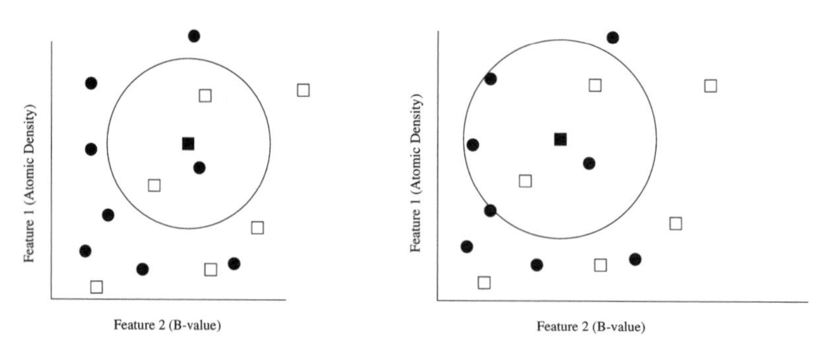

Figure 12.1 *Scaling of the x-axis (B-value) changes the classification of the water in question.*

System Structure

The GP system evolves functions that map the original values of the features into values allowing better separability of the patterns and, therefore, better classification of the water molecules. The entire system consisting of a GP module mapping the features into the Knn classifier module is depicted in Figure 12.2. Input to the system are the original features stored in the database as well as the correct classification results expected from the Knn classifier. These are then used as a quality measure for the GP system's mapping (scaling) of the features according to Figure 12.2.

GP Elements

Raymer et al. used the basic arithmetic functions $+, -, *, \%$ as the function set of the GP module. The terminals consisted of the original features to be mapped and of random ephemeral constants. A peculiarity of their work was that each tree of the genetic programming population consisted of 4 subtrees, corresponding to the four features to be mapped. In order to maintain those subtrees during evolution, they were coded as ADFs to be called by the main tree. The fitness measure was simply the degree of correct classification by a Knn with $K = 3$ among the 1700 water molecules, including 157 active-site binding water molecules.

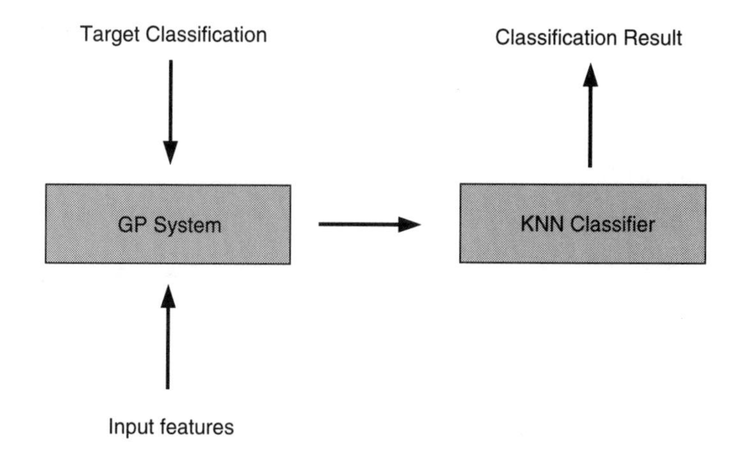

Figure 12.2
A system for improved Knn classification

The GP run was done using a ramped half-and-half initialization and the convergence termination criterion or the generation limit given below. The following parameters were chosen: $P = 100$; $G_{max} = 300$; $p_c = 0.9$; $MDP = 17$; $MDP_{init} = 6$. Raymer et al. report improved performance of the classifier and therefore improved prediction accuracy of the GP system over a comparable system using a genetic algorithm as feature-scaling device. The overall classification rate on all 157 active-site water molecules rose from 77% to 79%. The authors claim this to be a very good result given the difficulty of reaching more than 70% accuracy for protein secondary structure prediction from *ab initio* or knowledge-based methods [Mehta et al., 1995] [Rost and Sander, 1993].

Results

The four features were mapped differently, with an overall increase in importance given to the first and third. For illustration purposes, Figure 12.3 shows one of the developed non-linear mappings of features for atomic density. The scaling was done after the original features were normalized to the interval $[1, 10]$. The computation time per generation was about 15 minutes on a SUN SPARC 502.

12.3.2 Sequence Problems

Problem Domain

In many areas of science, there are problems with sequences of information carriers. These sequences might come about as the result of temporal processes or might be generated in the course of transforming spatial distributions into a form that can be subjected to sequence analysis. The very process by which you read these lines is such a case. During the process of reading text we are transforming

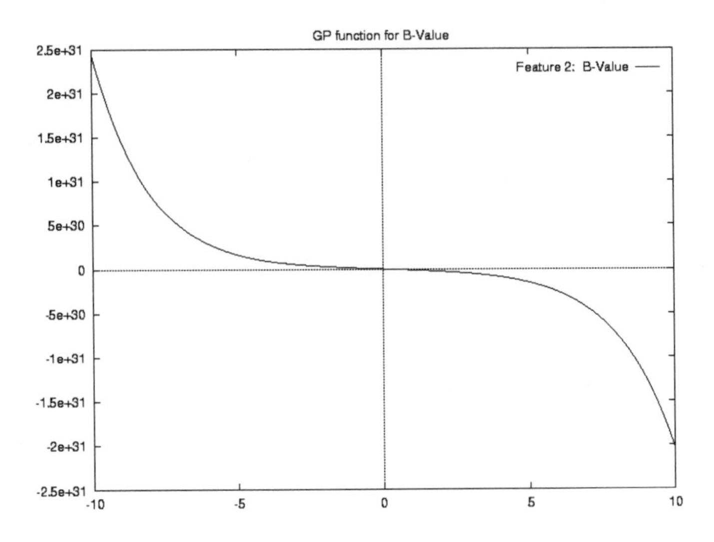

Figure 12.3
A GP-evolved function for the B-value with a Knn of $K = 3$. It approximately follows a $f(x) = -x^3$ function. ([Raymer et al., 1996], copyright MIT Press, reproduced with permission)

spatial patterns (dot pattern on paper) into a spatio-temporal form that our brain can process.

A glimpse of the ubiquity of sequence problems is given here:

- Speech processing
- Communication in general
- Language understanding and translation
- Analysis of economic problems
- DNA pattern recognition
- Time series analysis: prediction of weather, etc.
- Secondary and tertiary protein structure prediction

As we can see, sequence problems are widespread, and computer science and biology contain especially challenging instances of these problems. These problems have in common that the data type of sequence elements is usually the same over the entire sequence. Thus, whether an element is at the beginning of the sequence or at the end, it would have the same meaning in either case. In other words, there is positional independence or translational invariance as far as the sematics is concerned.

Handley has created a special set of functions for solving sequence problems [Handley, 1996a] and applied this set to two problems for demonstration purposes.

Task

The task given by the analysis of sequences is usually simply to recognize certain patterns in a sequence or to compare sequences for similarity when the number of elements that might express that pattern is not fixed. According to Handley, many machine learning techniques approach the problem by forcing the sequence into a fixed length pattern, that is, by sliding a window over the sequence. The window size thus determines the size of the pattern, which is now fixed in length irrespective of the entire length of the sequence. Handley, instead, proposes a more flexible approach using an arbitrary or adjustable window size, up to the extreme of taking the influence of the entire sequence into account for one specific pattern.

GP Elements

Handley's GP system is a traditional one, except that it is enhanced by his so-called statistical computing-zone function set. These functions may be divided into two classes: convolution functions and statistical functions.

Convolution functions are functions that compute values from parts of the sequence, independently of what the same convolution function has computed on other parts of the sequence. The situation is depicted in Figure 12.4.

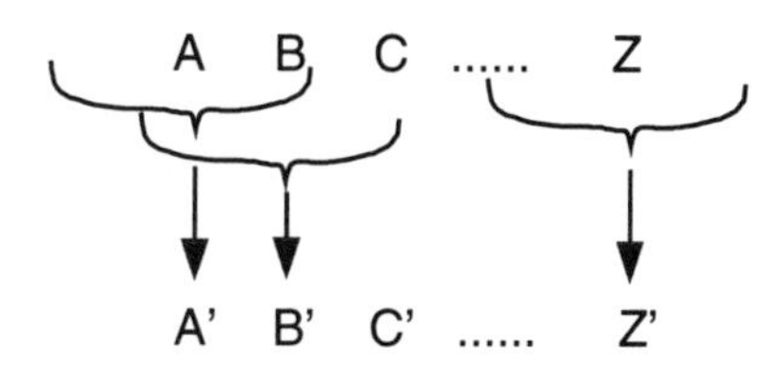

Figure 12.4
The effect of a convolution function on a sequence. Values are computed locally, independently of the computations for other parts of the sequence.

Typical representatives of those convolution functions are:

- arithmetic functions on single elements or on sequences
- conditional branches on single elements or on sequences
- sequence manipulation functions, like shift left or shift right

Statistical functions, on the other hand, do depend on the application of the same function earlier on in the sequence. As a typical instance, Handley mentions a summing operation that returns, at the end of the sequence, the sum of all values found on the sequence. Until the end, however, this function returns partial sums only, with each partial sum depending on partial sums computed earlier on. The situation is depicted in Figure 12.5.

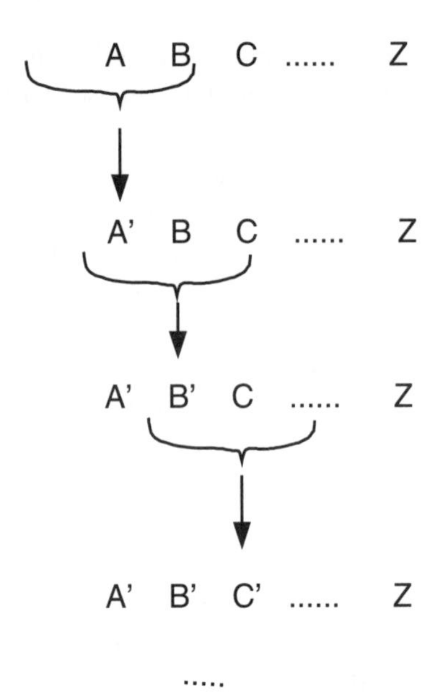

Figure 12.5
The effect of a statistical function on a sequence. Values are computed locally, but are dependent on computations of the same function earlier on in the sequence.

The statistical functions are based on one important function called *scz*, which inverts the order of execution usually valid in genetic programming. Instead of moving along a sequence and evaluating the elements in this order, this function moves over the sequence and evaluates in the opposite order. This allows it to take into account values that should be computed earlier in the sequence when evaluating later parts of the sequence. This is similar to storing values in registers in AIMGP when moving along the sequence.

Results

In one example, Handley applied this newly introduced function set to a problem of amino-acid classification in proteins. More specifically, the question was to predict the buriedness of a particular amino acid in the sequence, a problem closely related to tertiary structure prediction. Buriedness has to do with the tendency of an amino acid to try to hide from water within the protein fold or to try to be exposed to water at the surface.

Handley used 122 proteins from the Brookhaven Protein Data Bank [Bernstein et al., 1977] and separated them into training (60), test (30), and evalutation (32) sets. Runs were done with a 64-node parallel machine with demes of size 300, resulting in a total population of 19 200 individuals. Emigration of 5% was allowed per generation on a 2D toroidal mesh. Results reported by Handley [Handley, 1996b] compare very favorably with other methods on the same problem (although using another set of proteins). Prediction accuracy was 90% for the evaluation set, much better than the accuracy of 52% reached by Holbrook et al. using a neural network [Holbrook et al., 1990] [Holbrook et al., 1993].

12.3.3 Image Classification in Geoscience and Remote Sensing

Daida et al. [Daida et al., 1996a] have implemented an impressive GP-supported image processing system for the analysis of satellite radar images in a geoscience application. The objectives of their work are:

1. to describe an instance of a computer-assisted design of an image-processing algorithm where the computer assistance has a GP part,
2. to present one solution produced with the help of GP components, and
3. to present a special method for fitness specification using large data sets.

The images come from the ERS (European Remote Sensing Satellite) which scans the earth with a radar called SAR (Synthetic Aperture Radar). Figure 12.6 shows an example image (1024×1024 pixels).

The goal is to detect pressure ridges from images of ice in the Arctic Sea. A pressure ridge can be the result of first-year ice buckling under pressure from thicker, older ice. To a viewer on location the pressure ridge may look like a 5–10 meter high, long hill made of shattered ice blocks. The pressure ridges affect how the floating ice moves and drifts, which is of interest to meteorologists, for instance. On the radar images, pressure ridges appear, at best, as low-contrast brighter curves or blobs, which are very time consuming and tedious to extract by hand. The primary goal is, therefore, to find an automatic algorithm that can extract these diffuse features directly from satellite images.

Daida et al. use the *scaffolding* to describe a system that assists in algorithm design and that features GP as an essential component. The reason for using a system for algorithm design is partly that the goodness criterion or fitness function is not easy to define for pressure-ridge extraction. Experts may agree on where a pressure ridge is on a radar image, but they largely disagree on what defines such a feature in general. Hence, it is hard to just define a fitness criterion and then use a GP system for the algorithm design. Instead, the system uses an interactive cycle for designing the algorithm, illustrated in Figure 12.7.

Figure 12.6
Example target bitmap ([Daida et al., 1996a], copyright MIT Press, reproduced with permission)

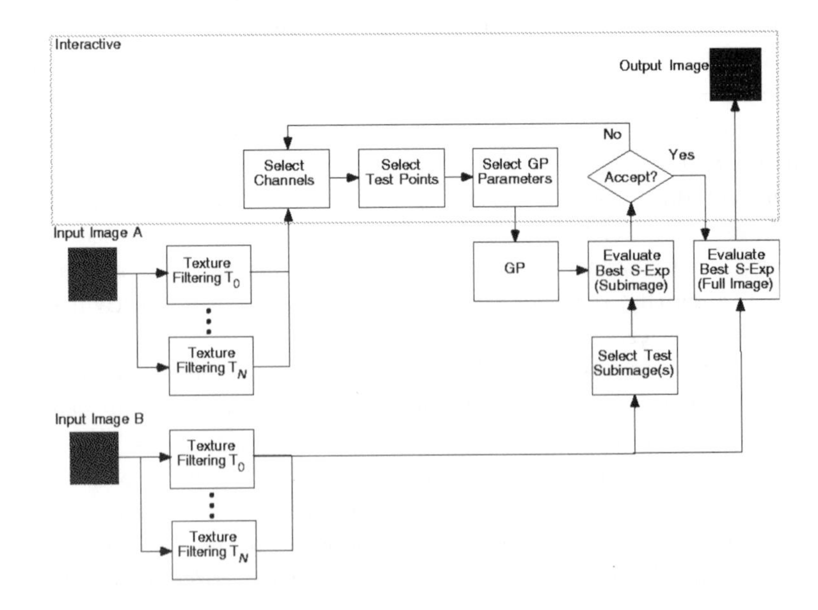

Figure 12.7
Diagram of scaffolded GP system ([Daida et al., 1996b], copyright MIT Press, reproduced with permission)

The user selects the terminal and function set together with test images, a fitness function, and GP parameters. When the GP system terminates, the best individual is tested on a full image, and an expert judges its performance. These steps are repeated until the GP component has found an acceptable image-processing algorithm.

The terminals that can be used in a terminal set represent image data and texture filters.

The function set components are the arithmetic operators and the conditional `If-Less-Than-or-Equal-to` (IFLTE). The operators are modified such that they are closed under 8-bit arithmetic.

The fitness calculation is based on manually classified single image points and on attributes of these points. The use of subimages turned out to be too computationally expensive even when subimages were as small as 8×8 pixels, and the results found did not generalize for full images. This is the reason for using manually classified single image points as fitness cases. A test point is a vector of values. The first value is a Boolean quantity simply giving the manually classified *ridge or non-ridge* property. Next comes the 8-bit pixel-intensity value of the test point followed by several intensity values for the image that has been processed by a selection of texture filters. The fitness is computed as the number of hits over the classification set.

The first work by Daida et al. in this field used a fixed training set. Later, better results were achieved using a dynamic training set that is changed during evolution. Daida et al. describe the method and its results:

> A GP system starts a run with a training set that is relatively small and contains test points that should, in theory, be easy for the algorithm to score well. When an individual scores a certain number of hits, a few points are added to the training set under evaluation. This process can continue until either an individual scores a maximum number of hits or maximum number of generations has been reached. Not only has this strategy resulted in a better individual than described in [Daida et al., 1995], but the overall process under this fitness function has been proven to be more controllable than when using a static training set.
>
> J. DAIDA ET AL., 1996

The method is inspired by the work of Goldberg [Goldberg, 1989] and Holland [Holland et al., 1986].

The results are encouraging when the best-of-runs individual is applied to two full test images. In a qualitative examination, it is shown that the extracted features are very well correlated with the pressure-ridge and rubble features identified by human experts; see Figure 12.8. These results constitute the first automatic extraction of pressure-ridge features as low-contrast curvilinear features from SAR imagery.

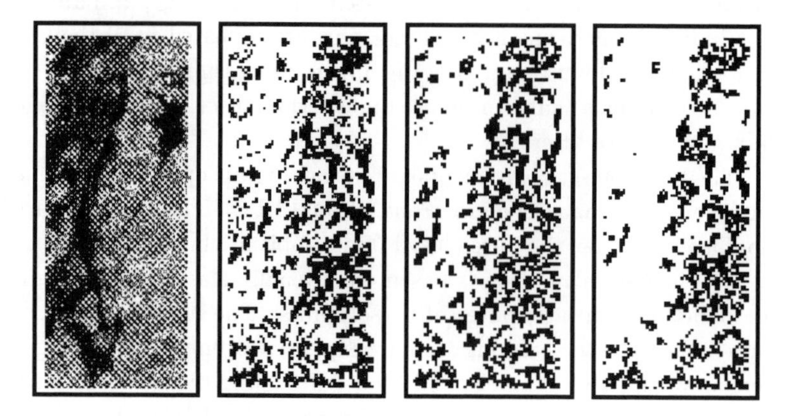

Figure 12.8
Filtered images ([Daida et al., 1996a], copyright MIT Press, reproduced with permission)

12.4 Computer Science–Oriented Applications

12.4.1 Cellular Encoding of Artificial Neural Networks

Problem Domain

Gruau has attacked the problem of automatic generation of neural networks using a developmental approach. He reasoned that for an encoding method to be compact and efficient, a modular approach must be employed. His idea was to use modular descriptions of parts of neural networks that could be used repeatedly in the course of construction (development) of a complete and presumably complex neural network.

Earlier work in the field of development of complex neural systems [Mjolsness et al., 1988] [Mjolsness et al., 1995] [Kitano, 1990] had already demonstrated the feasibility of grammars. In a series of papers Gruau proposed and later refined a developmental approach based on graph grammars [Gruau, 1992b] [Gruau and Whitley, 1993] [Gruau et al., 1994] [Gruau, 1995].

Task

The task Gruau considered in one demonstration of the feasibility of his approach is controlling a six-legged insect. Each leg has a number of neurons for control: three motor neurons and one sensor neuron recording the status of the leg. The task is to coordinate the different neurons on different legs so as to end up with coordinated motion in various gaits.

It is necessary to allow for recurrent connections in the network due to the problem of storing state information. Each artificial neural network cell consists of input and output connections to other cells, thus it is treated as a directed labeled graph. The author simplifies the concept as much as possible in order to be able to generate directed graphs.

The developmental approach comes in when Gruau specifies a list of graph rewriting rules to be applied to the cells. We have already seen a collection of graph rewriting rules in Figure 9.23. The basic idea is to allow for a division of cells under conservation of connections. Adding, removing, and changing weights are other possible rewriting rules. Originally, Gruau had only binary weights, but recently he has added more functionality to the weights.

The rewriting rules specified in this way are encoded as a tree and applied in a chosen order to arrive at a fully developed neural net. Figure 12.9 shows a sample with three different trees which are applied repeatedly.

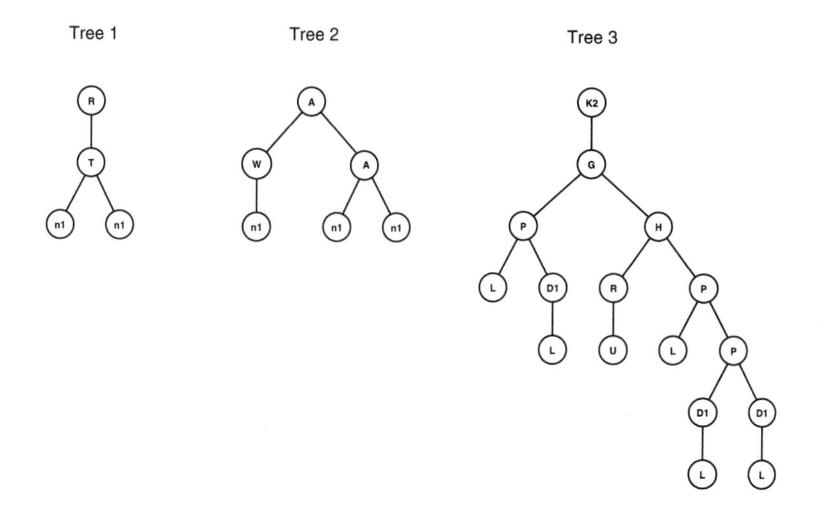

Figure 12.9
Sample cellular code, designed by hand. $n1$ means that the next tree in order from left to right is to be applied at that position. Capital letter nodes refer to rewriting rules of Figure 9.20.

The development of the neural network is shown in Figure 12.10. Keep in mind, however, that the rules are hand-designed for this demonstration, thus a highly symmetric solution results.

GP Elements

Genetic programming is now employed to evolve solutions similar to Figure 12.10 but starting from random rewriting rules and, hence, from randomly developed neural nets. Because the process is computationally expensive, Gruau implemented the algorithm on a MIMD machine (IPSC860 with 32 processors) as an island model with 2D torus-like topology. After the structure was developed, a stochastic hill climbing algorithm was applied to the weights of the neural net. Gruau used a population size of 32×64 individuals.

Results

Gruau found networks of the kind shown in Figure 12.11. The left example uses an additional refinement called *automatic definition*

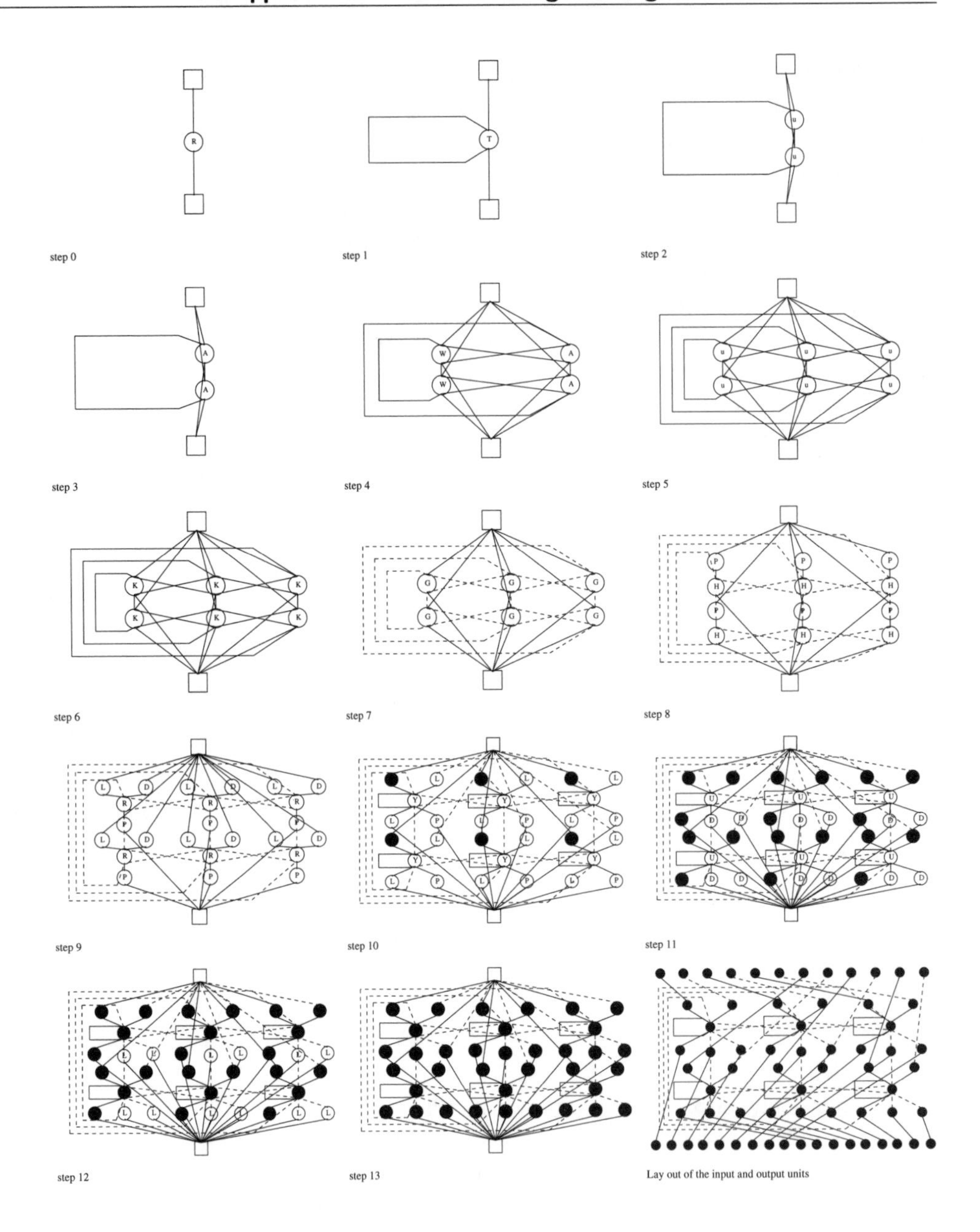

Figure 12.10
The development of a neural network specified through rules of Figure 12.9 ([Gruau, 1995], copyright MIT Press, reproduced with permission)

of subnetworks which is the cellular encoding version of ADFs. It can be seen clearly that such a network needs fewer nodes and has a more ordered structure than the simple approach of Figure 12.11 (right). The resulting networks are similar to the ones seen previously [Beer and Gallagher, 1992].

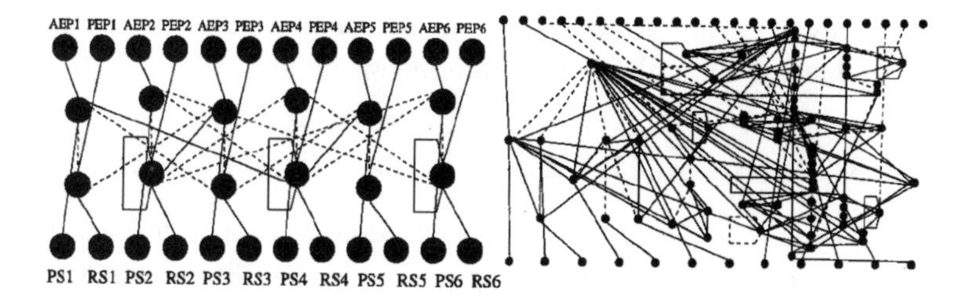

Figure 12.11
GP-evolved solutions: (left) including ADSN, (right) without ADSN ([Gruau, 1995], copyright MIT Press, reproduced with permission)

Gruau claims that his method is well suited for problems with regularities in them. Cellular encoding can discover those regularities and exploit them in the course of the development of a suitable solution.

12.4.2 Development and Evolution of Hardware Behaviors

Hemmi et al. describe an application from the domain of circuit synthesis [Hemmi et al., 1994a]. The authors do not report on a real-world application, but on a contribution – with GP-like means – to a field that will become highly practically relevant: hardware evolution. This work is a step toward hardware that may self-adapt its behavior with respect to an environment.

The authors present a process that consists of a developmental and an evolutionary phase. *Hardware description language* (HDL) programs get developed and evolved, which represent hardware specifications, which ultimately represent behavior of circuitry.

In the developmental phase, a program gets "grown" from a start symbol by using a rewriting system. Beginning with the start symbol, this system produces a tree structure by using production rules. The start symbol corresponds to a zygote while the sequence of leafs produced corresponds to the multi-cellular organism, in this case a program. The authors term this tree a chromosome since it represents the final program.

In the evolutionary phase, chromosomes may be subject to mutation and crossover, much like tree structures in common GP approaches. Fitness evaluation is performed by converting a chromosome into a programmable-logic device (PLD) program which modifies a PLD such that it represents the circuitry described by the chromosome. This circuitry can then be evaluated with respect to its behavior. This evaluation can be used to assign a fitness to the corresponding chromosome.

Finally, by using this method, the authors evolve a binary adder circuit that produces the binary sum of two input values.

12.4.3 Intrusion Detection

Crosbie et al. report on a prototype in the domain of computer system defense [Crosbie and Spafford, 1995]. Due to the high connectivity between systems worldwide, there is high potential for an intrusion into a system. In general, an intruder aims at getting access to and control over the system in order to perform an activity that is deleterious to the system's functionality or to a user, like deleting system files or accessing classified information.

A common intrusion detector is a central unit watching the complete system for intrusion. The authors propose a decentralized detection approach: autonomous software agents watch the system – each of them observing just a system *part* – and each of them learns to detect a potential intrusion related to the observed part. These agents are evolved by genetic programming.

The decentralized approach has several advantages over the central approach, according to the authors. Two important ones follow. First, a system change can be easily answered by a change in the number and potential speciation of agents. In contrast to this, a central approach might require a high effort in redesigning the detector.

Second, a central detector that gets conquered by an intruder is not only useless but actually very dangerous to the system, since it will provide information signaling non-existent system safety to the administrator. However, a single subverted software agent corrupts just a tiny part of the complete decentralized detector. Thus, the detector stays functional – although with decreased effectiveness.

Detector-Agent Design

The red line in the design of detector agents is concentration and inter-agent communication. A single agent focuses on one or just a few aspects of intrusive activity. If it detects a potential intrusion, it communicates this circumstance to all other agents, which will modify their operation accordingly. If the communication activity between agents reaches a certain degree – so that the probability of an actual intrusion is rather high – this may be communicated to a human agent, for instance, a system administrator. Of course, a single agent may wrongly assess a certain activity sequence – like two consecutive failed login trials to the same user account – as an intrusion trial. However, this single misinformation will not make it to the administrator. Thus, a single "paranoid" agent cannot disturb the system performance.

To detect intrusive activity, an agent scans system audit data, like login trials or ftp connections. For fitness evaluation, several scenarios get presented to each agent. Each scenario consists of potentially intrusive and legitimate activities, and it has a certain actual probability that it represents an intrusion. Based on this probability and the agent's assessment of this probability, the agent gets a certain fitness value.

Actually, the prototype application succeeds in evolving an agent that classifies two of three scenarios correctly. Certainly, the underlying concept has to be extended and tested more intensively, but the prototype results indicate a potential for future GP-based detection systems.

12.4.4 Autoparallelization

Walsh et al. report on an application from the domain of software engineering [Walsh and Ryan, 1996]. Considering the huge body of serial software, an automatic parallelization of serial programs is highly relevant. The authors present PARAGEN, a GP-based system for autoparallelization of serial software. The system's goal is to transform a serial program into a functionally equivalent highly parallel program. To that end, PARAGEN tries to reassemble a parallel program from the statements of the serial program such that the parallel program performs fast and correctly.

An individual is a parse tree. Each statement from a serial program represents a terminal. Fitness evaluation works in two ways: fitness reflects both a degree of functional equivalence and a degree of parallelism.

An evolved individual, representing a potentially functionally equivalent parallel version of a serial program, gets executed for several different initializations of the variables used in the serial program. The serial program gets executed using the same initializations. The smaller the differences between the corresponding results are, the better is the parallelizing individual with respect to functional equivalence.

Furthermore, the degree of parallelism gets evaluated. The faster the parallel version performs, the better is the parallelizing individual with respect to generating a high degree of parallelism. Thus, a balance between correctness and good parallelity gets established in a population.

PARAGEN was tested on common problems from the domain of parallelization. The authors report on a successful transformation of all corresponding serial programs, using a population size of 100 individuals over 10 generations. An example follows. When executing assignment statements that share variables, execution order is crucial. Consider the following sequence of assignment statements $S2, S5, S6, S8$.

```
for i:=1 to n do
  begin
    S2: f[i] := 57;
    S5: b[i] := a[i] + d[i];
    S6: a[i] := b[i] + 2;
    S8: c[i] := e[i-1]*2;
  end;
```

Obviously, the sequence $S5, S6$ is critical, since its statements share the array variables a and b with identical index i. Thus, in general, this sequence will result in different values for $a[i], b[i]$ than the sequence $S6, S5$. Therefore, when parallelizing the above code, PARAGEN must produce parallel code that ensures the execution order $S5, S6$. Indeed, the system comes up with

```
DoAcross i := 1 to n
  begin
    PAR-BEGIN
      S8: c[i] := e[i-1]*2;
      S5: b[i] := a[i] + d[i];
    PAR-END
    PAR-BEGIN
      S2: f[i] := 57;
      S6: a[i] := b[i] + 2;
    PAR-END
  end;
```

The semantics of `PAR-BEGIN..PAR-END` is to execute all statements between these two keywords in parallel. `DoAcross` loops over the *sequence* of PAR blocks. Note how this code ensures the execution of $S5$ prior to the execution of $S6$ yet also parallelizes execution.

The authors plan to focus further work on making PARAGEN generate OCCAM code – OCCAM is a prominent parallel language – and on having it generate parallel programs directly out of a problem description.

12.4.5 Confidence of Text Classification

The flood of information pouring over us is increasing by the day. We speak about the information society where information is the main commodity traded. One of the biggest problems in the information society is how to sort and discriminate information that is interesting and relevant to you and your work. Several companies offer services where information is classified and given keywords according to contents. The keywords can then be used to compile material of interest for different groups. One such provider is Dow Jones, which daily assigns keywords or codes to several thousand texts. The texts come from many different sources, like newspapers, magazines, news wires, and press releases. The large volume of texts makes it impractical or impossible for Dow Jones' editors to classify it all without computer support.

One such support system uses a *memory-based reasoning* technique (MBR) [Dasrathy, 1991] for assigning keywords automatically to articles. However, it is desirable that the automatic system can indicate which classification it is uncertain of and then call for manual assistance.

The system should, in other words, assign a confidence value to the classification that allows it to classify easy texts automatically while giving difficult cases to the editors. Brij Masand has successfully applied GP to the evolution of such confidence values for automatically classified news stories [Masand, 1994].

The coding of a text consists of assigning one or more keywords or codes to the document. There are about 350 different codes, such as *industrial, Far East Japan, technology, computers, automobile manufacturers, electrical components & equipment.* A single story typically will have a dozen codes assigned to it.

The automatic MBR classification system – which in itself has nothing to do with genetic programming – is trained on a database of 87 000 already classified stories. Performance is measured by the two concepts *recall* and *precision.* Recall is the proportion of codes that were assigned both by the system and the editors. Precision is the ratio of correct codes to total number of assigned codes.

The automatic system has a recall of about 82% and precision of about 71%. For each of the codes assigned to a text the system also produces a score which can be used as a measure of certainty for the particular code. The objective of the GP system is to evolve a general measure given the sorted list of confidence scores from the assigned codes.

The fitness of the GP system is the product of the recall, precision, and the proportion of texts accepted for automatic classification. Thus, the system simply tries to achieve highest possible quality in judgment together with automatic judgment of as many texts as possible.

The terminal set contains the scores of the assigned codes and five numerical constants $(1-5)$. The function set contains the four arithmetic operators $+, -, \times, /$ and the square-root function. The output of each individual is normalized, and all output above 0.8

is interpreted as an accept of the text, while output below 0.8 is interpreted as *give the text to manual classification.* The GP system is trained on 200 texts and validated on another 500 texts.

Results show the confidence formulas evolved by the GP system beat hand-constructed formulas by about 14%, measured by the number of correctly accepted or transfered texts.

12.4.6 Image Classification with the PADO System

Teller and Veloso [Teller and Veloso, 1996] use their PADO system for classification of images and sound. The PADO system is an evolutionary program-induction system that uses a graph representation of the individuals. It is a quite complex system which is briefly described in Section 12.12. Classifying images and sounds means dealing with large data sets and fuzzy relationships. The system has many different application areas, such as robotic control.

PADO has been used for both image and sound classification, but here we concentrate on an image classification experiment. The system is trained on one object image each time. After training, the classification performance is evaluated on an unseen image portraying the same object. This unseen image is taken from a validation set containing close to 100 images of the object. Examples of training and validation image pairs are shown in Figure 12.12. Objects in the images include book, bottle, cap, glasses, hammer, and shoe. Recognizing such objects in an image is an easy task for the human brain, but the problem is of extreme complexity, and it is very hard to find a general adaptive algorithm that can solve the task with a minimum of domain knowledge. Consequently, very few generic algorithms for this kind of image processing exist. The PADO system, however, has shown remarkable success in the limited domain it has been applied to.

Each program in the population has an indexed memory of 256 integers that are 8-bit values. The terminal set includes functions and procedures of many different types. There are arithmetic primitives (`ADD`, `SUB`, `MUL`, `DIV`, `NOT`, `MAX`, `MIN`) and conditional branches (`IF-THEN-ELSE` and `EITHER`). The terminal set also includes functions specific to the domain – `PIXEL`, `LEAST`, `MOST`, `AVERAGE`, `VARIANCE`, `DIFFERENCE` – which either give information on a single specific pixel value or perform a simple statistical function on an image area. The parameters for the functions reside on the system stack.

The population size is 2800 individuals, and the system is run over 80 generations for each of the 98 training images. The results show that the classification program is correct between 70% and 90%

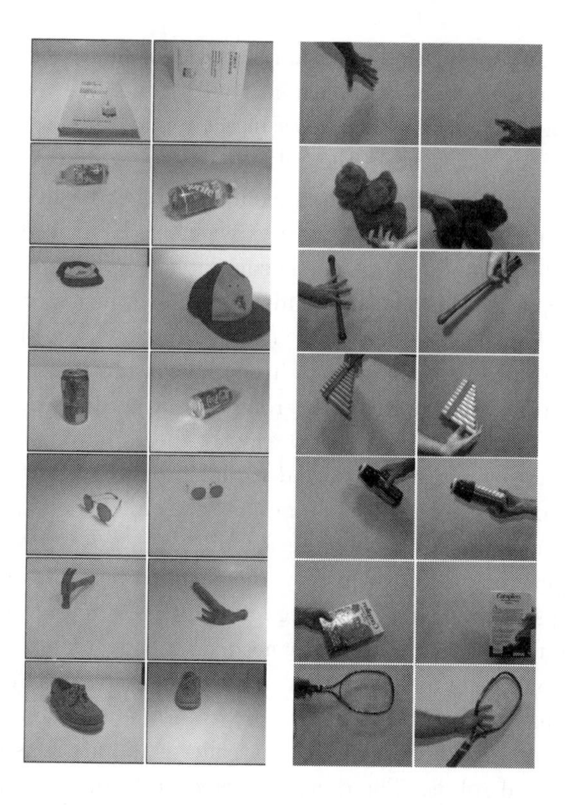

Figure 12.12
The training and test images classified with PADO ([Teller and Veloso, 1996], copyright Oxford University Press, Inc., reproduced with permission)

of the test cases, depending on the difficulty of the respective image. Similar results are achieved in a sound classification task.

12.5 Engineering-Oriented Applications of GP

12.5.1 Online Control of Real Robot

Genetic programming has great potential for control applications. Many control applications demand solutions that are hard or impossible to derive with analytical methods. Nevertheless, nature has proved to us that there exist ingenious solutions to many advanced control problems.

Consider the miracle of controlling a house fly (*Musca domestica*), for instance. The fly shows significant precision in its movements, controlling several wings in the context of the extremely non-linear dynamics of small-scale aeronautics. In addition, we know that the hardware elements controlling the flight – the neurones – are fewer than one million, with a clock frequency thousands of times slower than that of the processing elements of a computer. The control algorithm is also run on a fault tolerant massively parallel tiny hardware base without problems. This example – one of many – is indeed an impressive achievement of evolution. If GP could evolve solutions to problems a fraction as complex as insect flying, this would already have enormous potential in the field of control.

Problem Domain

Robot control is an area that has interested several GP researchers (see Table 12.2). However, most experiments have been performed with a simulation of a robot in the computer. A problem with many of these approaches is that it is difficult to move the experiment from a simulated robot to a real one. The commonest way of doing GP for a control problem is to create a population of random programs. Each program then controls the robot for a predefined number of time steps. The robot is reset to exactly the same point before every program evaluation. The individual program's performance is judged according to the fitness function as usual. When all programs have been evaluated, the best performing ones are reproduced into the next generation. Subsequently, the genetic operators are applied. These steps are repeated until a good solution is found. Let us say approximately 100 generations are needed for convergence. If we have a population size of 500 programs and each individual is evolved during 100 time steps then we need $100 \times 500 \times 100 = 5\,000\,000$ time steps of evaluation before a good controller is found. This is impractical for a real robot. The dynamic response from the environment generally requires at least 300 ms for meaningful feedback – the result of too short a control action will drown in the different kinds of mechanical, electronic, and algorithmic noise the robot produces. This means an evolution time of 1 500 000 seconds, or two weeks. The robot also needs to be reset to its initial position before each program evaluation, in this case 50 000 times. A computer *simulation*, on the other hand, could be run much faster than a real-time simulation, and resetting the position would not be a problem either.

In this section, we present our own work on an online version of a GP control architecture which allows for efficient use of GP to control a real robot.[1] The experiments are performed with a genetic programming system (AIMGP) evolving machine code control programs. This system – described in Section 9.2.4 – is well suited for low-level control, since it gives acceptable performance even on very weak architectures while also allowing for a compact representation of the population.

Task

The objective of the control system is to evolve obstacle-avoiding behavior given data from eight infrared proximity sensors. The exper-

[1]For details, see the following papers: [Nordin and Banzhaf, 1995c] [Olmer et al., 1996] [Banzhaf et al., 1997b] [Nordin and Banzhaf, 1997b] [Nordin and Banzhaf, 1997c].

iments were performed with a standard autonomous miniature robot, the Swiss mobile robot platform Khepera [Mondada et al., 1993], shown in Figure 12.13. The mobile robot has a circular shape, a diameter of 6 cm, and a height of 5 cm. It possesses two motors and an on-board power supply. The motors can be independently controlled by a PID controller. The robot has a microcontroller onboard which runs both a simple operating system and the GP-based online learning system.

Figure 12.13
The Khepera robot

GP Elements

The fitness in the obstacle-avoiding task has a pain and a pleasure part. The negative contribution to fitness – pain – is simply the sum of all proximity sensor values. The closer the robot's sensors are to an object, the higher the pain. In order to keep the robot from standing still or gyrating, it has a positive contribution to fitness – pleasure – as well. It receives pleasure from going straight and fast. Both motor speed values minus the absolute value of their difference is thus added to the fitness.

The online GP is based on a probabilistic sampling of the environment. Different solution candidates (programs) are evaluated in different situations. This could result in unfair comparison because a good individual dealing with a hard situation can be rejected in favor of a bad individual dealing with a very easy situation. The conclusion of these experiments, however, is that a good overall individual tends to survive and reproduce in the long term. The somewhat paradoxical fact is that sparse training data sets or probabilistic sampling in evolutionary algorithms often both increase convergence speed and keep diversity high enough to escape local optima during the search.

The remarkable fact that evolutionary algorithms might prefer noisy fitness functions is also illustrated in [Fitzpatrick et al., 1984]. Here, a genetic algorithm is used to match two digital pictures, each consisting of 10 000 pixels. The most efficient sample size in the fitness evaluation turned out to be 10 (out of 10 000) pixels. In other words, the fastest run to reach a good solution only looked at 1/1000 of the available data in each individual evaluation.

The online method is based on a similar assumption, and each individual is evaluated against only a single event during the robot's learning process. An event includes one reading of the sensors, the execution of one action, and the feedback from the fitness function. Figure 12.14 gives a schematic view of the control architecture and the GP system.

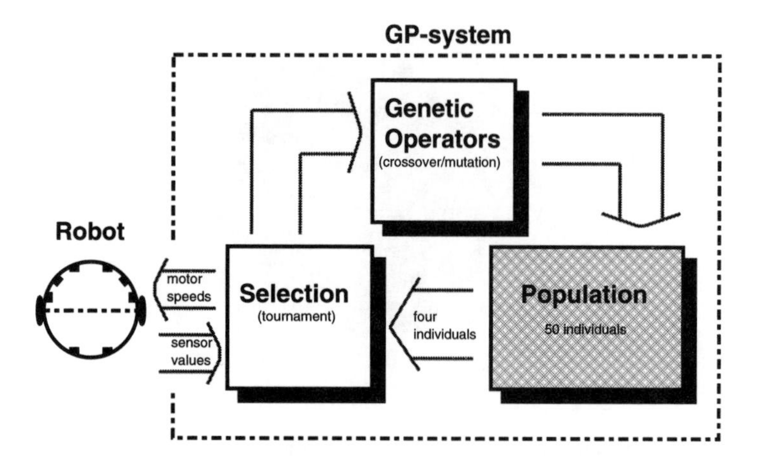

Figure 12.14
A schematic view of the control system

The GP system has the following execution cycle:

1. Select four members for tournament.
2. For all members in the tournament:
    (a) Read out the eight proximity sensors and feed the values to one individual in the tournament.
    (b) Execute the individual and store the resulting robot motor speeds.
    (c) Send motor speeds to the robot.
    (d) Sleep for 400 ms to await the results of the action.
    (e) Read the proximity sensors again and compute fitnes; see below.
3. Replace the two worst performing individuals with copies of the two best ones.
4. Do mutation and crossover on the offspring (copies).
5. Go to step 1.

Results

The robot shows exploratory behavior from the beginning on. This is a result of the diversity in behavior residing in the first generation of programs which has been generated randomly. Naturally, the behavior is erratic at the outset of a run. During the first minutes, the robot keeps colliding with different objects, but, as time goes on, the collisions become less and less frequent. The first intelligent behavior usually emerging is some kind of backing up after a collision. Then, the robot gradually learns to steer away in an increasingly sophisticated manner.

On average, in 90% of the experiments, the robot learns how to reduce the collision frequency to less than 2 collisions per minute. The convergence time is about one hour. It takes about 40–60 minutes, or 200–300 generation equivalents, to evolve good obstacle-avoiding behavior. Figure 12.15 shows how the number of collisions per minute diminishes as the robot learns and as the population becomes dominated by good control strategies.

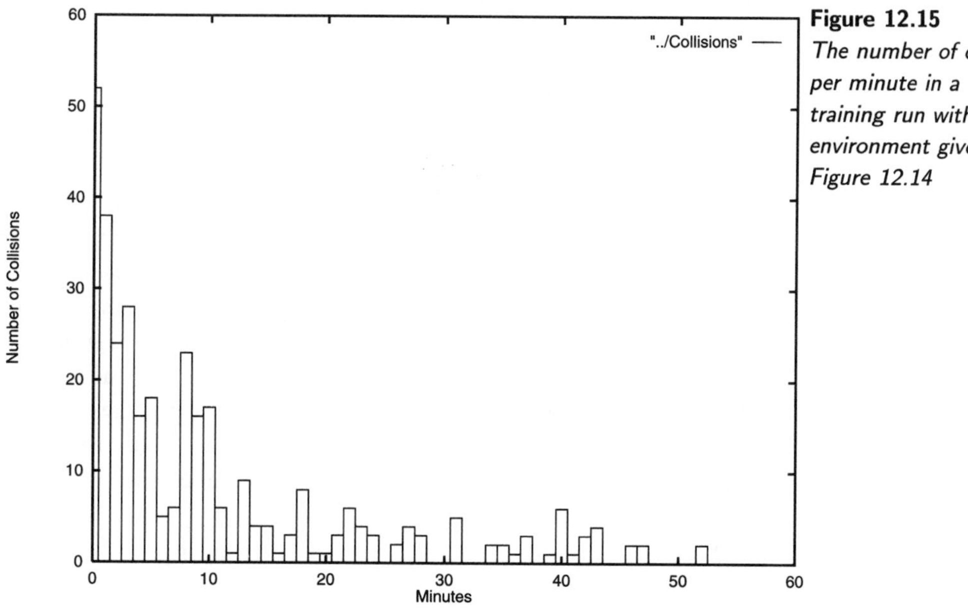

Figure 12.15
The number of collisions per minute in a typical training run with the environment given in Figure 12.14

These experiments indicate the potential of GP for online control and how genetic programming combined with a technique for evolving machine code can make the evolution of low-end architectures possible.

12.5.2 Spacecraft Attitude Maneuvers

Howley describes a nice application from the domain of optimal control [Howley, 1996]. The object to be controlled is a spacecraft. The problem is to find a control program that performs a 3D craft re-orientation in minimal time. This means that, given an initial craft attitude and a desired final attitude, the program is required to rotate the craft into the final attitude in minimal time by rotating the craft around one or more of its three body axes. These axes are pairwisely orthogonal.

Problem Domain

The craft components affecting the rotation are called actuators. For each body axis, there is an actuator, which generates a positive or a negative torque that lets the craft rotate – depending on the definition of *positive* and *negative* – anticlockwise or clockwise, say, about the respective axis. Each actuator is assumed to have a bang-bang characteristic, so that it generates either a *maximal* positive or *maximal* negative torque.

This problem is practically relevant in the area of satellite-based data transmission and observation, for instance. Typically, an observation satellite has to keep an optical system in focus on a planetary target area. If the satellite is moving relative to the planetary surface, it has to reorient continuously – or, at least, at sufficiently short time intervals – in order to stay focused.

Task

The application concentrates on two maneuver types: rest-to-rest and rate-limited non-zero (RLNZ) terminal velocity. A rest-to-rest maneuver begins and ends with zero angular rates – rate is a synonym for velocity here – so there is no rotation about any axis before or after the maneuver. Often, however, a RLNZ maneuver is needed: before or after the maneuver, the craft is rotating about one or more axes with certain angular rates. For instance, when the craft is supposed to track a moving and maneuvering target with an optical system, RLNZ maneuvers are needed. The maximal angular rates are limited by the maximal forces the actuators can generate. In particular, a rest-to-rest maneuver can be viewed as a RLNZ maneuver with zero initial and final rate.

Note that, according to a theorem of Euler, a rigid body can get from an arbitrary attitude into another arbitrary attitude by a rotation through a certain angle about a certain axis, called the *eigenaxis*. Thus, instead of doing a rotation sequence about one or more body axes, the craft may just rotate through a certain angle about the corresponding eigenaxis. This rotation is implemented by operating one or more actuators in parallel.

If you have trouble imagining this, take a rubber eraser and a needle. Hold the eraser in some initial attitude, then move it into some final attitude by a sequence of body axis rotations. For each pair of initial and final attitude, you can always stick the needle through the eraser such that, when rotating the needle appropriately, the eraser will move from the initial to the final attitude: the needle represents the eigenaxis.

Thus, a maneuver means changing an initial eigenaxis/rate into a final eigenaxis/rate. The control problem is to do this as fast as possible.

Control Law

Since an actuator has a bang-bang characteristic, the complete maneuver consists of a sequence of actuator-switching commands. Each command switches one or more actuators, each to positive or negative torque. For instance, such a command may switch actuators one and two to positive and three to negative. Thus, a command can be represented by a 3D torque vector u_1, u_2, u_3 with each u_i designating maximal positive or negative torque. This vector is the output of a control law that takes as input the final eigenaxis/rate and the current eigenaxis/rate.

Thus, for the control law implementation at the start of the maneuver, the initial eigenaxis/rate are the *current* parameters. The control law becomes active for the first time, computing a torque vector. The corresponding actuator activities lead to a new current eigenaxis and rate. The control law becomes active again, and so on, until the desired eigenaxis/rate are reached. Then the control loop terminates.

The control problem for rest-to-rest maneuvers has a known numerical solution. However, the computation of this solution takes some time, whereas the problem must be solved in real time: there is no sense in computing an optimal solution that, when finally found, can no longer be used since it is out of date. An approximate but real-time solution is required. It is realized by the control loop that makes the craft move incrementally into the final eigenaxis/rate.

GP Elements

For the genetic programming search, an individual is a control law: GP evolves expressions that are used as control law within the control loop. The terminal set reflects the described input and output parameters. The function set contains $+, -, \times$, a protected division operator, sign inversion, the absolute-value function, an if-*a*-less-*b*-then-action-else-action construct, and three ADF symbols.

The fitness cases consist of the initial eigenaxis/rate, the final eigenaxis/rate, and a time-out limit. Fitness evaluation considers an individual as successful if its application by the control loop results in the final eigenaxis/rate within certain error bounds and before time-out.

Results

For rest-to-rest maneuvers, runs went over 51 generations and used population size 5000. For RLNZ maneuvers, the values were

74 and 10 000. As genetic operators, 80% node crossover, 10% leaf crossover, and 10% copying were employed. As for hardware and software, Andre's DGPC was used on an IBM RS6000.

Genetic programming produced a result for the rest-to-rest maneuver within plus or minus 2% of the numerical solution. Additionally, this solution generalized to solve further randomly generated maneuvers. As for the RLNZ maneuver, GP produced a solution that solved all fitness cases but did not generalize.

12.5.3 Hexapodal Robot

Spencer reports on an application of genetic programming to robotics [Spencer, 1994]. A simulated hexapodal (six-legged) robot is to be controlled by an evolved program such that the robot walks as far as possible in an environment before a time-out. Each leg of the robot can be lifted up to a final position or put down until the foot touches the ground. Each leg can also be moved forward and backward – parallel to the ground – with varying force. Thus, if the leg is down and moves, this action results in a robot-body movement. The robot is unstable if its center of gravity is outside the polygon defined by the feet that are on the ground. After a certain time, an unstable robot falls.

Task

A subgoal of this application is having the robot learn to walk with minimal *a priori* knowledge about this function. Thus, three experimental setups are presented, each providing the robot with less knowledge.

GP Elements

The genotype is a tree expression. The terminal set contains random floating-point constants, and – depending on the experimental setup – `Oscillator` and `Get-Leg-Force-n`. `Oscillator` represents $10\sin(t)$, t being the number of elapsed time units since the start of an experiment. `Get-Leg-Force-n` returns the force leg n moves with.

The function set consists of unary negation ˜ , $+, -, \times$, protected division, `min` and `max` functions, and the `fmod` function (floating-point modulo). A ternary `if` function answers its second argument if the first argument is positive, and its third argument if the first argument is non-positive.

For leg control, there are two functions. Unary `Set-Leg-Force-n` sets a certain force for leg n. The leg will be moved with this force. Unary `Set-Leg-State-n` lifts leg n up or puts it down dependent on the argument's sign.

The fitness of an individual corresponds to the distance – between the start and end points of the walked path – the robot walks under the individual's control until a time-out. Distance is measured in space units. The time-out occurs 500 time units after the simulation has started. After each elapsed time unit, an individual gets evaluated. The evaluation – the execution – may cause leg actions. As a consequence, the individual may fall. In this case, the simulation stops at once.

A population size of 1000 individuals is used, with 50, 65, or 100 generations, depending on the experimental setup. The selection scheme is a 6-tournament, and 0.5% mutation and 75% crossover are employed as genetic operators.

Spencer introduces constant perturbation, which gets used with 25% probability. This operator, applied to an individual, changes each constant in the expression by plus or minus 10%. Spencer argues that this kind of mutation is more beneficial than standard mutation since it is just the coefficients that get changed, not the structure, which might disrupt the program logic.

The leg-related terminals and functions may give rise to interesting side effects. For instance, certain legs can be controlled immediately by the state of other legs. For example, the statement `Set-Leg-Force-1(Get-Leg-Force-2(..))` makes leg 1 move with the force of leg 2. This kind of inter-leg communication is actually a software analog of a similar concept implemented in the neural leg-motion control system of certain insects.

Results

Spencer reports that, in all three experiments, individuals emerge that can control the robot such that it performs "efficient, sustained locomotion," and believes that the application can be scaled up to three-dimensional robots.

This application raises a very interesting possibility: real-time GP control of robotic systems. Obviously, evolving a program for a real-time application is usually infeasible *during* run time, since the evolution takes too long. So, the naive approach is to evolve a very good solution offline and to transfer this program to the control unit of a robot where it performs in real time. However, a big disadvantage with this approach is that, no matter how good the solution is, once it has been evolved and transferred, it cannot evolve further to adapt to a potentially changing environment. Spencer briefly considers combining genetic programming with classifier systems to overcome this disadvantage.

12.5.4 Design of Electrical Circuits

The most obvious application area for GP is evolving programs. However, variants of the technique can be used to evolve structures representing objects that are not immediately identified as conventional programs. One such application is the automatic design of electrical circuits.

Koza et al. have used genetic programming successfully to evolve a large number of different circuits with good results [Koza et al., 1996b] [Koza et al., 1996c]. Here, a recipe for how to construct a circuit is evolved by the GP system. Each individual contains nodes with operations that manipulate an electrical circuit. At the beginning of an individual evaluation a – predefined – embryonic circuit is created. This small but consistent circuit is then changed by the operators in the GP individual, and the resulting circuit is evaluated for its performance by a simulator for electrical circuits using the original task. The simulator is often very complex and the fitness function thus takes a very long time to compute. Nevertheless, it is possible to evolve well-performing electrical circuit designs faster and better than a human could design them manually.

The technique is an example of a method where the phenotype – the circuit – differs from the genotype – the GP individual – and is partly based on the work of Gruau described in Section 12.4.1.

The circuits that have been successfully synthesized include both passive components (wires, resistors, capacitors, etc.) and active components (transistors). Figures 12.16 and 12.17 show examples of evolved circuits.

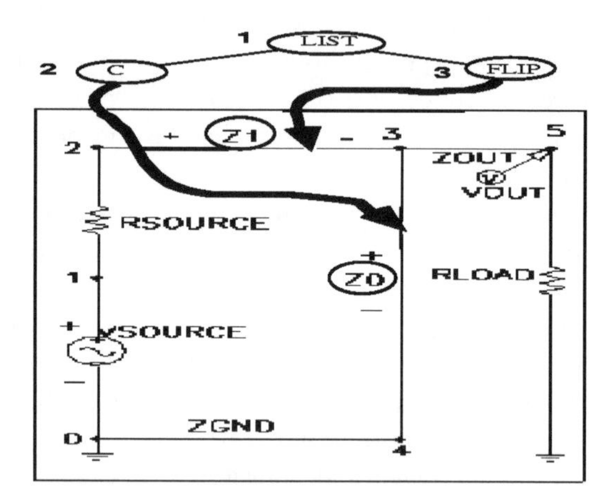

Figure 12.16
An evolved circuit ([Koza et al., 1996b], copyright MIT Press, reproduced with permission)

The function set contains functions of three basic types:

- Connection-modifying functions
- Component-creating functions
- Automatically defined functions

At each moment, there are several *writing heads* in the circuit. The writing heads point to components and wires in the circuit that will be changed by the next function to be evaluated in the program

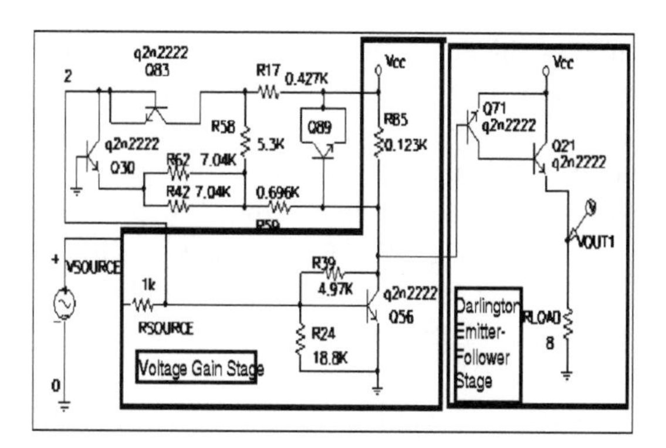

Figure 12.17
An evolved circuit ([Koza et al., 1996b], copyright MIT Press, reproduced with permission)

tree. A writing head moves and/or spawns new writing heads after each operation.

Connection-modifying functions can flip a component upside down or duplicate a component in series or in parallel with it. Component-creating functions insert a new component into the circuit at the location of a writing head.

The program tree is typed (see Section 10.2.10), and crossover exchanges only subtrees of the same type. Initialization is guaranteed to create trees obeying syntactic constraints, that is, a tree has a correct number of subtrees. There is also a mutation operator that complies with the syntactic rules of the circuit.

The SPICE package from the University of California at Berkeley is used to simulate electronic circuits for fitness evaluation. Simulating electronic circuits is very time consuming, thus the system is run on a 64-processor (PowerPC) parallel machine with a parallel GP implementation using demes (see Section 10.1.2).

The results show that, using a population size of 640 000 over about 200 generations, it is possible to solve problems that are regarded as difficult for a human designer.

12.5.5 Articulated Figure Motion Animation

Problem Domain

Computer-assisted character animation became a multimillion-dollar business in the 1990s. With the advent of powerful graphic workstations and parallel-processor farms for simulating lighting conditions and other aspects of animated scenes, the automated generation of articulated figure motion is a growing need.

Professional animators have until recently been able to keep up with the demand for frames, but the manual generation of frames is becoming more and more outdated. Gritz and Hahn have proposed using GP for animating figures [Gritz and Hahn, 1995]. They observe that, although it is difficult for humans to generate character motion, it is easy for humans to judge generated motion for its quality. Their suggestion is to let a GP system generate motion and to have the human provide the judgment as to whether it is good or bad.

The agents to be animated are controlled by programs which will be subject to evolution via GP. The hope is that GP is able to find good controller programs, starting from a population of randomly generated motion control programs.

Task

The task Gritz and Hahn consider is to generate control programs for artificial agents consisting of a figure model with fixed geometry and dynamical features. A figure is treated as a tree of rigid links with damped angular springs, and a simulation of its dynamics is performed using established methods [Armstrong and Green, 1985] [Hahn, 1988] [Wilhelms, 1990].

Figure 12.18 shows one of the joints with the corresponding quantities to be used for the dynamics simulation. The joint has a desired orientation that can be achieved by integrating, with a proportional-derivative (PD) controller, the forces of the spring.

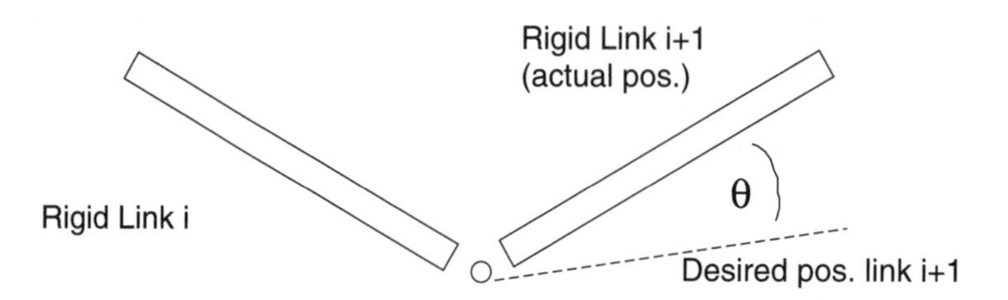

Figure 12.18
The torque at the joint is proportional to the angular difference between actual and desired orientation of the links. The desired angle is the output of the controller program ([Gritz and Hahn, 1995]).

The GP system evolves controller programs that output desired orientations for all the joints which are then used by the dynamics model to generate motion by integrating the dynamics equations for a number of time steps. The dynamics simulation also performs collision detection and collision response, yielding a physically realistic motion of the figure.

System Structure

The entire system consisting of a GP module communicating with the dynamics module is depicted in Figure 12.19.

The figure model of the particular character to be animated is considered the input, as is the fitness function specifying what kind of motion is good or bad. The control programs giving commands as to how to move the figure are the output of the system.

GP Elements

The minimal set of functions Gritz and Hahn consider are the basic arithmetic functions $+, -, *, \%$ and a function `ifltz`, which needs three arguments. More elaborate functions, such as `cos` or `abs`, or `while`-control functions or special functions like `release--grip` or `distance--to--nearest--neighbor`, are mentioned but not used.

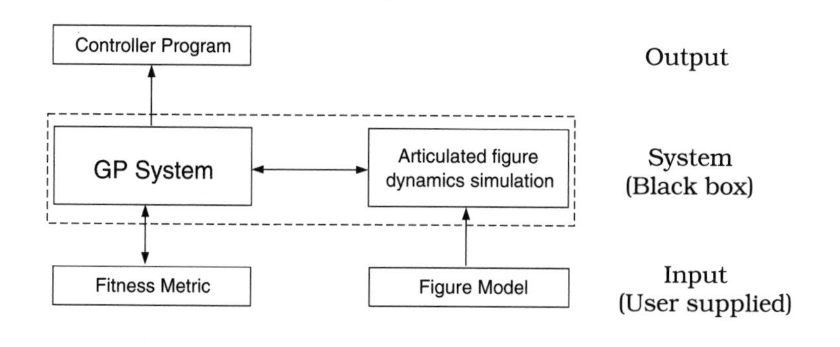

Figure 12.19
A system for rigid body animation ([Gritz and Hahn, 1995])

The terminals consist of the internal state variables of the figure and outputs of sensors accessible to the figure. These quantities are read out at the time of evaluation of the control program and used as its input. Gritz and Hahn also use random ephemeral constants as terminals for their simulations.

Because there is usually more than one joint in a figure, the control programs for each of the joints are evolved simultaneously but separately.

The fitness measure consists of two terms one evaluates the main goal of the movement, and the other one judges the style of movement. In one example used by Gritz and Hahn, the main goal is to move a lamp – as the animated figure – to a certain place. Parameters used were: $P = 250, G_{max} = 50, p_c = 0.9$.

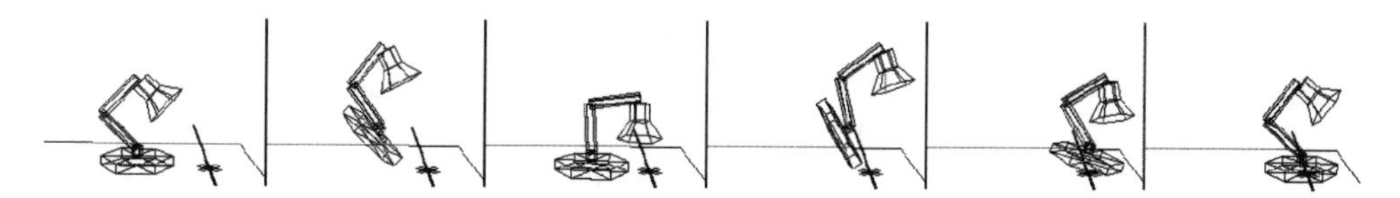

Figure 12.20
The lamp's jumping motion: the main goal is to move the lamp to the target point X ([Gritz and Hahn, 1995], copyright John Wiley, reproduced with permission)

The lamp figure has four links and three internally controllable 3D degrees of freedom. The resulting lamp motions (see Figure 12.20) consisted of jumps and looked smooth, physically realistic, efficient and surprisingly organic [Gritz and Hahn, 1995]. Gritz and Hahn also use a humanoid figure with a total of 28 degrees of freedom, between 4 and 10 of which are controlled by the GP system.

As to the style of the motion, Gritz and Hahn played a nice trick applicable anywhere. Initially, the fitness function consisted of the distance to the goal only. The movement-style terms were gradually phased in after some generations. "In this manner, we allowed the motion to start crudely and get progressively more stable over several generations. This was easier for the system than requiring optimal motion at the very start" [Gritz and Hahn, 1995].

Style was judged by a weighted sum of the following:

- bonus for completing the motion early
- penalty for excess movement after the goal was met
- penalty for hitting its head or falling over
- bonus for ending with joints at neutral angles

The computation time per generation was a few minutes on a MIPS R4000 processor. This performance appears good when compared to that of a human animator drawing a sequence of frames by hand.

12.6 Summary

In this chapter, we have seen an amazing variety of applications. Genetic programming has shown its worth in a broad spectrum of real-life problem domains with remarkable flexibility as a machine learning technique. GP is also unique in its combination of symbolic and subsymbolic application areas. In contrast to neural networks, for instance, it is possible – but not always easy – to *interpret* the GP output individual and thus to generate potentially new insights into the solved problem. In this way, we have seen or mentioned how GP can be applied to highly symbolic tasks such as natural language processing and even theorem proving. At the other end of the spectrum, it can be used for low-level signal processing.

Genetic programming has shown its value in hybrid techniques together with other machine learning paradigms such as neural networks. In other experiments, it compared well to other such machine learning paradigms in terms both of speed and learning capability. GP techniques have been used to evolve binary machine code, which gives very efficient evolved programs.

Some GP applications learn with only a few fitness cases while others work in domains with huge data sets such as image processing. In other examples, for instance, in Koza's circuit design approach, GP has been shown to match human expert performance both in time expenditure and in solution quality.

Some of this flexibility may be attributed to the freedom in choosing an arbitrary function set. By changing the function set, it is possible to adapt GP toward any problem domain where a suitable fitness function can be defined. Another reason for the broad spectrum of GP application domains is the generally very robust evolutionary search. Evolutionary search is not always the most efficient search in specialized problem domains but it is known to be very robust in its applicability – at worst degrading to a hill climbing algorithm.

The achievements in GP research over only five years or so are truly encouraging for the future.

13 Summary and Perspectives

Contents

> The engineering process doesn't work very well when it gets complicated. We are beginning to depend on computers that use a process very different from engineering – a process that allows us to produce things of much more complexity than we could with normal engineering. Yet we don't quite understand the possibilities of that process, so in a sense it's getting ahead of us. We are now using those programs to make much faster computers so that we will be able to run this process much faster. The process is feeding on itself. It's becoming faster. It's autocatalytic. We are analogous to the single-celled organisms when they were turning into multicellular organisms. We are the amoebas, and we can't quite figure out what the hell this thing is we are creating. We are right at that point of transition, and there is something coming along after us.
>
> D. HILLIS, 1991

13.1 Summary

In this book we have set out to introduce the reader to genetic programming, an exciting new field in computer science. GP is part of the machine learning branch of artificial intelligence. Therefore, it seemed to us appropriate to start this book with a look at machine learning in Chapter 1. Our main point here was that GP is able to evolve computer programs and, as such, GP's problem representation is a superset of the representations of all other ML methods, all of which work on computers, too.

GP has also drawn heavily on ideas from biology. In particular, it is an ML paradigm built on the theory of evolution. Chapter 2 thus presented some biological background, and we looked at how natural evolution proceeds in the living world. Throughout this book, we have attempted to return to the biological metaphor in explaining GP or in posing possible answers to unresolved questions. In the judgment of the authors, much more is to be learned by studying this analogy further and into more depth.

GP can be said to be the "offspring" resulting from a crossover of biology with computer science. Accordingly, Chapter 3 presented some mathematics and computer science background. In particular, we looked at the question of generating randomness, and randomness with different features for algorithms, and into the computer science methods available for GP.

There are other offspring from such a crossover of biology with computer science, notably EP, GA, and ES, which were discussed in more detail in Chapter 4. The main differences between GP and these other algorithms were mentioned.

The chapters treating GP directly started with an introduction to the basics in Chapter 5. Elementary representations for breeding programs were presented along with the commonest genetic operators and selection schemes that work on these representations.

One of the most heavily used genetic operators in GP, crossover, was the subject of Chapter 6. Our point of view was that crossover, in its simplest embodiment, has some flaws that need improvement. We argued that a homologous crossover operator or other means to decrease the rate of destructive crossover would be important steps in that direction.

In looking at the run dynamics of GP, we learned in Chapter 7 how introns can spoil GP runs and why they emerge in the first place. The mixed blessing of this emergent phenomenon was the main theme in the chapter. We tried to make clear that understanding emergent effects helps in understanding evolution.

Validation of results and understanding complex run dynamics are essential to advancing the discipline. Accordingly, Chapter 8 describes the toolkit available to practitioners to measure important observables during GP runs and to validate GP results.

After these general considerations, Chapter 9 presented a set of GP variants. Roughly, they can be classified by their genotypic structure into three groups: sequence, tree, and graph structures.

Chapter 10 was devoted to discussing a large variety of advanced methods for GP which were organized around how they improved the features of GP: speed, power of evolution, and the power of the evolved programs.

Chapter 11 then dealt with implementation issues – the techniques necessary to make GP work on the computers available today. We discussed systems based on list processing, on arrays and stacks, and on machine code.

Chapter 12, finally, discussed a large variety of subjectively selected application problems that have been tackled using GP.

13.2 The Future of Genetic Programming

> But the only way of discovering the limits of the possible is to venture a little way past them into the impossible.
>
> A.C. Clarke

Some of the ideas presented in this section will seem to be just around the corner, while others may appear far-fetched. However, we should not forget that dynamic systems – and GP is part of such a system, science and technology – sometimes show non-linear behavior which may result in accelerated development during critical phases.

Let us remind ourselves of the ultimate goal. We want to be able to tell computers *what* we want them to do – and computers should learn *how* to do it automatically. Computers should be able to program themselves, and GP is one approach that might take us toward that goal.

However, if *we* can tell a computer what to do, why should not a computer tell another computer what to do? Consider the increasing connectivity of computers worldwide. It is intriguing to think about computers delegating subtasks of a complex overall task to their peers. In July 1996 there were close to 13 million hosts in the Internet. A certain percentage may often be down, slow, or busy, but this still leaves us with an impressive number of CPUs that could solve delegated GP subtasks instead of being idle.

Structure Evolution

Genetic programming requires substantial computing resources in order to perform the task of breeding generations of complex structures. Programs can be considered to be just special cases of such structures. It does not actually matter to GP whether crossover works with parse trees, strings, graphs, or, say, *bridge components.* The generation of general structures may become a major topic of GP in the future.

The idea of generating all kinds of structures once more illustrates the flexibility of GP. At least one strength of GP compared to other evolutionary paradigms might be mentioned here: its power to handle symbolic expressions. GP also has its weaknesses, like depending on explicit fitness measures. Thus, there is a great potential for combining genetic programming with other paradigms into hybrids appropriate for certain problem domains.

GP and Software

Many real-world situations may well be so complex that explicit fitness measures cannot be defined. For instance, what is the fitness function for evolving a program telling you when to sell and when to buy stocks and shares, for controlling traffic in a complex highway system, or for simulating tissue growth in an organism? Typically, we are in trouble when trying to define a meaningful fitness measure for problems that involve many interacting entities with different strategies in a dynamic environment. The trouble arises because, in such situations, the underlying fitness landscape becomes dynamic and cannot be described by a static fitness measure.

Such situations are very similar to what happens in organic evolution. Organisms change their environment, which then changes the organisms. Hence, it is interesting to combine genetic programming and artificial life: evolved programs should represent behavior implicitly evaluated by program–program and program–environment interactions. It is typical for artificial life environments that there is no explicit fitness measure. A certain entity continues its existence or vanishes due to its specific interaction with its environment.

Some research has already gone into combining genetic programming and artificial life, such as ERUNTICLAB[1] or PHILIA,[2] the latter a project implemented by students at Dortmund University. GP for co-evolving agents in an environment is becoming a more widespread technique [Luke and Spector, 1996] [Qureshi, 1996].

Hardware Evolution

We referred to work on making electronic hardware evolve "like an organism." May we, in the future, expect GP to grow circuits more complex than a binary adder or an asymmetric bandpass filter [Koza et al., 1996b]? For instance, imagine a circuit that rewires in order to replace the lost functionality of a damaged subsystem, like certain brain parts may take over tasks from damaged parts.

GP and Robotics

Work on combining genetic programming with robotics has also been presented. Usually, such approaches focus on evolving a control program for a mechanical device that does not develop or evolve, like a little mobile robot. May we, in the future, expect GP to *grow* and *evolve* hardware in general, not just electronic hardware? A blending of GP with nanotechnology [Drexler, 1992] could be interesting.

Meta-GP

Meta-GP is another area where we can expect dramatic progress in the coming years. Self-adaptation of parameters of GP runs and evolution of operators through a GP system are issues that have been mentioned in this book. With the acceleration of GP systems, these areas will become more and more accessible for researchers.

13.3 Conclusion

It is our strong belief that, over the next decade, GP will make progress on the challenges we have presented and on other issues mentioned. The question is: How far can we go? The evolution of GP might itself follow an improvement law similar to the ones we can observe in numerous evolutionary algorithm runs: very fast improvement over the first few years of its evolution, then a gradual decline in speed until phases of stagnation are only occasionally interrupted by incremental improvements.

GP must be able to scale up and generate solutions for problems demanding larger solution programs. There should be clearer guidelines for the application engineer, so as to make GP application less of an art and more routine. Off-the-shelf software packages would help in the emergence of commercial applications. GP must also prove its

[1] http://hamp.hampshire.edu/CCL/Projects/ErunticLab/

[2] http://ls11-www.informatik.uni-dortmund.de/gp/philia.html

worth in more real-life applications and generate money in large-scale industrial applications.

We are still in the first phase of rapid progress as more and more researchers enter the field of GP. It is impossible to predict where the field will be in ten years. All the indications are that, as the speed of software and hardware improves, the learning domains that may be addressed successfully with GP will grow. We see two main factors that will determine the rate and longevity of that growth.

GP Needs Speed

Returning one last time to the biological metaphor, greater speed means that GP populations may become much larger and a GP run may be able to conduct a much broader search of the space of possible programs. The importance of the *massive* parallelism seen in biological evolution may be the key to better GP algorithms. The remarkable results of the $Q\beta$ replicase experiments involve literally billions of RNA molecules, all evolving in parallel. The same is true for the elegant and promising SELEX algorithm (Chapter 2). Today, even very large GP runs involve, perhaps, no more than one million individuals because of the speed of present-day hardware and GP systems. Accordingly, increases in the speed of GP software and hardware will be one key factor determining GP's growth over the next few years.

GP Needs Efficiency

Speed is not the only factor. It is also important to return again to our discussion in Chapter 1 about GP's place among machine learning systems. The problem representation and the efficiency of the search algorithm are very important. Important research and innovation lies ahead in improving the GP search algorithm.

If increases in GP speed are also accompanied by increases in the efficiency of the GP search algorithm, it is possible that we may already have begun a historic move away from the "guild" era of writing computer programs toward an era where we can, as Friedberg said in 1958, tell a computer "precisely how to learn" – how to program itself and other computers. Even in the new era, however, it would remain *our* burden to specify tasks worthy of being learned.

A Printed and Recorded Resources

The following URLs are also available from the homepage of this book.

http://www.mkp.com/GP-Intro

The reader is advised to check with the URL to find the most up-to-date information.

A.1 Books on Genetic Programming

- Koza, J.R. (1992). *Genetic Programming: On Programming Computers by Means of Natural Selection.* MIT Press, Cambridge, MA.

  http://www-leland.stanford.edu/~phred/jaws1.html

- Kinnear, K.E. Jr. (ed.) (1994). *Advances in Genetic Programming.* MIT Press, Cambridge, MA.

  http://www-cs-faculty.stanford.edu/~koza/aigp.html

- Koza, J.R. (1994). *Genetic Programming II: Automatic Discovery of Reusable Programs.* MIT Press, Cambridge, MA.

  http://www-leland.stanford.edu/~phred/jaws2.html

- Angeline, P.J. and Kinnear, K.E. Jr. (eds.) (1996). *Advances in Genetic Programming 2.* MIT Press, Cambridge, MA.

  http://www-dept.cs.ucl.ac.uk/staff/w.langdon/aigp2.html

- Koza, J.R. and Goldberg, D. E. and Fogel, D. B. and Riolo, R. L. (eds.) (1996). *Genetic Programming 1996: Proceedings of the First Annual Conference.* Stanford University, Stanford, CA. MIT Press, Cambridge, MA.

- Koza, J.R. and Deb, K. and Dorigo, M. and Fogel, D. B. and Garzon, M. and Iba, H. and Riolo, R. L. (eds.) (1997). *Genetic Programming 1997: Proceedings of the Second Annual Conference.* Stanford University, Stanford, CA. Morgan Kaufmann, San Francisco, CA.

A.2 GP Video Tapes

- Koza, J.R. and Rice, J. P. (1992). *Genetic Programming: The Movie.* MIT Press, Cambridge, MA.
- Koza, J.R (1994). *Genetic Programming II Videotape: The Next Generation.* MIT Press, Cambridge, MA.
- (1996). *Genetic Programming 1996: Video Proceedings of the First Annual Conference.* Sound Photo Synthesis, CA.

A.3 Books on Evolutionary Algorithms

- Holland, J.H. (1975 and 1995). *Adaptation in Natural and Artificial Systems.* University of Michigan Press, Ann Arbor, MI.
- Schwefel, H.-P. (1981 and 1995). *Evolution and Optimum Seeking.* John Wiley & Sons, New York.
- Davis, L. (ed.) (1987). *Genetic Algorithms and Simulated Annealing.* Pitman, London.
- Goldberg, D. E. (1989). *Genetic Algorithms in Search, Optimization, and Machine Learning.* Addison-Wesley, Reading, MA.
- Davis, L. (1991). *Handbook of Genetic Algorithms.* Van Nostrand Reinhold, New York.
- Fogel, D.B. (1995). *Evolutionary Computation.* IEEE Press, New York.
- Michalewicz, Z. (1992). *Genetic Algorithms + Data Structures = Evolution Programs.* Springer-Verlag, Berlin. 1996: 3rd edition.
- Mitchell, M. (1996). *An Introduction to Genetic Algorithms.* MIT Press, Cambridge, MA.

Some dissertations have also been published in book form.

A.4 Selected Journals

- *Adaptive Behavior*, MIT Press
- *Artificial Intelligence*, Kluwer Academic
- *Artificial Life*, MIT Press
- *Biological Cybernetics*, Springer-Verlag
- *BioSystems*, Elsevier Science
- *Complexity*, Academic Press
- *Complex Systems*, Complex Systems Publications
- *Evolutionary Computation*, MIT Press
- *IEEE Transactions on Evolutionary Computation*, IEEE
- *IEEE Transactions on Systems, Man, and Cybernetics*, IEEE
- *Machine Learning*, Kluwer Academic

B Information Available on the Internet

The following URLs are also available from the homepage of this book.

http://www.mkp.com/GP-Intro

The reader is advised to check with the URL to find the most up-to-date information.

B.1 GP Tutorials

http://metricanet.com/people/jjf/gp/Tutorial/tutorial.html

http://research.germany.eu.net:8080/encore/www/Q1_5.htm

http://alphard.ethz.ch/gerber/approx/default.html

http://www.byte.com/art/9402/sec10/art1.htm

http://aif.wu-wien.ac.at/~geyers/archive/ga/gp/gp/node2.html

http://www.geneticprogramming.com

B.2 GP Frequently Asked Questions

http://www.salford.ac.uk/docs/depts/eee/gp2faq.html

http://www.salford.ac.uk/docs/depts/eee/gpfaq.html

B.3 GP Bibliographies

This always close-to-complete bibliography is being maintained by Bill Langdon:

ftp://ftp.cs.bham.ac.uk/pub/authors/W.B.Langdon/biblio/gp-bibliography.bib

http://liinwww.ira.uka.de/bibliography/Ai/genetic.programming.html

Jarmo Alander's GP Bibliography:

http://reimari.uwasa.fi/~jal/gaGPbib/gaGPlist.html

B.4 GP Researchers

Lists of researchers and their homepages can be found on

http://www-cs-faculty.stanford.edu/~koza/gpers.html

http://www.cs.ucl.ac.uk/research/genprog/

http://metricanet.com/people/jjf/gp/GPpages/misc.html

This resource offers various GP-oriented links and links to Koza's papers:

http://www-leland.standford.edu/~phred/john.html

B.5 General Evolutionary Computation

GGAA

http://www.aic.nrl.navy.mil/galist/

ENCORE

http://research.germany.eu.net:8080/encore/www/top.htm

Evolutionary Computation Page

http://rodin.cs.uh.edu/~twryu/genetic.html

B.6 Mailing Lists

There are two genetic programming e-mailing lists: one global list with more than 1000 researchers participating, and one local list for the San Francisco bay area. You may subscribe to the global list by sending a subscription request consisting of the message

subscribe genetic-programming

to

genetic programming-REQUEST@cs.stanford.edu

If you would like to unsubscribe, send a request consisting of the message

```
unsubscribe genetic-programming
```

to

```
genetic programming-REQUEST@cs.stanford.edu
```

The local list announces the periodic genetic programming lunches held at Stanford to people in the San Francisco bay area. It is occasionally also used to announce conference events and jobs in the bay area. You can subscribe to the local list by sending a subscription request consisting of the message

```
subscribe ba-gp
```

to

```
ba-gp-REQUEST@cs.stanford.edu
```

If you wish to unsubscribe, send a request consisting of the message

```
unsubscribe ba-gp
```

to

```
ba-gp-REQUEST@cs.stanford.edu
```

Related Mailing Lists and News Groups

- Genetic algorithm mailing list

  `ga-list-request@aic.nrl.navy.mil`

- Genetic algorithms and neural networks mailing list

  `gann-request@cs.iastate.edu`

- Genetic algorithms news group
  USENET news group:

  `comp.ai.genetic`

- Artificial life news group
  USENET news group:

  `alife.bbs.ga`

C GP Software

The following URLs are also available from the homepage of this book.

http://www.mkp.com/GP-Intro

The reader is advised to check with the URL to find the most up-to-date information.

C.1 Public Domain GP Systems

- GP in C++; author: Adam Fraser

  ftp://ftp.salford.ac.uk/pub/gp/

- lilGP; source: GARAGe

  http://garage.cps.msu.edu/software/lil-gp/index.html

- GP-QUICK; author: Andy Singleton
- GP-QUICK with data structures (Bill Langdon)

  ftp://ftp.io.com/pub/genetic-programming/GPdata-20-aug-95.tar.Z

- DGPC; author: David Andre

  http://www-leland.stanford.edu/~phred/gp.tar.gz

- Genetic Programming Kernel, C++ class library;
  author: Helmut Horner

  http://aif.wu-wien.ac.at/~geyers/archive/gpk/Dok/kurz/kurz.html

- SGPC–Simple Genetic Programming in C;
  authors: Walter Alden Tackett, Aviram Carmi
  available at the genetic programming FTP site
- YAGPLIC–Yet Another Genetic Programming Library In C
  contact Tobias Blickle: blickle@tik.ee.ethz.ch
- Common LISP implementation;

ftp://ftp.io.com/pub/genetic-programming/code/
koza-book-gp-implementation.lisp

This is a LISP implementation of genetic programming as described Koza's first book. There is also a file containing source from Koza's second book which includes ADFs.

- GPX/Abstractica
  Interactive evolution à la Karl Sims

  ftp://ftp.io.com/pub/genetic-programming/code/
  abs.tar.Z

- Symbolic regression using genetic programming in MATHEMATICA

  ftp://ftp.io.com/pub/genetic-programming/code/
  GPSRegress.m

- A framework for the genetic programming of neural networks

  ftp://ftp.io.com/pub/genetic-programming/code/
  cerebrum.tar.Z

C.2 Commercial GP Software

Discipulus™, genetic programming software for desktop PCs. This tool evolves machine code directly and is fast and efficient. Nice user interface. Free version is available at

http://www.aimlearning.com

C.3 Related Software Packages

DAVINCI tree drawing tool

http://www.informatik.uni-bremen.de/~inform/forschung/
daVinci/daVinci.html

C.4 C++ Implementation Issues

http://www.frc.ri.cmu.edu/~mcm/chapt.html

D Events

The following URLs are also available from the homepage of this book.

http://www.mkp.com/GP-Intro

The reader is advised to check with the URL to find the most up-to-date information.

D.1 GP Conferences

The annual genetic programming conference series started in 1996. Information about it can be found at

http://www.cs.brandeis.edu/~zippy/gp-96.html

The Genetic Programming 1997 conference is presented at

http://www-cs-faculty.stanford.edu/~koza/gp97.html

The Genetic Programming 1998 conference is presented at

http://www.genetic-programming.org/

D.2 Related Conferences and Workshops

ICGA International Conference on Genetic Algorithms (ICGA) conference series

- Grefenstette, John J. (ed.). Proceedings of the First International Conference on Genetic Algorithms and Their Applications. Hillsdale, NJ. Lawrence Erlbaum Associates. 1985.
- Grefenstette, John J.(ed.). Proceedings of the Second International Conference on Genetic Algorithms. Hillsdale, NJ. Lawrence Erlbaum Associates. 1987.
- Schaffer, J. David (ed.). Proceedings of the Third International Conference on Genetic Algorithms. San Mateo, CA. Morgan Kaufmann. 1989.
- Belew, Richard and Booker, Lashon (eds.). Proceedings of the Fourth International Conference on Genetic Algorithms. San Mateo, CA. Morgan Kaufmann. 1991.

- Forrest, Stephanie (ed.). Proceedings of the Fifth International Conference on Genetic Algorithms. San Mateo, CA. Morgan Kaufmann. 1993.

- Eshelman, Larry (ed.). Proceedings of the Sixth International Conference on Genetic Algorithms. San Francisco, CA. Morgan Kaufmann. 1995.

Parallel Problem Solving from Nature (PPSN) conference series

- Schwefel, Hans-Paul and Männer, Reinhard (eds.). Parallel Problem Solving from Nature I. Volume 496 of Lecture Notes in Computer Science. Berlin. Springer-Verlag. 1991.

- Männer, Reinhard and Manderick, Bernard (eds.). Parallel Problem Solving from Nature II. Amsterdam. North-Holland. 1992.

- Davidor, Yuval and Schwefel, Hans-Paul and Männer, Reinhard (eds.). Parallel Problem Solving from Nature III. Volume 866 of Lecture Notes in Computer Science. Berlin. Springer-Verlag. 1994.

- Ebeling, Werner and Rechenberg, Ingo and Schwefel, Hans-Paul and Voigt, Hans-Michael (eds.). Parallel Problem Solving from Nature IV. Volume 1141 of Lecture Notes in Computer Science. Berlin. Springer-Verlag. 1996.

Evolutionary Programming (EP) conference series

- Fogel, David B. and Atmar, Wirt (eds.). Proceedings of the First Annual Conference on Evolutionary Programming. San Diego, CA. Evolutionary Programming Society. 1992.

- Fogel, David B. and Atmar, Wirt (eds.). Proceedings of the Second Annual Conference on Evolutionary Programming. San Diego, CA. Evolutionary Programming Society. 1993.

- Sebald, Anthony V. and Fogel, Lawrence J. (eds.). Proceedings of the Third Annual Conference on Evolutionary Programming. River Edge, NJ. World Scientific. 1994.

- McDonnell, John R. and Reynolds, Robert G. and Fogel, David (eds.). Proceedings of the Fourth Annual Conference on Evolutionary Programming. Cambridge, MA. MIT Press. 1995.

- Fogel, Lawrence J. and Angeline, Peter J. and Bäck, Thomas (eds.). Proceedings of the Fifth Annual Conference on Evolutionary Programming. Cambridge, MA. MIT Press. 1996.

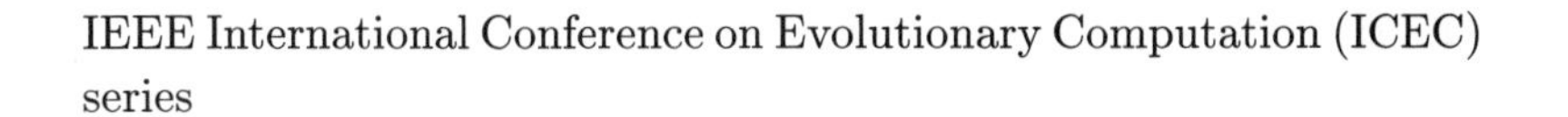

IEEE International Conference on Evolutionary Computation (ICEC) series

- Proceedings of the First IEEE Conference on Evolutionary Computation. New York. IEEE Press. 1994.
- Proceedings of the Second IEEE Conference on Evolutionary Computation. New York. IEEE Press. 1995.
- Proceedings of the Third IEEE Conference on Evolutionary Computation. New York. IEEE Press. 1996.

Foundations Of Genetic Algorithms (FOGA) series

- Rawlins, Gregory (ed.). Foundations of Genetic Algorithms. San Mateo, CA. Morgan Kaufmann. 1991.
- Whitley, Darrell (ed.). Proceedings of the Workshop on the Foundations of Genetic Algorithms and Classifier Systems. Vail, CO. Morgan Kaufmann. 1992.
- Whitley, Darrell and Vose, Michael (eds.). Proceedings of the Third Workshop on the Foundations of Genetic Algorithms. San Mateo, CA. Morgan Kaufmann. 1995.
- Belew, Richard and Vose, Michael (eds.). Proceedings of the Fourth Workshop on the Foundations of Genetic Algorithms. San Mateo, CA. Morgan Kaufmann. 1997.

Artificial Life (AL) series

- Brooks, Rodney and Maes, Pattie (eds.). Proceedings of the Fourth International Workshop on the Synthesis and Simulation of Living Systems. Cambridge, MA. MIT Press. 1994.

European Conference on Artificial Life (ECAL) series

- Varela, Francisco J. and Bourgine, Paul (eds.). Toward a Practice of Autonomous Systems: Proceedings of the First European Conference on Artificial Life. Cambridge, MA. MIT Press. 1992.
- Moran, Federico and Moreno, Alvaro and Merelo, Juan J. and Chacon, Pablo (eds.). Advances in Artificial Life. Berlin. Springer-Verlag. 1995.

Others

- Cliff, Dave and Husbands, Philip and Meyer, Jean-Arcady and Wilson, Stewart W. (eds.). Proceedings of the Third International Conference on the Simulation of Adaptive Behavior. Cambridge, MA. MIT Press. 1994.

- Altman, Russ and Brutlag, Douglas and Karp, Peter and Lathrop, Richard and Searls, David (eds.). Proceedings of the Second International Conference on Intelligent Systems for Molecular Biology. Menlo Park, CA. AAAI Press. 1994.

Bibliography

[Abramson and Hunter, 1996] Abramson, M. and Hunter, L. (1996). Classification using cultural co-evolution and genetic programming. In Koza, J. R., Goldberg, D. E., Fogel, D. B., and Riolo, R. L., editors, *Genetic Programming 1996: Proceedings of the First Annual Conference*, pages 249–254, Stanford University, CA. MIT Press, Cambridge, MA.

[Aho, 1986] Aho, A. (1986). *Compilers.* Addison-Wesley, London.

[Alander, 1995] Alander, J. T. (1995). An indexed bibliography of genetic programming. Report Series no 94-1-GP, Department of Information Technology and Industrial Management, University of Vaasa, Finland.

[Alba et al., 1996] Alba, E., Cotta, C., and Troyo, J. J. (1996). Type-constrained genetic programming for rule-base definition in fuzzy logic controllers. In Koza, J. R., Goldberg, D. E., Fogel, D. B., and Riolo, R. L., editors, *Genetic Programming 1996: Proceedings of the First Annual Conference*, pages 255–260, Stanford University, CA. MIT Press, Cambridge, MA.

[Altenberg, 1994a] Altenberg, L. (1994a). Emergent phenomena in genetic programming. In Sebald, A. V. and Fogel, L. J., editors, *Evolutionary Programming — Proceedings of the Third Annual Conference*, pages 233–241. World Scientific, Singapore.

[Altenberg, 1994b] Altenberg, L. (1994b). The evolution of evolvability in genetic programming. In Kinnear, Jr., K. E., editor, *Advances in Genetic Programming*, chapter 3, pages 47–74. MIT Press, Cambridge, MA.

[Altenberg, 1995] Altenberg, L. (1995). Genome growth and the evolution of the genotype-phenotype map. In Banzhaf, W. and Eeckman, F. H., editors, *Evolution as a Computational Process.* Springer-Verlag, Berlin, Germany.

[Andre, 1994a] Andre, D. (1994a). Automatically defined features: The simultaneous evolution of 2-dimensional feature detectors and an algorithm for using them. In Kinnear, Jr., K. E., editor, *Advances in Genetic Programming*, chapter 23, pages 477–494. MIT Press, Cambridge, MA.

[Andre, 1994b] Andre, D. (1994b). Learning and upgrading rules for an OCR system using genetic programming. In *Proceedings of the 1994 IEEE World Congress on Computational Intelligence*, Orlando, FL. IEEE Press, New York.

[Andre et al., 1996a] Andre, D., Bennett III, F. H., and Koza, J. R. (1996a). Discovery by genetic programming of a cellular automata rule that is better than any known rule for the majority classification problem. In Koza, J. R., Goldberg, D. E., Fogel, D. B., and Riolo, R. L., editors, *Genetic Programming 1996: Proceedings of the First Annual Conference*, pages 3–11, Stanford University, CA. MIT Press, Cambridge, MA.

[Andre et al., 1996b] Andre, D., Bennett III, F. H., and Koza, J. R. (1996b). Evolution of intricate long-distance communication signals in cellular automata using genetic programming. In *Artificial Life V: Proceedings of the Fifth International Workshop on the Synthesis and Simulation of Living Systems*, volume 1, Nara, Japan. MIT Press, Cambridge, MA.

[Andre and Koza, 1996a] Andre, D. and Koza, J. (1996a). A parallel implementation of genetic programming that achieves superlinear performance. In Arabnia, H. R., editor, *Proceedings of the International Conference on Parallel and Distributed Processing Techniques and Applications*, volume Volume III, pages 1163–1174, Athens, GA. CSREA.

[Andre and Koza, 1996b] Andre, D. and Koza, J. R. (1996b). Parallel genetic programming: A scalable implementation using the transputer network architecture. In Angeline, P. J. and Kinnear, Jr., K. E., editors, *Advances in Genetic Programming 2*, chapter 16, pages 317–338. MIT Press, Cambridge, MA.

[Andre and Teller, 1996] Andre, D. and Teller, A. (1996). A study in program response and the negative effects of introns in genetic programming. In Koza, J. R., Goldberg, D. E., Fogel, D. B., and Riolo, R. L., editors, *Genetic Programming 1996: Proceedings of the First Annual Conference*, pages 12–20, Stanford University, CA. MIT Press, Cambridge, MA.

[Andrews and Prager, 1994] Andrews, M. and Prager, R. (1994). Genetic programming for the acquisition of double auction market strategies. In Kinnear, Jr., K. E., editor, *Advances in Genetic*

Programming, chapter 16, pages 355–368. MIT Press, Cambridge, MA.

[Angeline, 1997] Angeline, P. (1997). Subtree crossover: Building block engine or macromutation. In Koza, J., Deb, K., Dorigo, M., Fogel, D., Garzon, M., Iba, H., and Riolo, R., editors, *Genetic Programming 1997: Proceedings of the Second Annual Conference, July 13–16, 1997*, pages 9–17, Stanford University, Stanford, CA. Morgan Kaufmann, San Francisco, CA.

[Angeline, 1993] Angeline, P. J. (1993). *Evolutionary Algorithms and Emergent Intelligence.* PhD thesis, Ohio State University.

[Angeline, 1994] Angeline, P. J. (1994). Genetic programming and emergent intelligence. In Kinnear, Jr., K. E., editor, *Advances in Genetic Programming*, chapter 4, pages 75–98. MIT Press, Cambridge, MA.

[Angeline, 1996] Angeline, P. J. (1996). Two self-adaptive crossover operators for genetic programming. In Angeline, P. J. and Kinnear, Jr., K. E., editors, *Advances in Genetic Programming 2*, chapter 5, pages 89–110. MIT Press, Cambridge, MA.

[Angeline and Pollack, 1992] Angeline, P. J. and Pollack, J. B. (1992). The evolutionary induction of subroutines. In *Proceedings of the Fourteenth Annual Conference of the Cognitive Science Society*, Bloomington, Indiana. Lawrence Erlbaum Associates, Hillsdale, NJ.

[Angeline and Pollack, 1993] Angeline, P. J. and Pollack, J. B. (1993). Competitive environments evolve better solutions for complex tasks. In Forrest, S., editor, *Proceedings of the 5th International Conference on Genetic Algorithms, ICGA-93*, pages 264–270, University of Illinois at Urbana-Champaign. Morgan Kaufmann, San Mateo, CA.

[Armstrong and Gecsei, 1979] Armstrong, W. W. and Gecsei, J. (1979). Adaptation algorithms for binary tree networks. *IEEE Transactions on Systems, Man and Cybernetics (SMC)*, 9:276–285.

[Armstrong and Green, 1985] Armstrong, W. W. and Green, M. W. (1985). The dynamics of articulated rigid bodies for purposes of animation. volume 1, pages 231–240. Springer-Verlag, Berlin.

[Atkin and Cohen, 1993] Atkin, M. and Cohen, P. R. (1993). Genetic programming to learn an agent's monitoring strategy. In *Proceedings of the AAAI-93 Workshop on Learning Action Models.* AAAI, Menlo Park, CA.

[Atkin and Cohen, 1994] Atkin, M. S. and Cohen, P. R. (1994). Learning monitoring strategies: A difficult genetic programming application. In *Proceedings of the 1994 IEEE World Congress on Computational Intelligence*, pages 328–332a, Orlando, FL. IEEE Press, New York.

[Atlan et al., 1994] Atlan, L., Bonnet, J., and Naillon, M. (1994). Learning distributed reactive strategies by genetic programming for the general job shop problem. In *Proceedings of the 7th annual Florida Artificial Intelligence Research Symposium*, Pensacola, FL. IEEE Press, New York.

[Babovic, 1995] Babovic, V. (1995). Genetic model induction based on experimental data. In *Proceedings of the XXVIth Congress of International Association for Hydraulics Research.*

[Balakrishnan and Honavar, 1996] Balakrishnan, K. and Honavar, V. (1996). On sensor evolution in robotics. In Koza, J. R., Goldberg, D. E., Fogel, D. B., and Riolo, R. L., editors, *Genetic Programming 1996: Proceedings of the First Annual Conference*, pages 455–460, Stanford University, CA. MIT Press, Cambridge, MA.

[Banzhaf, 1993a] Banzhaf, W. (1993a). Genetic programming for pedestrians. MERL Technical Report 93-03, Mitsubishi Electric Research Labs, Cambridge, MA.

[Banzhaf, 1993b] Banzhaf, W. (1993b). Genetic programming for pedestrians. In Forrest, S., editor, *Proceedings of the 5th International Conference on Genetic Algorithms, ICGA-93*, page 628, University of Illinois at Urbana-Champaign. Morgan Kaufmann, San Francisco, CA.

[Banzhaf, 1994] Banzhaf, W. (1994). Genotype-phenotype-mapping and neutral variation – A case study in genetic programming. In Davidor, Y., Schwefel, H.-P., and M R., editors, *Parallel Problem Solving from Nature III.*

[Banzhaf et al., 1996] Banzhaf, W., Francone, F., and Nordin, P. (1996). The effect of extensive use of the mutation operator on generalization in genetic programming using sparse data sets. In Voigt, H., Ebeling, W., Rechenberg, I., and Schwefel, H.-P., editors, *Parallel Problem Solving From Nature IV. Proceedings of the International Conference on Evolutionary Computation*, volume 1141 of *Lecture Notes in Computer Science*, pages 300–310, Berlin. Springer-Verlag, Berlin.

[Banzhaf and Friedrich, 1994] Banzhaf, W. and Friedrich, J. (1994). Efficient implementation of GP in C++ and assembler. (In German).

[Banzhaf et al., 1997a] Banzhaf, W., Nordin, P., and Olmer, M. (1997a). Generating adaptive behavior for a real robot using function regression within genetic programming. In Koza, J. R., Deb, K., Dorigo, M., Fogel, D. B., Garzon, M., Iba, H., and Riolo, R. L., editors, *Genetic Programming 1997: Proceedings of the Second Annual Conference*, pages 35–43, Stanford University, CA, USA. Morgan Kaufmann.

[Banzhaf et al., 1997b] Banzhaf, W., Nordin, P., and Olmer, M. (1997b). Generating adaptive behavior for a real robot using function regression within genetic programming. In Koza, J. R., Deb, K., Dorigo, M., Fogel, D. B., Garzon, M., Iba, H., and Riolo, R. L., editors, *Genetic Programming 1997: Proceedings of the Second Annual Conference*, pages 35–43, Stanford University, CA. Morgan Kaufmann, San Francisco, CA.

[Barnes, 1982] Barnes, J. G. (1982). *Programming in Ada.* Addison-Wesley, London.

[Bartel and Szostak, 1993] Bartel, D. and Szostak, J. (1993). Isolation of new ribozymes from a large pool of random sequences. *Science*, 261:1411–1418.

[Barto et al., 1983] Barto, A., Sutton, R., and Anderson, C. (1983). Neuronlike elements that can solve difficult learning control problems. *IEEE Transactions on Systems, Man, and Cybernetics*, 13:835–846.

[Beer and Gallagher, 1992] Beer, R. and Gallagher, J. (1992). Evolving dynamical neural netsworks for adaptive behavior. *Adaptive Behavior*, 1:92–122.

[Belady and Lehman, 1985] Belady, L. and Lehman, M. (1985). *Software Evolution.* Addison-Wesley, Reading, MA.

[Bengio et al., 1994] Bengio, S., Bengio, Y., and Cloutier, J. (1994). Use of genetic programming for the search of a new learning rule for neutral networks. In *Proceedings of the 1994 IEEE World Congress on Computational Intelligence*, pages 324–327, Orlando, FL. IEEE Press, New York.

[Bennett III, 1997] Bennett III, F. H. (1997). A multi-skilled robot that recognizes and responds to different problem environments. In Koza, J. R., Deb, K., Dorigo, M., Fogel, D. B., Garzon, M., Iba,

H., and Riolo, R. L., editors, *Genetic Programming 1997: Proceedings of the Second Annual Conference*, pages 44–51, Stanford University, CA. Morgan Kaufmann, San Francisco, CA.

[Bernstein et al., 1977] Bernstein, F., Koetzle, T., Williams, G., Meyer, J., Brice, M., J.R., R., Kennard, O., Shimaouchi, T., and Tasumi, M. (1977). The protein data bank: A computer based archival file for macromolecular structures. *J. Mol.Biol.*, 112:535–542.

[Bersano-Begey et al., 1996] Bersano-Begey, T. F., Daida, J. M., Vesecky, J. F., and Ludwig, F. L. (1996). A platform-independent collaborative interface for genetic programming applications: Image analysis for scientific inquiry. In Koza, J. R., editor, *Late Breaking Papers at the Genetic Programming 1996 Conference*, pages 1–8, Stanford University, CA. Stanford Bookstore, Stanford, CA.

[Bettenhausen et al., 1995a] Bettenhausen, K. D., Gehlen, S., Marenbach, P., and Tolle, H. (1995a). BioX++ – New results and conceptions concerning the intelligent control of biotechnological processes. In Munack, A. and Sch K., editors, *6th International Conference on Computer Applications in Biotechnology.*

[Bettenhausen et al., 1995b] Bettenhausen, K. D., Marenbach, P., Freyer, S., Rettenmaier, H., and Nieken, U. (1995b). Self-organizing structured modeling of a biotechnological fed-batch fermentation by means of genetic programming. In Zalzala, A. M. S., editor, *First International Conference on Genetic Algorithms in Engineering Systems: Innovations and Applications, GALESIA*, volume 414, pages 481–486, Sheffield, UK. IEE, London, UK.

[Beyer, 1995] Beyer, H. (1995). Toward a theory of evolution strategies: On the benefit of sex – the $(\mu/\mu, \lambda)$-strategy. *Evolutionary Computation*, 3:81–111.

[Biethahn and Nissen, 1995] Biethahn, J. and Nissen, V. E. (1995). *Evolutionary Algorithms in Management Applications.* Springer-Verlag, Berlin.

[Blickle, 1996] Blickle, T. (1996). Evolving compact solutions in genetic programming: A case study. In Voigt, H.-M., Ebeling, W., Rechenberg, I., and Schwefel, H.-P., editors, *Parallel Problem Solving From Nature IV. Proceedings of the International Conference on Evolutionary Computation*, volume 1141 of *Lecture Notes in Computer Science*, pages 564–573, Berlin. Springer-Verlag, Berlin, Germany.

[Blickle and Thiele, 1995] Blickle, T. and Thiele, L. (1995). A comparison of selection schemes used in genetic algorithms. TIK-Report 11, TIK Institut fur Technische Informatik und Kommunikationsnetze, Computer Engineering and Networks Laboratory, ETH, Swiss Federal Institute of Technology, Gloriastrasse 35, 8092 Zurich, Switzerland.

[Brave, 1994] Brave, S. (1994). Evolution of planning: Using recursive techniques in genetic planning. In Koza, J. R., editor, *Artificial Life at Stanford 1994*, pages 1–10. Stanford Bookstore, Stanford, CA.

[Brave, 1995] Brave, S. (1995). Using genetic programming to evolve recursive programs for tree search. In Louis, S., editor, *Fourth Golden West Conference on Intelligent Systems*, pages 60–65. International Society for Computers and their Applications, ISCA.

[Brave, 1996] Brave, S. (1996). Evolving deterministic finite automata using cellular encoding. In Koza, J. R., Goldberg, D. E., Fogel, D. B., and Riolo, R. L., editors, *Genetic Programming 1996: Proceedings of the First Annual Conference*, pages 39–44, Stanford University, CA. MIT Press, Cambridge, MA.

[Bremermann, 1962] Bremermann, H. (1962). Optimization through evolution and recombination. In Yovits, M., Jacobi, G., and Goldstein, G., editors, *Self-Organizing Systems*, pages 93–106. Spartan Books, New York.

[Bruce, 1996] Bruce, W. S. (1996). Automatic generation of object-oriented programs using genetic programming. In Koza, J. R., Goldberg, D. E., Fogel, D. B., and Riolo, R. L., editors, *Genetic Programming 1996: Proceedings of the First Annual Conference*, pages 267–272, Stanford University, CA. MIT Press, Cambridge, MA.

[Bulmer, 1980] Bulmer, M. G. (1980). *The Mathematical Theory of Quantitative Genetics.* Clarendon Press, Oxford, UK.

[Cacoullos, 1966] Cacoullos, T. (1966). Estimation of a multivariate density. *Annals of the Institute of Statistical Mathematics* (Tokyo), 18:179–189.

[Celko, 1995] Celko, J. (1995). *SQL for Smarties: Advanced SQL Programming.* Morgan Kaufmann, San Francisco, CA.

[Chen and Yeh, 1996] Chen, S.-H. and Yeh, C.-H. (1996). Genetic programming and the efficient market hypothesis. In Koza, J. R., Goldberg, D. E., Fogel, D. B., and Riolo, R. L., editors, *Genetic*

Programming 1996: Proceedings of the First Annual Conference, pages 45–53, Stanford University, CA. MIT Press, Cambridge, MA.

[Chen and Yeh, 1997] Chen, S.-H. and Yeh, C.-H. (1997). Using genetic programming to model volatility in financial time series. In Koza, J. R., Deb, K., Dorigo, M., Fogel, D. B., Garzon, M., Iba, H., and Riolo, R. L., editors, *Genetic Programming 1997: Proceedings of the Second Annual Conference*, pages 58–63, Stanford University, CA. Morgan Kaufmann, San Francisco, CA.

[Choi, 1996] Choi, A. (1996). Optimizing local area networks using genetic algorithms. In Koza, J. R., Goldberg, D. E., Fogel, D. B., and Riolo, R. L., editors, *Genetic Programming 1996: Proceedings of the First Annual Conference*, pages 467–472, Stanford University, CA. MIT Press, Cambridge, MA.

[Clack et al., 1996] Clack, C., Farringdon, J., Lidwell, P., and Yu, T. (1996). An adaptive document classification agent. Research Note RN/96/45, University College London, University College London, Dept. of Computer Science, London, UK. Submitted to BCS-ES96.

[Coon, 1994] Coon, B. W. (1994). Circuit synthesis through genetic programming. In Koza, J. R., editor, *Genetic Algorithms at Stanford 1994*, pages 11–20. Stanford Bookstore, Stanford, CA.

[Cramer, 1985] Cramer, N. L. (1985). A representation for the adaptive generation of simple sequential programs. In Grefenstette, J. J., editor, *Proceedings of an International Conference on Genetic Algorithms and the Applications*, pages 183–187, Carnegie-Mellon University, Pittsburgh, PA.

[Crepeau, 1995] Crepeau, R. L. (1995). Genetic evolution of machine language software. In Rosca, J. P., editor, *Proceedings of the Workshop on Genetic Programming: From Theory to Real-World Applications*, pages 121–134, Tahoe City, CA.

[Crosbie and Spafford, 1995] Crosbie, M. and Spafford, E. H. (1995). Applying genetic programming to intrusion detection. In Siegel, E. V. and Koza, J. R., editors, *Working Notes for the AAAI Symposium on Genetic Programming*, pages 1–8, MIT, Cambridge, MA. AAAI, Menlo Park, CA.

[Crow and Kimura, 1970] Crow, J. F. and Kimura, M. (1970). *An introduction to population genetics theory*. Burgess Publ. Co., Minneapolis, MN.

[Daida et al., 1996a] Daida, J. M., Bersano-Begey, T. F., Ross, S. J., and Vesecky, J. F. (1996a). Computer-assisted design of image classification algorithms: Dynamic and static fitness evaluations in a scaffolded genetic programming environment. In Koza, J. R., Goldberg, D. E., Fogel, D. B., and Riolo, R. L., editors, *Genetic Programming 1996: Proceedings of the First Annual Conference*, pages 279–284, Stanford University, CA. MIT Press, Cambridge, MA.

[Daida et al., 1996b] Daida, J. M., Hommes, J. D., Bersano-Begey, T. F., Ross, S. J., and Vesecky, J. F. (1996b). Algorithm discovery using the genetic programming paradigm: Extracting low-contrast curvilinear features from SAR images of arctic ice. In Angeline, P. J. and Kinnear, Jr., K. E., editors, *Advances in Genetic Programming 2*, chapter 21, pages 417–442. MIT Press, Cambridge, MA.

[Daida et al., 1995] Daida, J. M., Hommes, J. D., Ross, S. J., and Vesecky, J. F. (1995). Extracting curvilinear features from SAR images of arctic ice: Algorithm discovery using the genetic programming paradigm. In Stein, T., editor, *Proceedings of IEEE International Geoscience and Remote Sensing*, pages 673–675, Florence, Italy. IEEE Press, New York.

[Dain, 1997] Dain, R. A. (1997). Genetic programming for mobile robot wall-following algorithms. In Koza, J. R., Deb, K., Dorigo, M., Fogel, D. B., Garzon, M., Iba, H., and Riolo, R. L., editors, *Genetic Programming 1997: Proceedings of the Second Annual Conference*, pages 70–76, Stanford University, CA. Morgan Kaufmann, San Francisco, CA.

[Darwin, 1859] Darwin, C. (1859). *On the Origin of Species by Means of Natural Selection or the Preservation of Favoured Races in the Struggle for Life.* Murray, London, UK.

[Das et al., 1994] Das, S., Franguidakis, T., Papka, M., DeFanti, T. A., and Sandin, D. J. (1994). A genetic programming application in virtual reality. In *Proceedings of the first IEEE Conference on Evolutionary Computation*, volume 1, pages 480–484, Orlando, FL. IEEE Press, New York. Part of 1994 IEEE World Congress on Computational Intelligence, Orlando, FL.

[Dasrathy, 1991] Dasrathy (1991). *Nearest Neighbor (NN) Norms: NN Pattern Classification Techniques.* IEEE Computer Society Press, Los Alamitos, CA.

[Dawkins, 1987] Dawkins, R. (1987). *The Blind Watchmaker.* Penguin Books, London, UK.

[Dawkins, 1989] Dawkins, R. (1989). *The Selfish Gene.* Oxford University Press, Oxford, UK.

[DeGaris, 1993] DeGaris, H. (1993). Evolvable hardware: Genetic programming of a Darwin machine. In Albrecht, R. F., Reeves, C. R., and Steele, N. C., editors, *Proc. of Artificial Neural Nets and Genetic Algorithms, Innsbruck, Austria*, pages 443–449. Springer-Verlag, Berlin.

[D'haeseleer, 1994] D'haeseleer, P. (1994). Context preserving crossover in genetic programming. In *Proceedings of the 1994 IEEE World Congress on Computational Intelligence*, volume 1, pages 256–261, Orlando, FL. IEEE Press, New York.

[D'haeseleer and Bluming, 1994] D'haeseleer, P. and Bluming, J. (1994). Effects of locality in individual and population evolution. In Kinnear, Jr., K. E., editor, *Advances in Genetic Programming*, chapter 8, pages 177–198. MIT Press, Cambridge, MA.

[Dickinson, 1994] Dickinson, A. (1994). Evolution of damage-immune programs using genetic programming. In Koza, J. R., editor, *Genetic Algorithms at Stanford 1994*, pages 21–30. Stanford Bookstore, Stanford, CA.

[Dickmanns et al., 1987] Dickmanns, D., Schmidhuber, J., and Winklhofer, A. (1987). Der genetische Algorithmus: Eine Implementierung in Prolog. Project, Technical University Munich, Munich.

[Dracopoulos, 1997] Dracopoulos, D. C. (1997). Evolutionary control of a satellite. In Koza, J. R., Deb, K., Dorigo, M., Fogel, D. B., Garzon, M., Iba, H., and Riolo, R. L., editors, *Genetic Programming 1997: Proceedings of the Second Annual Conference*, pages 77–81, Stanford University, CA. Morgan Kaufmann, San Francisco, CA.

[Drechsler et al., 1996] Drechsler, R., Becker, B., and Gockel, N. (1996). A genetic algorithm for the construction of small and highly testable OKFDD circuits. In Koza, J. R., Goldberg, D. E., Fogel, D. B., and Riolo, R. L., editors, *Genetic Programming 1996: Proceedings of the First Annual Conference*, pages 473–478, Stanford University, CA. MIT Press, Cambridge, MA.

[Drexler, 1992] Drexler, K. E. (1992). *Nanosystems: Molecular Machinery, Manufacturing, and Computation.* Wiley Interscience, New York.

[Dunay et al., 1994] Dunay, B. D., Petry, F. E., and Buckles, W. P. (1994). Regular language induction with genetic programming. In *Proceedings of the 1994 IEEE World Congress on Computational Intelligence*, pages 396–400, Orlando, FL. IEEE Press, New York.

[Dunning and Davis, 1996] Dunning, T. E. and Davis, M. W. (1996). Evolutionary algorithms for natural language processing. In Koza, J. R., editor, *Late Breaking Papers at the Genetic Programming 1996 Conference*, pages 16–23, Stanford University, CA. Stanford Bookstore, Stanford, CA.

[Duprè, 1987] Duprè, J. (1987). *The latest on the best. Essays on Evolution an Optimality.* MIT Press, Cambridge, MA.

[Dworman et al., 1996] Dworman, G., Kimbrough, S. O., and Laing, J. D. (1996). Bargaining by artificial agents in two coalition games: A study in genetic programming for electronic commerce. In Koza, J. R., Goldberg, D. E., Fogel, D. B., and Riolo, R. L., editors, *Genetic Programming 1996: Proceedings of the First Annual Conference*, pages 54–62, Stanford University, CA. MIT Press, Cambridge, MA.

[Dzeroski et al., 1995] Dzeroski, S., Todorovski, L., and Petrovski, I. (1995). Dynamical system identification with machine learning. In Rosca, J. P., editor, *Proceedings of the Workshop on Genetic Programming: From Theory to Real-World Applications*, pages 50–63, Tahoe City, CA.

[Ehlers, 1992] Ehlers, J. (1992). Seminar on cycles. Compuserve Information Service, INVFORUM, Library.

[Ehrenburg, 1996] Ehrenburg, H. (1996). Improved direct acyclic graph handling and the combine operator in genetic programming. In Koza, J. R., Goldberg, D. E., Fogel, D. B., and Riolo, R. L., editors, *Genetic Programming 1996: Proceedings of the First Annual Conference*, pages 285–291, Stanford University, CA. MIT Press, Cambridge, MA.

[Ehrenburg and van Maanen, 1994] Ehrenburg, H. H. and van Maanen, H. A. N. (1994). A finite automaton learning system using genetic programming. NeuroColt Tech Rep CS-R9458, Dept. of Computer Science, CWI, Centrum voor Wiskunde en Informatica, CWI, P.O. Box 94079, 1090 GB Amsterdam, The Netherlands.

[Eigen, 1992] Eigen, M. (1992). *Steps Toward Life.* Oxford University Press, Oxford, UK.

[Esparcia-Alcazar and Sharman, 1996] Esparcia-Alcazar, A. I. and Sharman, K. (1996). Some applications of genetic programming in digital signal processing. In Koza, J. R., editor, *Late Breaking Papers at the Genetic Programming 1996 Conference*, pages 24–31, Stanford University, CA. Stanford Bookstore, Stanford, CA.

[Esparcia-Alcazar and Sharman, 1997] Esparcia-Alcazar, A. I. and Sharman, K. (1997). Evolving recurrent neural network architectures by genetic programming. In Koza, J. R., Deb, K., Dorigo, M., Fogel, D. B., Garzon, M., Iba, H., and Riolo, R. L., editors, *Genetic Programming 1997: Proceedings of the Second Annual Conference*, pages 89–94, Stanford University, CA. Morgan Kaufmann, San Francisco, CA.

[Faglia and Vetturi, 1996] Faglia, R. and Vetturi, D. (1996). Motion planning and design of CAM mechanisms by means of a genetic algorithm. In Koza, J. R., Goldberg, D. E., Fogel, D. B., and Riolo, R. L., editors, *Genetic Programming 1996: Proceedings of the First Annual Conference*, pages 479–484, Stanford University, CA. MIT Press, Cambridge, MA.

[Feller, 1968] Feller, W. (1968). *An Introduction to Probability Theory and Its Applications, Volume 1.* John Wiley & Sons, New York.

[Feller, 1971] Feller, W. (1971). *An Introduction to Probability Theory and Its Applications, Volume 2.* John Wiley & Sons, New York.

[Fernandez et al., 1996] Fernandez, J. J., Farry, K. A., and Cheatham, J. B. (1996). Waveform recognition using genetic programming: The myoelectric signal recognition problem. In Koza, J. R., Goldberg, D. E., Fogel, D. B., and Riolo, R. L., editors, *Genetic Programming 1996: Proceedings of the First Annual Conference*, pages 63–71, Stanford University, CA. MIT Press, Cambridge, MA.

[Ferrer and Martin, 1995] Ferrer, G. J. and Martin, W. N. (1995). Using genetic programming to evolve board evaluation functions for a boardgame. In *1995 IEEE Conference on Evolutionary Computation*, volume 2, page 747, Perth, Australia. IEEE Press, New York.

[Fitzpatrick et al., 1984] Fitzpatrick, J. M., Grefenstette, J., and Van Gucht, D. (1984). Image registration by genetic search. In *Proceedings of IEEE Southeast Conference*, Piscataway, NY. IEEE Press, New York.

[Fogel, 1995] Fogel, D. B. (1995). *Evolutionary Computation.* IEEE Press, New York.

[Fogel et al., 1965] Fogel, L., Owens, A., and Walsh, M. (1965). Artificial intelligence through a simulation of evolution. In Maxfield,

M., Callahan, A., and Fogel, L., editors, *Biophysics and Cybernetic Systems*, pages 131–155.

[Fogel et al., 1966] Fogel, L., Owens, A., and Walsh, M. (1966). *Artificial Intelligence through Simulated Evolution.* John Wiley & Sons, New York.

[Francone et al., 1996] Francone, F. D., Nordin, P., and Banzhaf, W. (1996). Benchmarking the generalization capabilities of a compiling genetic programming system using sparse data sets. In Koza, J. R., Goldberg, D. E., Fogel, D. B., and Riolo, R. L., editors, *Genetic Programming 1996: Proceedings of the First Annual Conference*, pages 72–80, Stanford University, CA. MIT Press, Cambridge, MA.

[Fraser and Rush, 1994] Fraser, A. P. and Rush, J. R. (1994). Putting INK into a BIRo: A discussion of problem domain knowledge for evolutionary robotics. In Fogarty, T. C., editor, *Evolutionary Computing*, volume 865 of *Lecture Notes in Computer Science.* Springer-Verlag, Berlin, Germany.

[Freitas, 1997] Freitas, A. A. (1997). A genetic programming framework for two data mining tasks: Classification and generalized rule induction. In Koza, J. R., Deb, K., Dorigo, M., Fogel, D. B., Garzon, M., Iba, H., and Riolo, R. L., editors, *Genetic Programming 1997: Proceedings of the Second Annual Conference*, pages 96–101, Stanford University, CA. Morgan Kaufmann, San Francisco, CA.

[Friedberg, 1958] Friedberg, R. (1958). A learning machine, part I. *IBM J. Research and Development*, 2:2–13.

[Friedberg et al., 1959] Friedberg, R., Dunham, B., and North, J. (1959). A learning machine, part II. *IBM J. of Research and Development*, 3:282–287.

[Friedrich and Banzhaf, 1997] Friedrich, J. and Banzhaf, W. (1997). A simple genetic compiler system for efficient genetic programming on CISC-architectures. Unpublished.

[Fujiki and Dickinson, 1987] Fujiki, C. and Dickinson, J. (1987). Using the genetic algorithm to generate LISP source code to solve the prisoner's dilemma. In Grefenstette, J. J., editor, *Genetic Algorithms and their Applications: Proceedings of the second international conference on Genetic Algorithms*, pages 236–240, MIT, Cambridge, MA. Lawrence Erlbaum Associates, Hillsdale, NJ.

[Garces-Perez et al., 1996] Garces-Perez, J., Schoenefeld, D. A., and Wainwright, R. L. (1996). Solving facility layout problems using genetic programming. In Koza, J. R., Goldberg, D. E., Fogel,

D. B., and Riolo, R. L., editors, *Genetic Programming 1996: Proceedings of the First Annual Conference*, pages 182–190, Stanford University, CA. MIT Press, Cambridge, MA.

[Gathercole and Ross, 1994] Gathercole, C. and Ross, P. (1994). Dynamic training subset selection for supervised learning in genetic programming. In Davidor, Y., Schwefel, H.-P., and M R., editors, *Parallel Problem Solving from Nature III*, volume 866 of *Lecture Notes in Computer Science*.

[Gathercole and Ross, 1997a] Gathercole, C. and Ross, P. (1997a). Small populations over many generations can beat large populations over few generations in genetic programming. In Koza, J. R., Deb, K., Dorigo, M., Fogel, D. B., Garzon, M., Iba, H., and Riolo, R. L., editors, *Genetic Programming 1997: Proceedings of the Second Annual Conference*, pages 111–118, Stanford University, CA. Morgan Kaufmann, San Francisco, CA.

[Gathercole and Ross, 1997b] Gathercole, C. and Ross, P. (1997b). Tackling the boolean even N parity problem with genetic programming and limited-error fitness. In Koza, J. R., Deb, K., Dorigo, M., Fogel, D. B., Garzon, M., Iba, H., and Riolo, R. L., editors, *Genetic Programming 1997: Proceedings of the Second Annual Conference*, pages 119–127, Stanford University, CA. Morgan Kaufmann, San Francisco, CA.

[Ghanea-Hercock and Fraser, 1994] Ghanea-Hercock, R. and Fraser, A. P. (1994). Evolution of autonomous robot control architectures. In Fogarty, T. C., editor, *Evolutionary Computing*, volume 865 of *Lecture Notes in Computer Science*. Springer-Verlag, Berlin, Germany.

[Goldberg, 1989] Goldberg, D. (1989). *Genetic Algorithms in Search, Optimization and Machine Learning*. Addison-Wesley, Reading, MA.

[Gordon, 1994] Gordon, B. M. (1994). Exploring the underlying structure of natural images through genetic programming. In Koza, J. R., editor, *Genetic Algorithms at Stanford 1994*, pages 49–56. Stanford Bookstore, Stanford, CA.

[Gray et al., 1996a] Gray, G. J., Murray-Smith, D. J., Li, Y., and Sharman, K. C. (1996a). Nonlinear model structure identification using genetic programming. In Koza, J. R., editor, *Late Breaking Papers at the Genetic Programming 1996 Conference*, pages 32–37, Stanford University, CA. Stanford Bookstore, Stanford, CA.

[Gray and Maxwell, 1997] Gray, H. F. and Maxwell, R. J. (1997). Genetic programming for multi-class classification of magnetic resonance spectroscopy data. In Koza, J. R., Deb, K., Dorigo, M., Fogel, D. B., Garzon, M., Iba, H., and Riolo, R. L., editors, *Genetic Programming 1997: Proceedings of the Second Annual Conference*, page 137, Stanford University, CA. Morgan Kaufmann, San Francisco, CA.

[Gray et al., 1996b] Gray, H. F., Maxwell, R. J., Martinez-Perez, I., Arus, C., and Cerdan, S. (1996b). Genetic programming for classification of brain tumours from nuclear magnetic resonance biopsy spectra. In Koza, J. R., Goldberg, D. E., Fogel, D. B., and Riolo, R. L., editors, *Genetic Programming 1996: Proceedings of the First Annual Conference*, page 424, Stanford University, CA. MIT Press, Cambridge, MA.

[Greeff and Aldrich, 1997] Greeff, D. J. and Aldrich, C. (1997). Evolution of empirical models for metallurgical process systems. In Koza, J. R., Deb, K., Dorigo, M., Fogel, D. B., Garzon, M., Iba, H., and Riolo, R. L., editors, *Genetic Programming 1997: Proceedings of the Second Annual Conference*, page 138, Stanford University, CA. Morgan Kaufmann, San Francisco, CA.

[Grefenstette and Baker, 1989] Grefenstette, J. J. and Baker, J. E. (1989). How genetic algorithms work: A critical look at implicit parallelism. In *Proc. 3rd International Conference on Genetic Algorithms*, pages 20–27, San Mateo, CA. Morgan Kaufmann, San Francisco, CA.

[Grimes, 1995] Grimes, C. A. (1995). Application of genetic techniques to the planning of railway track maintenance work. In Zalzala, A. M. S., editor, *First International Conference on Genetic Algorithms in Engineering Systems: Innovations and Applications, GALESIA*, volume 414, pages 467–472.

[Gritz and Hahn, 1995] Gritz, L. and Hahn, J. (1995). Genetic programming for articulated figure motion. *J. of Visualization and Computer Animation*, 6:129–142.

[Gritz and Hahn, 1997] Gritz, L. and Hahn, J. K. (1997). Genetic programming evolution of controllers for 3-D character animation. In Koza, J. R., Deb, K., Dorigo, M., Fogel, D. B., Garzon, M., Iba, H., and Riolo, R. L., editors, *Genetic Programming 1997: Proceedings of the Second Annual Conference*, pages 139–146, Stanford University, CA. Morgan Kaufmann, San Francisco, CA.

[Gruau, 1992a] Gruau, F. (1992a). Cellular encoding of genetic neural networks. Technical report 92-21, Laboratoire de l'Informatique du Parallelisme. Ecole Normale Superieure de Lyon, France.

[Gruau, 1992b] Gruau, F. (1992b). Genetic synthesis of boolean neural networks with a cell rewriting developmental process. In Schaffer, J. D. and Whitley, D., editors, *Proceedings of the Workshop on Combinations of Genetic Algorithms and Neural Networks (COGANN92)*, pages 55–74. IEEE Computer Society Press.

[Gruau, 1993] Gruau, F. (1993). Genetic synthesis of modular neural networks. In Forrest, S., editor, *Proceedings of the 5th International Conference on Genetic Algorithms, ICGA-93*, pages 318–325, University of Illinois at Urbana-Champaign. Morgan Kaufmann, San Francisco, CA.

[Gruau, 1994a] Gruau, F. (1994a). Genetic micro programming of neural networks. In Kinnear, Jr., K. E., editor, *Advances in Genetic Programming*, chapter 24, pages 495–518. MIT Press, Cambridge, MA.

[Gruau, 1994b] Gruau, F. (1994b). *Neural Network Synthesis using Cellular Encoding and the Genetic Algorithm.* PhD thesis, Laboratoire de l'Informatique du Parallelisme, Ecole Normale Superieure de Lyon, France.

[Gruau, 1995] Gruau, F. (1995). Automatic definition of modular neural networks. *Adaptive Behaviour*, 3(2):151–183.

[Gruau et al., 1994] Gruau, F., Ratajszczak, J., and Wiber, G. (1994). A neural compiler. *Journal of Theoretical Computer Science*, 1834:1–52.

[Gruau and Whitley, 1993] Gruau, F. and Whitley, D. (1993). Adding learning to the cellular development process: a comparative study. *Evolutionary Computation*, 1(3):213–233.

[Gustafson et al., 1986] Gustafson, D., Barrett, W., Bates, R., and Couch, J. D. (1986). *Compiler Construction: Theory and Practice.* Science Research Assoc.

[Hahn, 1988] Hahn, J. (1988). Realistic animation of rigid bodies. *Computer Graphics*, 22:299–308.

[Haigh, 1978] Haigh, J. (1978). The accumulation of deleterious genes in a population — Muller's rachet. *Theoretical Population Biology*, 14:251–267.

[Hampo et al., 1994] Hampo, R. J., Bryant, B. D., and Marko, K. A. (1994). IC engine misfire detection algorithm generation using genetic programming. In *EUFIT'94*, pages 1674–1678, Aachen, Germany. ELITE Foundation.

[Hampo and Marko, 1992] Hampo, R. J. and Marko, K. A. (1992). Application of genetic programming to control of vehicle systems. In *Proceedings of the Intelligent Vehicles '92 Symposium*. June 29–July 1, 1992, Detroit, MI.

[Handley, 1993a] Handley, S. (1993a). Automatic learning of a detector for alpha-helices in protein sequences via genetic programming. In Forrest, S., editor, *Proceedings of the 5th International Conference on Genetic Algorithms, ICGA-93*, pages 271–278, University of Illinois at Urbana-Champaign. Morgan Kaufmann, San Francisco, CA.

[Handley, 1993b] Handley, S. (1993b). The genetic planner: The automatic generation of plans for a mobile robot via genetic programming. In *Proceedings of the Eighth IEEE International Symposium on Intelligent Control*, Chicago, IL.

[Handley, 1994a] Handley, S. (1994a). Automated learning of a detector for the cores of a-helices in protein sequences via genetic programming. In *Proceedings of the 1994 IEEE World Congress on Computational Intelligence*, volume 1, pages 474–479, Orlando, FL. IEEE Press, New York.

[Handley, 1994b] Handley, S. (1994b). The automatic generations of plans for a mobile robot via genetic programming with automatically defined functions. In Kinnear, Jr., K. E., editor, *Advances in Genetic Programming*, chapter 18, pages 391–407. MIT Press, Cambridge, MA.

[Handley, 1995] Handley, S. (1995). Predicting whether or not a 60-base DNA sequence contains a centrally-located splice site using genetic programming. In Rosca, J. P., editor, *Proceedings of the Workshop on Genetic Programming: From Theory to Real-World Applications*, pages 98–103, Tahoe City, CA.

[Handley, 1996a] Handley, S. (1996a). A new class of function sets for solving sequence problems. In Koza, J. R., Goldberg, D. E., Fogel, D. B., and Riolo, R. L., editors, *Genetic Programming 1996: Proceedings of the First Annual Conference*, pages 301–308, Stanford University, CA. MIT Press, Cambridge, MA.

[Handley, 1996b] Handley, S. (1996b). The prediction of the degree of exposure to solvent of amino acid residues via genetic programming. In Koza, J. R., Goldberg, D. E., Fogel, D. B., and Riolo, R. L., editors, *Genetic Programming 1996: Proceedings of the First*

Annual Conference, pages 297–300, Stanford University, CA. MIT Press, Cambridge, MA.

[Harris and Buxton, 1996] Harris, C. and Buxton, B. (1996). Evolving edge detectors. Research Note RN/96/3, UCL, University College London, Dept. of Computer Science, London, UK.

[Haynes et al., 1996] Haynes, T., Gamble, R., Knight, L., and Wainwright, R. (1996). Entailment for specification refinement. In Koza, J. R., Goldberg, D. E., Fogel, D. B., and Riolo, R. L., editors, *Genetic Programming 1996: Proceedings of the First Annual Conference*, pages 90–97, Stanford University, CA. MIT Press, Cambridge, MA.

[Haynes et al., 1995] Haynes, T. D., Wainwright, R. L., and Sen, S. (1995). Evolving cooperating strategies. In Lesser, V., editor, *Proceedings of the first International Conference on Multiple Agent Systems*, page 450, San Francisco, CA. AAAI, Menlo Park, CA/MIT Press, Cambridge, MA. Poster.

[Hecht-Nielsen, 1988] Hecht-Nielsen, R. (1988). *Neurocomputing.* Addison-Wesley, Reading, MA.

[Hemmi et al., 1994a] Hemmi, H., Mizoguchi, J., and Shimohara, K. (1994a). Development and evolution of hardware behaviours. In Brooks, R. and Maes, P., editors, *Artificial Life IV*, pages 371–376. MIT Press.

[Hemmi et al., 1994b] Hemmi, H., Mizoguchi, J., and Shimohara, K. (1994b). Hardware evolution — an HDL approach. In *Proc. of the Japan–US Symposium on Flexible Automation.* The Institute of Systems, Control and Information Engineers.

[Hicklin, 1986] Hicklin, J. F. (1986). Application of the genetic algorithm to automatic program generation. Master's thesis, Dept. of Computer Science, University of Idaho.

[Hillis, 1989] Hillis, D. (1989). Co-evolving parasites. In Forrest, S., editor, *Emergent Computation.* MIT Press, Cambridge, MA.

[Hillis, 1992] Hillis, W. D. (1992). Co-evolving parasites improve simulated evolution as an optimization procedure. In Langton, C. G., Taylor, C., Farmer, J. D., and Rasmussen, S., editors, *Artificial Life II*, volume X of *Sante Fe Institute Studies in the Sciences of Complexity*, pages 313–324. Addison-Wesley, Reading, MA.

[Hinchliffe et al., 1996] Hinchliffe, M., Hiden, H., McKay, B., Willis, M., Tham, M., and Barton, G. (1996). Modelling chemical process systems using a multi-gene genetic programming algorithm.

In Koza, J. R., editor, *Late Breaking Papers at the Genetic Programming 1996 Conference*, pages 56–65, Stanford University, CA. Stanford Bookstore, Stanford, CA.

[Hirst, 1996] Hirst, A. (1996). Notes on the evolution of adaptive hardware. In *Proc. 2nd Int. Conf. on Adaptive Computing in Engineering Design and Control (ACEDC-96)*, Plymouth, UK, http://kmi.open.ac.uk/ monty/evoladaphwpaper.html.

[Holbrook et al., 1990] Holbrook, S., Muskal, S., and Kim, S. (1990). Predicting surface exposure of amino acids from proteins sequence. *Protein Engineering*, 3:659–665.

[Holbrook et al., 1993] Holbrook, S., Muskal, S., and Kim, S. (1993). Predicting protein structural features with artificial neural networks. In Hunter, L., editor, *Artificial Intelligence and Molecular Biology*, pages 161–194. AAAI, Menlo Park, CA.

[Holland, 1975] Holland, J. (1975). *Adaptation in natural and artificial systems.* MIT Press, Cambridge, MA.

[Holland, 1992] Holland, J. (1992). *Adaptation in natural and artificial systems.* MIT Press, Cambridge, MA.

[Holland et al., 1986] Holland, J., Holyak, R., and Thagard, P. (1986). *Induction: Processes of Inference, Learning and Discovery.* MIT Press, Cambridge, MA.

[Holland and Reitman, 1978] Holland, J. and Reitman, J. (1978). Cognitive systems based on adaptive algorithms. In Waterman, D. and Hayes-Roth, F., editors, *Pattern-Directed Inference Systems.* Academic Press, New York.

[Howley, 1996] Howley, B. (1996). Genetic programming of near-minimum-time spacecraft attitude maneuvers. In Koza, J. R., Goldberg, D. E., Fogel, D. B., and Riolo, R. L., editors, *Genetic Programming 1996: Proceedings of the First Annual Conference*, pages 98–106, Stanford University, CA. MIT Press, Cambridge, MA.

[Howley, 1997] Howley, B. (1997). Genetic programming and parametric sensitivity: a case study in dynamic control of a two link manipulator. In Koza, J. R., Deb, K., Dorigo, M., Fogel, D. B., Garzon, M., Iba, H., and Riolo, R. L., editors, *Genetic Programming 1997: Proceedings of the Second Annual Conference*, pages 180–185, Stanford University, CA. Morgan Kaufmann, San Francisco, CA.

[Huelsbergen, 1996] Huelsbergen, L. (1996). Toward simulated evolution of machine language iteration. In Koza, J. R., Goldberg,

D. E., Fogel, D. B., and Riolo, R. L., editors, *Genetic Programming 1996: Proceedings of the First Annual Conference*, pages 315–320, Stanford University, CA. MIT Press, Cambridge, MA.

[Iba, 1997] Iba, H. (1997). Multiple-agent learning for a robot navigation task by genetic programming. In Koza, J. R., Deb, K., Dorigo, M., Fogel, D. B., Garzon, M., Iba, H., and Riolo, R. L., editors, *Genetic Programming 1997: Proceedings of the Second Annual Conference*, pages 195–200, Stanford University, CA. Morgan Kaufmann, San Francisco, CA.

[Iba and de Garis, 1996] Iba, H. and de Garis, H. (1996). Extending genetic programming with recombinative guidance. In Angeline, P. J. and Kinnear, Jr., K. E., editors, *Advances in Genetic Programming 2*, chapter 4, pages 69–88. MIT Press, Cambridge, MA.

[Iba et al., 1994] Iba, H., de Garis, H., and Sato, T. (1994). Genetic programming with local hill-climbing. In Davidor, Y., Schwefel, H.-P., and M R., editors, *Parallel Problem Solving from Nature III*, volume 866 of *Lecture Notes in Computer Science.*

[Iba et al., 1995a] Iba, H., de Garis, H., and Sato, T. (1995a). Temporal data processing using genetic programming. In Eshelman, L., editor, *Genetic Algorithms: Proceedings of the Sixth International Conference (ICGA95)*, pages 279–286, Pittsburgh, PA. Morgan Kaufmann, San Francisco, CA.

[Iba et al., 1993] Iba, H., Niwa, H., de Garis, H., and Sato, T. (1993). Evolutionary learning of boolean concepts: An empirical study. Technical Report ETL-TR-93-25, Electrotechnical Laboratory, Tsukuba, Japan.

[Iba et al., 1995b] Iba, H., Sato, T., and de Garis, H. (1995b). Numerical genetic programming for system identification. In Rosca, J. P., editor, *Proceedings of the Workshop on Genetic Programming: From Theory to Real-World Applications*, pages 64–75, Tahoe City, CA.

[Ivakhnenko, 1971] Ivakhnenko, A. (1971). Polymonial theory of complex systems. *IEEE Transactions on Systems, Man and Cybernetics*, 1:364–378.

[Jacob, 1994] Jacob, C. (1994). Genetic L-system programming. In Davidor, Y., Schwefel, H.-P., and M R., editors, *Parallel Problem Solving from Nature III*, volume 866 of *Lecture Notes in Computer Science.*

[Jacob, 1996a] Jacob, C. (1996a). Evolving evolution programs: Genetic programming and L-systems. In Voigt, H., Ebeling, W.,

Rechenberg, I., and Schwefel, H., editors, *Parallel Problem Solving From Nature IV. Proceedings of the International Conference on Evolutionary Computation*, pages 42–51, Berlin. Springer-Verlag.

[Jacob, 1996b] Jacob, C. (1996b). Evolving evolution programs: Genetic programming and L-systems. In Koza, J. R., Goldberg, D. E., Fogel, D. B., and Riolo, R. L., editors, *Genetic Programming 1996: Proceedings of the First Annual Conference*, pages 107–115, Stanford University, CA. MIT Press, Cambridge, MA.

[Jannink, 1994] Jannink, J. (1994). Cracking and co-evolving randomizers. In Kinnear, Jr., K. E., editor, *Advances in Genetic Programming*, chapter 20, pages 425–443. MIT Press, Cambridge, MA.

[Jiang and Butler, 1996] Jiang, J. and Butler, D. (1996). An adaptive genetic algorithm for image data compression. In Koza, J. R., editor, *Late Breaking Papers at the Genetic Programming 1996 Conference*, pages 83–87, Stanford University, CA. Stanford Bookstore, Stanford, CA.

[Johannsen, 1911] Johannsen, W. (1911). The genotype conception of heredity. *The American Naturalist*, 45:129–159.

[Johnson et al., 1994] Johnson, M. P., Maes, P., and Darrell, T. (1994). Evolving visual routines. In Brooks, R. A. and Maes, P., editors, *Artificial Life IV, Proceedings of the fourth International Workshop on the Synthesis and Simulation of Living Systems*, pages 198–209, MIT, Cambridge, MA. MIT Press, Cambridge, MA.

[Jones, 1995] Jones, T. (1995). Crossover, macromutation, and population-based search. In Eshelman, L., editor, *Genetic Algorithms: Proceedings of the Sixth International Conference (ICGA95)*, pages 310–317, Pittsburgh, PA. Morgan Kaufmann, San Francisco, CA.

[Juels and Wattenberg, 1995] Juels, A. and Wattenberg, M. (1995). Stochastic hillclimbing as a baseline method for evaluating genetic algorithms. Technical Report CSD-94-834, Department of Computer Science, University of California at Berkeley, Berkeley, CA.

[Juille and Pollack, 1996] Juille, H. and Pollack, J. B. (1996). Massively parallel genetic programming. In Angeline, P. J. and Kinnear, Jr., K. E., editors, *Advances in Genetic Programming 2*, chapter 17, pages 339–358. MIT Press, Cambridge, MA.

[Keith and Martin, 1994] Keith, M. J. and Martin, M. C. (1994). Genetic programming in C++: Implementation issues. In Kinnear,

Jr., K. E., editor, *Advances in Genetic Programming*, chapter 13, pages 285–310. MIT Press, Cambridge, MA.

[Keller and Banzhaf, 1994] Keller, R. and Banzhaf, W. (1994). Explicit maintenance of genetic diversity on genospaces. http://ls11-www.cs.uni-dortmund.de/people/banzhaf/publications.html.

[Keller and Banzhaf, 1996] Keller, R. E. and Banzhaf, W. (1996). Genetic programming using genotype-phenotype mapping from linear genomes into linear phenotypes. In Koza, J. R., Goldberg, D. E., Fogel, D. B., and Riolo, R. L., editors, *Genetic Programming 1996: Proceedings of the First Annual Conference*, pages 116–122, Stanford University, CA. MIT Press, Cambridge, MA.

[Kernighan et al., 1988] Kernighan, B. W., Tondo, C. L., and Ritchie, D. M. (1988). *The C Programming Language.* Prentice Hall, Englewood Cliffs, NJ.

[Kimura, 1983] Kimura, M. (1983). *The Neutral Theory of Molecular Evolution.* Cambridge University Press, Cambridge, UK.

[Kinnear, Jr., 1993a] Kinnear, Jr., K. E. (1993a). Evolving a sort: Lessons in genetic programming. In *Proceedings of the 1993 International Conference on Neural Networks*, volume 2, San Francisco, CA. IEEE Press, New York.

[Kinnear, Jr., 1993b] Kinnear, Jr., K. E. (1993b). Generality and difficulty in genetic programming: Evolving a sort. In Forrest, S., editor, *Proceedings of the 5th International Conference on Genetic Algorithms, ICGA-93*, pages 287–294, University of Illinois at Urbana-Champaign. Morgan Kaufmann, San Francisco, CA.

[Kinnear, Jr., 1994] Kinnear, Jr., K. E. (1994). Alternatives in automatic function definition: A comparison of performance. In Kinnear, Jr., K. E., editor, *Advances in Genetic Programming*, chapter 6, pages 119–141. MIT Press, Cambridge, MA.

[Kirkerud, 1989] Kirkerud, B. (1989). *Object-Oriented Programming with Simula.* Addison-Wesley, London.

[Kirkpatrick et al., 1983] Kirkpatrick, S., Gelatt, C., and Vecchi, M. (1983). Optimization by simulated annealing. *Science*, 220:671–680.

[Kitano, 1990] Kitano, H. (1990). Designing neural networks using genetic algorithms with graph generation system. *Complex Systems*, 4:461–476.

[Knuth, 1981] Knuth, D. (1981). *The Art of Computer Programming. Vol. 2: Seminumerical Algorithms.* Addison-Wesley, Reading, MA, 2nd edition.

[Kohonen, 1989] Kohonen, T. (1989). *Self-organization and Associative Memory.* Springer-Verlag, Berlin, 2nd edition.

[Kolmogorov, 1950] Kolmogorov, A. (1950). *Foundations of the Theory of Probability.* Chelsea Publishing Company, New York.

[Koza, 1989] Koza, J. R. (1989). Hierarchical genetic algorithms operating on populations of computer programs. In Sridharan, N. S., editor, *Proceedings of the Eleventh International Joint Conference on Artificial Intelligence IJCAI-89*, volume 1, pages 768–774. Morgan Kaufmann, San Francisco, CA.

[Koza, 1991] Koza, J. R. (1991). Concept formation and decision tree induction using the genetic programming paradigm. In Schwefel, H.-P. and Männer, R., editors, *Parallel Problem Solving from Nature–Proceedings of 1st Workshop, PPSN 1*, volume 496 of *Lecture Notes in Computer Science*, pages 124–128. Springer-Verlag, Berlin, Germany.

[Koza, 1992a] Koza, J. R. (1992a). Evolution of subsumption using genetic programming. In Varela, F. J. and Bourgine, P., editors, *Proceedings of the First European Conference on Artificial Life. Towards a Practice of Autonomous Systems*, pages 110–119, Paris, France. MIT Press, Cambridge, MA.

[Koza, 1992b] Koza, J. R. (1992b). A genetic approach to finding a controller to back up a tractor-trailer truck. In *Proceedings of the 1992 American Control Conference*, volume III, pages 2307–2311, Evanston, IL. American Automatic Control Council.

[Koza, 1992c] Koza, J. R. (1992c). A genetic approach to the truck backer upper problem and the inter-twined spiral problem. In *Proceedings of IJCNN International Joint Conference on Neural Networks*, volume IV, pages 310–318. IEEE Press, New York.

[Koza, 1992d] Koza, J. R. (1992d). *Genetic Programming: On the Programming of Computers by Natural Selection.* MIT Press, Cambridge, MA.

[Koza, 1993a] Koza, J. R. (1993a). Discovery of rewrite rules in Lindenmayer systems and state transition rules in cellular automata via genetic programming. In *Symposium on Pattern Formation (SPF-93)*, Claremont, CA.

[Koza, 1993b] Koza, J. R. (1993b). Simultaneous discovery of detectors and a way of using the detectors via genetic programming. In

1993 IEEE International Conference on Neural Networks, volume III, pages 1794–1801, San Francisco, CA. IEEE, New York.

[Koza, 1994a] Koza, J. R. (1994a). *Genetic Programming II: Automatic Discovery of Reusable Programs.* MIT Press, Cambridge, MA.

[Koza, 1994b] Koza, J. R. (1994b). Recognizing patterns in protein sequences using iteration-performing calculations in genetic programming. In *1994 IEEE World Congress on Computational Intelligence*, Orlando, FL. IEEE Press, New York.

[Koza, 1995a] Koza, J. R. (1995a). Evolving the architecture of a multi-part program in genetic programming using architecture-altering operations. In McDonnell, J. R., Reynolds, R. G., and Fogel, D. B., editors, *Evolutionary Programming IV Proceedings of the Fourth Annual Conference on Evolutionary Programming*, pages 695–717, San Diego, CA. MIT Press, Cambridge, MA.

[Koza, 1995b] Koza, J. R. (1995b). Gene duplication to enable genetic programming to concurrently evolve both the architecture and work-performing steps of a computer program. In *IJCAI-95 Proceedings of the Fourteenth International Joint Conference on Artificial Intelligence*, volume 1, pages 734–740, Montreal, Quebec, Canada. Morgan Kaufmann, San Francisco, CA.

[Koza and Andre, 1996a] Koza, J. R. and Andre, D. (1996a). Classifying protein segments as transmembrane domains using architecture-altering operations in genetic programming. In Angeline, P. J. and Kinnear, Jr., K. E., editors, *Advances in Genetic Programming 2*, chapter 8, pages 155–176. MIT Press, Cambridge, MA.

[Koza and Andre, 1996b] Koza, J. R. and Andre, D. (1996b). Evolution of iteration in genetic programming. In Fogel, L. J., Angeline, P. J., and Baeck, T., editors, *Evolutionary Programming V: Proceedings of the Fifth Annual Conference on Evolutionary Programming.* MIT Press, Cambridge, MA.

[Koza et al., 1996a] Koza, J. R., Andre, D., Bennett III, F. H., and Keane, M. A. (1996a). Use of automatically defined functions and architecture-altering operations in automated circuit synthesis using genetic programming. In Koza, J. R., Goldberg, D. E., Fogel, D. B., and Riolo, R. L., editors, *Genetic Programming 1996: Proceedings of the First Annual Conference*, pages 132–149, Stanford University, CA. MIT Press, Cambridge, MA.

[Koza et al., 1996b] Koza, J. R., Bennett III, F. H., Andre, D., and Keane, M. A. (1996b). Automated WYWIWYG design of both the topology and component values of electrical circuits using genetic programming. In Koza, J. R., Goldberg, D. E., Fogel, D. B., and Riolo, R. L., editors, *Genetic Programming 1996: Proceedings of the First Annual Conference*, pages 123–131, Stanford University, CA. MIT Press, Cambridge, MA.

[Koza et al., 1996c] Koza, J. R., Bennett III, F. H., Andre, D., and Keane, M. A. (1996c). Four problems for which a computer program evolved by genetic programming is competitive with human performance. In *Proceedings of the 1996 IEEE International Conference on Evolutionary Computation*, volume 1, pages 1–10. IEEE Press, New York.

[Koza et al., 1997a] Koza, J. R., Bennett III, F. H., Keane, M. A., and Andre, D. (1997a). Evolution of a time-optimal fly-to controller circuit using genetic programming. In Koza, J. R., Deb, K., Dorigo, M., Fogel, D. B., Garzon, M., Iba, H., and Riolo, R. L., editors, *Genetic Programming 1997: Proceedings of the Second Annual Conference*, pages 207–212, Stanford University, CA. Morgan Kaufmann, San Francisco, CA.

[Koza et al., 1997b] Koza, J. R., Bennett III, F. H., Lohn, J., Dunlap, F., Keane, M. A., and Andre, D. (1997b). Use of architecture-altering operations to dynamically adapt a three-way analog source identification circuit to accommodate a new source. In Koza, J. R., Deb, K., Dorigo, M., Fogel, D. B., Garzon, M., Iba, H., and Riolo, R. L., editors, *Genetic Programming 1997: Proceedings of the Second Annual Conference*, pages 213–221, Stanford University, CA. Morgan Kaufmann, San Francisco, CA.

[Koza and Keane, 1990] Koza, J. R. and Keane, M. A. (1990). Cart centering and broom balancing by genetically breeding populations of control strategy programs. In *Proceedings of International Joint Conference on Neural Networks*, volume I, pages 198–201, Washington, DC. Lawrence Erlbaum Associates, Hillsdale, NJ.

[Kraft et al., 1994] Kraft, D. H., Petry, F. E., Buckles, W. P., and Sadasivan, T. (1994). The use of genetic programming to build queries for information retrieval. In *Proceedings of the 1994 IEEE World Congress on Computational Intelligence*, pages 468–473, Orlando, FL. IEEE Press, New York.

[Lang, 1995] Lang, K. J. (1995). Hill climbing beats genetic search on a boolean circuit synthesis of Koza's. In *Proceedings of the Twelfth*

International Conference on Machine Learning, Tahoe City, CA. Morgan Kaufmann, San Francisco, CA.

[Langdon, 1995a] Langdon, W. B. (1995a). Data structures and genetic programming. Research Note RN/95/70, University College London, University College London, Dept. of Computer Science, London, UK.

[Langdon, 1995b] Langdon, W. B. (1995b). Evolving data structures using genetic programming. In Eshelman, L., editor, *Genetic Algorithms: Proceedings of the Sixth International Conference (ICGA95)*, pages 295–302, Pittsburgh, PA. Morgan Kaufmann, San Francisco, CA.

[Langdon, 1995c] Langdon, W. B. (1995c). Evolving data structures using genetic programming. Research Note RN/95/1, UCL, University College London, Dept. of Computer Science, London, UK.

[Langdon, 1996a] Langdon, W. B. (1996a). Scheduling maintenance of electrical power transmission networks using genetic programming. Research Note RN/96/49, University College London, University College London, Dept. of Computer Science, London, UK.

[Langdon, 1996b] Langdon, W. B. (1996b). Using data structures within genetic programming. In Koza, J. R., Goldberg, D. E., Fogel, D. B., and Riolo, R. L., editors, *Genetic Programming 1996: Proceedings of the First Annual Conference*, pages 141–148, Stanford University, CA. MIT Press, Cambridge, MA.

[Langdon and Poli, 1997] Langdon, W. B. and Poli, R. (1997). An analysis of the MAX problem in genetic programming. In Koza, J. R., Deb, K., Dorigo, M., Fogel, D. B., Garzon, M., Iba, H., and Riolo, R. L., editors, *Genetic Programming 1997: Proceedings of the Second Annual Conference*, pages 222–230, Stanford University, CA. Morgan Kaufmann, San Francisco, CA.

[Langdon and Qureshi, 1995] Langdon, W. B. and Qureshi, A. (1995). Genetic programming – computers using "natural selection" to generate programs. Research Note RN/95/76, University College London, University College London, Dept. of Computer Science, London, UK.

[Langley, 1996] Langley, P. (1996). *Elements of Machine Learning*. Morgan Kaufmann, San Francisco, CA.

[Lay, 1994] Lay, M.-Y. (1994). Application of genetic programming in analyzing multiple steady states of dynamical systems. In *Proceedings of the 1994 IEEE World Congress on Computational Intelligence*, pages 333–336b, Orlando, FL. IEEE Press, New York.

[Lee, 1995] Lee, G. Y. (1995). Explicit models for chaotic and noisy time series through the genetic recursive regression. Unpublished.

[Lehmer, 1951] Lehmer, D. (1951). In *Proc. 2nd Symp. on Large-Scale Digital Calculating Machinery Cambridge*, pages 141–146. Harvard University Press, Cambridge, MA.

[Lensberg, 1997] Lensberg, T. (1997). A genetic programming experiment on investment behavior under knightian uncertainty. In Koza, J. R., Deb, K., Dorigo, M., Fogel, D. B., Garzon, M., Iba, H., and Riolo, R. L., editors, *Genetic Programming 1997: Proceedings of the Second Annual Conference*, pages 231–239, Stanford University, CA. Morgan Kaufmann, San Francisco, CA.

[Lent, 1994] Lent, B. (1994). Evolution of trade strategies using genetic algorithms and genetic programming. In Koza, J. R., editor, *Genetic Algorithms at Stanford 1994*, pages 87–98. Stanford Bookstore, Stanford, CA.

[Levenshtein, 1966] Levenshtein, A. (1966). Binary codes capable of correcting deletions, insertions, and reversals. *Soviet Physics-Doklady*, 10:703–710.

[Li and Vitanyi, 1997] Li, M. and Vitanyi, P. (1997). *An Introduction to Kolmogorov Complexity and its Applications.* Springer-Verlag, Berlin, 2nd revised and expanded edition.

[Lindenmayer, 1968] Lindenmayer, A. (1968). Mathematical models for cellular interaction in development. part I and II. *Journal of Theoretical Biology*, 18:280–315.

[Lohmann, 1992] Lohmann, R. (1992). Structure evolution and incomplete induction. In Männer, R. and Manderick, B., editors, *Proceedings PPSN II*, pages 175–185. North-Holland, Amsterdam.

[Lott, 1994] Lott, C. G. (1994). Terrain flattening by autonomous robot: A genetic programming application. In Koza, J. R., editor, *Genetic Algorithms at Stanford 1994*, pages 99–109. Stanford Bookstore, Stanford, CA.

[Ludvikson, 1995] Ludvikson, A. (1995). Private communication.

[Luke and Spector, 1996] Luke, S. and Spector, L. (1996). Evolving teamwork and coordination with genetic programming. In Koza, J. R., Goldberg, D. E., Fogel, D. B., and Riolo, R. L., editors, *Genetic Programming 1996: Proceedings of the First Annual Conference*, pages 150–156, Stanford University, CA. MIT Press, Cambridge, MA.

[Maniatis, 1991] Maniatis, T. (1991). Review. *Science*, 251:33–34.

[Masand, 1994] Masand, B. (1994). Optimising confidence of text classification by evolution of symbolic expressions. In Kinnear, Jr., K. E., editor, *Advances in Genetic Programming*, chapter 21, pages 445–458. MIT Press, Cambridge, MA.

[Masand and Piatesky-Shapiro, 1996] Masand, B. and Piatesky-Shapiro, G. (1996). Discovering time oriented abstractions in historical data to optimize decision tree classification. In Angeline, P. J. and Kinnear, Jr., K. E., editors, *Advances in Genetic Programming 2*, chapter 24, pages 489–498. MIT Press, Cambridge, MA.

[Masters, 1995a] Masters, T. (1995a). *Advanced Algorithms for Neural Networks.* John Wiley & Sons, New York.

[Masters, 1995b] Masters, T. (1995b). *Neural, Novel and Hybrid Algorithms for Time Series Prediction.* John Wiley & Sons, New York.

[Maynard-Smith, 1994] Maynard-Smith, J. (1994). *Evolutionary Genetics.* Oxford University Press, Oxford, UK.

[McKay et al., 1995] McKay, B., Willis, M., and Barton, G. W. (1995). On the application of genetic programming to chemical process systems. In *1995 IEEE Conference on Evolutionary Computation*, volume 2, pages 701–706, Perth, Australia. IEEE Press, New York.

[McKay et al., 1996] McKay, B., Willis, M., Montague, G., and Barton, G. W. (1996). Using genetic programming to develop inferential estimation algorithms. In Koza, J. R., Goldberg, D. E., Fogel, D. B., and Riolo, R. L., editors, *Genetic Programming 1996: Proceedings of the First Annual Conference*, pages 157–165, Stanford University, CA. MIT Press, Cambridge, MA.

[McKnight et al., 1994] McKnight, R., Wall, R., and Hennighausen, L. (1994). Expressions of genomic and CDNA transgenes after co-integration in transgenic mice. Technical Report ARS Report Number: 0000031671, United States Department of Agriculture, National Agricultural Library.

[McPhee and Miller, 1995] McPhee, N. F. and Miller, J. D. (1995). Accurate replication in genetic programming. In Eshelman, L., editor, *Genetic Algorithms: Proceedings of the Sixth International Conference (ICGA95)*, pages 303–309, Pittsburgh, PA. Morgan Kaufmann, San Francisco, CA.

[Mehta et al., 1995] Mehta, P., Heringa, J., and Argos, P. (1995). A simple and fast approach to prediction of protein secondary structure from multiply aligned sequences with accuracy above 70%. *Protein Science*, 4:2517–2525.

[Meisel, 1972] Meisel, W. (1972). *Computer-Oriented Approaches to Pattern Recognition.* Academic Press, New York.

[Michalewicz, 1994] Michalewicz, Z. (1994). *Genetic Algorithms + Data Structures = Evolution Programs.* Springer-Verlag, Berlin, 1996: 3rd edition.

[Mitchell, 1996] Mitchell, T. (1996). *Machine Learning.* McGraw-Hill, New York.

[Mjolsness et al., 1995] Mjolsness, E., Garret, C., Reinitz, J., and Sharp, D. (1995). Modelling the connection between development and evolution: Preliminary report. In Banzhaf, W. and Eeckman, F., editors, *Evolution and Biocomputation.* Springer-Verlag, Berlin.

[Mjolsness et al., 1988] Mjolsness, E., Sharp, D., and Albert, B. (1988). Scaling, machine learning and genetic neural nets. Technical Report LA-UR-88-142, Los Alamos National Laboratories.

[Mondada et al., 1993] Mondada, F., Franzi, E., and Ienne, P. (1993). Mobile robot miniaturization. In Yoshikawa, T. and Miyazaki, F., editors, *Experimental Robotics III: The 3rd International Symposium,* Lecture Notes in Control and Information Sciences, Vol, 200., Kyoto, Japan. Springer-Verlag, Berlin.

[Montana, 1994] Montana, D. J. (1994). Strongly typed genetic programming. BBN Technical Report #7866, Bolt Beranek and Newman, Cambridge, MA.

[Morgan and Schonfelder, 1993] Morgan, J. and Schonfelder, J. (1993). *Programming in Fortran 90.* Alfred Waller, Oxfordshire, UK.

[Mühlenbein and Schlierkamp-Voosen, 1994] Mühlenbein, H. and Schlierkamp-Voosen, D. (1994). The science of breeding and its application to the breeder genetic algorithm. *Evolutionary Computation*, 1:335–360.

[Muller, 1932] Muller, H. J. (1932). Some genetic aspects of sex. *The American Naturalist*, 66:118–138.

[Mulloy et al., 1996] Mulloy, B. S., Riolo, R. L., and Savit, R. S. (1996). Dynamics of genetic programming and chaotic time series prediction. In Koza, J. R., Goldberg, D. E., Fogel, D. B., and Riolo, R. L., editors, *Genetic Programming 1996: Proceedings of the First Annual Conference*, pages 166–174, Stanford University, CA. MIT Press, Cambridge, MA.

[Ngo and Marks, 1993] Ngo, J. and Marks, J. (1993). Physically realistic motion synthesis in animation. *Evolutionary Computation*, 1(3):235–268.

[Nguyen and Huang, 1994] Nguyen, T. and Huang, T. (1994). Evolvable 3D modeling for model-based object recognition systems. In Kinnear, Jr., K. E., editor, *Advances in Genetic Programming*, chapter 22, pages 459–475. MIT Press, Cambridge, MA.

[Nguyen et al., 1993] Nguyen, T. C., Goldberg, D. S., and Huang, T. S. (1993). Evolvable modeling: structural adaptation through hierarchical evolution for 3-D model-based vision. Technical report, Beckman Institute and Coordinated Science Laboratory, University of Illinois, Urbana, IL.

[Nilsson, 1971] Nilsson, N. (1971). *Problem-Solving Methods in Artificial Intelligence.* McGraw-Hill, New York.

[Nordin, 1997] Nordin, J. (1997). *Evolutionary Program Induction of Binary Machine Code and its Application.* Krehl-Verlag, Münster, Germany.

[Nordin, 1994] Nordin, P. (1994). A compiling genetic programming system that directly manipulates the machine code. In Kinnear, Jr., K. E., editor, *Advances in Genetic Programming*, chapter 14, pages 311–331. MIT Press, Cambridge, MA.

[Nordin and Banzhaf, 1995a] Nordin, P. and Banzhaf, W. (1995a). Complexity compression and evolution. In Eshelman, L., editor, *Genetic Algorithms: Proceedings of the Sixth International Conference (ICGA95)*, pages 310–317, Pittsburgh, PA. Morgan Kaufmann, San Francisco, CA.

[Nordin and Banzhaf, 1995b] Nordin, P. and Banzhaf, W. (1995b). Evolving Turing-complete programs for a register machine with self-modifying code. In Eshelman, L., editor, *Genetic Algorithms: Proceedings of the Sixth International Conference (ICGA95)*, pages 318–325, Pittsburgh, PA. Morgan Kaufmann, San Francisco, CA.

[Nordin and Banzhaf, 1995c] Nordin, P. and Banzhaf, W. (1995c). Genetic programming controlling a miniature robot. In Siegel, E. V. and Koza, J. R., editors, *Working Notes for the AAAI Symposium on Genetic Programming*, pages 61–67, MIT, Cambridge, MA. AAAI, Menlo Park, CA.

[Nordin and Banzhaf, 1996] Nordin, P. and Banzhaf, W. (1996). Programmatic compression of images and sound. In Koza, J. R., Goldberg, D. E., Fogel, D. B., and Riolo, R. L., editors, *Genetic Pro-*

gramming 1996: Proceedings of the First Annual Conference, pages 345–350, Stanford University, CA. MIT Press, Cambridge, MA.

[Nordin and Banzhaf, 1997a] Nordin, P. and Banzhaf, W. (1997a). Genetic reasoning – evolving proofs with genetic search. In Koza, J. R., Deb, K., Dorigo, M., Fogel, D. B., Garzon, M., Iba, H., and Riolo, R. L., editors, *Genetic Programming 1997: Proceedings of the Second Annual Conference*, July 13–16, Stanford University, Stanford, CA. Morgan Kaufmann, San Francisco, CA.

[Nordin and Banzhaf, 1997b] Nordin, P. and Banzhaf, W. (1997b). An on-line method to evolve behavior and to control a miniature robot in real time with genetic programming. *Adaptive Behavior*, 5:107–140.

[Nordin and Banzhaf, 1997c] Nordin, P. and Banzhaf, W. (1997c). Real time control of a khepera robot using genetic programming. *Control and Cybernetics*, 26 (3).

[Nordin et al., 1995] Nordin, P., Francone, F., and Banzhaf, W. (1995). Explicitly defined introns and destructive crossover in genetic programming. In Rosca, J. P., editor, *Proceedings of the Workshop on Genetic Programming: From Theory to Real-World Applications*, pages 6–22, Tahoe City, CA.

[Nordin et al., 1996] Nordin, P., Francone, F., and Banzhaf, W. (1996). Explicitly defined introns and destructive crossover in genetic programming. In Angeline, P. J. and Kinnear, Jr., K. E., editors, *Advances in Genetic Programming 2*, chapter 6, pages 111–134. MIT Press, Cambridge, MA.

[Oakley, 1993] Oakley, E. H. N. (1993). Signal filtering and data processing for laser rheometry. Technical report, Institute of Naval Medicine, Portsmouth, UK.

[Oakley, 1994a] Oakley, E. H. N. (1994a). The application of genetic programming to the investigation of short, noisy, chaotic data series. In Fogarty, T. C., editor, *Evolutionary Computing*, volume 865 of *Lecture Notes in Computer Science*. Springer-Verlag, Berlin, Germany.

[Oakley, 1994b] Oakley, E. H. N. (1994b). Two scientific applications of genetic programming: Stack filters and non-linear equation fitting to chaotic data. In Kinnear, Jr., K. E., editor, *Advances in Genetic Programming*, chapter 17, pages 369–389. MIT Press, Cambridge, MA.

[Oakley, 1996] Oakley, E. H. N. (1996). Genetic programming, the reflection of chaos, and the bootstrap: Towards a useful test for chaos. In Koza, J. R., Goldberg, D. E., Fogel, D. B., and Riolo, R. L., editors, *Genetic Programming 1996: Proceedings of the First Annual Conference*, pages 175–181, Stanford University, CA. MIT Press, Cambridge, MA.

[Ohno, 1970] Ohno, S. (1970). *Evolution by Gene Duplication.* Springer-Verlag, New York.

[Olmer et al., 1996] Olmer, M., Banzhaf, W., and Nordin, P. (1996). Evolving real-time behavior modules for a real robot with genetic programming. In *Proceedings of the international symposium on robotics and manufacturing*, Montpellier, France.

[Openshaw and Turton, 1994] Openshaw, S. and Turton, I. (1994). Building new spatial interaction models using genetic programming. In Fogarty, T. C., editor, *Evolutionary Computing*, volume 865 of *Lecture Notes in Computer Science.* Springer-Verlag, Berlin, Germany.

[O'Reilly, 1995] O'Reilly, U.-M. (1995). *An Analysis of Genetic Programming.* PhD thesis, Carleton University, Ottawa-Carleton Institute for Computer Science, Ottawa, Ontario, Canada.

[O'Reilly and Oppacher, 1992] O'Reilly, U. M. and Oppacher, F. (1992). The troubling aspects of a building block hypothesis for genetic programming. Working Paper 94-02-001, Santa Fe Institute, Santa Fe, NM.

[O'Reilly and Oppacher, 1994a] O'Reilly, U.-M. and Oppacher, F. (1994a). Program search with a hierarchical variable length representation: Genetic programming, simulated annealing and hill climbing. In Davidor, Y., Schwefel, H.-P., and M R., editors, *Parallel Problem Solving from Nature – PPSN III.*

[O'Reilly and Oppacher, 1994b] O'Reilly, U.-M. and Oppacher, F. (1994b). Using building block functions to investigate a building block hypothesis for genetic programming. Working Paper 94-02-029, Santa Fe Institute, Santa Fe, NM.

[O'Reilly and Oppacher, 1995a] O'Reilly, U.-M. and Oppacher, F. (1995a). Hybridized crossover-based search techniques for program discovery. In *Proceedings of the 1995 World Conference on Evolutionary Computation*, volume 2, page 573, Perth, Australia.

[O'Reilly and Oppacher, 1995b] O'Reilly, U.-M. and Oppacher, F. (1995b). The troubling aspects of a building block hypothesis for genetic programming. In Whitley, L. D. and Vose, M. D., editors, *Foundations of Genetic Algorithms 3*, pages 73–88, Estes Park, CO. Morgan Kaufmann, San Francisco, CA.

[O'Reilly and Oppacher, 1996] O'Reilly, U.-M. and Oppacher, F. (1996). A comparative analysis of GP. In Angeline, P. J. and Kinnear, Jr., K. E., editors, *Advances in Genetic Programming 2*, chapter 2, pages 23–44. MIT Press, Cambridge, MA.

[Orgel, 1979] Orgel, L. (1979). Selection in vitro. *Proc. Royal Soc. of London.*

[Osborn et al., 1995] Osborn, T. R., Charif, A., Lamas, R., and Dubossarsky, E. (1995). Genetic logic programming. In *1995 IEEE Conference on Evolutionary Computation*, volume 2, page 728, Perth, Australia. IEEE Press, New York.

[Oussaidene et al., 1996] Oussaidene, M., Chopard, B., Pictet, O. V., and Tomassini, M. (1996). Parallel genetic programming: An application to trading models evolution. In Koza, J. R., Goldberg, D. E., Fogel, D. B., and Riolo, R. L., editors, *Genetic Programming 1996: Proceedings of the First Annual Conference*, pages 357–380, Stanford University, CA. MIT Press, Cambridge, MA.

[Park and Song, 1997] Park, Y. and Song, M. (1997). Genetic programming approach to sense clustering in natural language processing. In Koza, J. R., Deb, K., Dorigo, M., Fogel, D. B., Garzon, M., Iba, H., and Riolo, R. L., editors, *Genetic Programming 1997: Proceedings of the Second Annual Conference*, page 261, Stanford University, CA. Morgan Kaufmann, San Francisco, CA.

[Parzen, 1962] Parzen, E. (1962). On estimation of a probability density function and mode. *Annals of Mathematical Statistics*, 33:1065–1076.

[Paterson and Livesey, 1997] Paterson, N. and Livesey, M. (1997). Evolving caching algorithms in C by genetic programming. In Koza, J. R., Deb, K., Dorigo, M., Fogel, D. B., Garzon, M., Iba, H., and Riolo, R. L., editors, *Genetic Programming 1997: Proceedings of the Second Annual Conference*, pages 262–267, Stanford University, CA. Morgan Kaufmann, San Francisco, CA.

[Pei et al., 1995] Pei, M., Goodman, E., Punch, W., and Ding, Y. (1995). Further research on feature selection and classification using genetic algorithms for classification and feature extraction. In *Proc. of the Annual Meeting of the Classification Society of North America*, Denver, CO.

[Perry, 1994] Perry, J. E. (1994). The effect of population enrichment in genetic programming. In *Proceedings of the 1994 IEEE World Congress on Computational Intelligence*, pages 456–461, Orlando, FL. IEEE Press, New York.

[Poli and Cagnoni, 1997] Poli, R. and Cagnoni, S. (1997). Genetic programming with user-driven selection: Experiments on the evolution of algorithms for image enhancement. In Koza, J. R., Deb, K., Dorigo, M., Fogel, D. B., Garzon, M., Iba, H., and Riolo, R. L., editors, *Genetic Programming 1997: Proceedings of the Second Annual Conference*, pages 269–277, Stanford University, CA. Morgan Kaufmann, San Francisco, CA.

[Poli and Langdon, 1997a] Poli, R. and Langdon, W. B. (1997a). An experimental analysis of schema creation, propagation and disruption in genetic programming. In Goodman, E., editor, *Genetic Algorithms: Proceedings of the Seventh International Conference*, Michigan State University, East Lansing, MI. Morgan Kaufmann, San Francisco, CA.

[Poli and Langdon, 1997b] Poli, R. and Langdon, W. B. (1997b). A new schema theory for genetic programming with one-point crossover and point mutation. In Koza, J. R., Deb, K., Dorigo, M., Fogel, D. B., Garzon, M., Iba, H., and Riolo, R. L., editors, *Genetic Programming 1997: Proceedings of the Second Annual Conference*, pages 278–285, Stanford University, CA. Morgan Kaufmann, San Francisco, CA.

[Prusinkiewicz and Lindenmayer, 1990] Prusinkiewicz, P. and Lindenmayer, A. (1990). *The Algorithmic Beauty of Plants.* Springer-Verlag, New York.

[Punch et al., 1993] Punch, W., Goodman, E., and Pei, M. (1993). Further research on feature selection and classification using genetic algorithms. In *Proceedings of the Fifth International Conference on Genetic Algorithms, ICGA-93*, page 557. Morgan Kaufmann, San Mateo, CA.

[Quinlan, 1979] Quinlan, J. (1979). Discovering rules by induction from large collections of examples. In Michie, D., editor, *In Expert Systems in the Micro-electronic Age*, pages 168–201. Edinburgh University Press, Edinburgh, UK.

[Quinlan, 1993] Quinlan, J. (1993). *C4.5: Programs for Machine Learning.* Morgan Kaufmann, San Francisco, CA.

[Qureshi, 1996] Qureshi, A. (1996). Evolving agents. Research Note RN/96/4, UCL, University College London, Dept. of Computer Science, London, UK.

[Rao and Rao, 1995] Rao, V. and Rao, H. (1995). *C++, Neural Networks and Fuzzy Logic.* MIS:Press/M&T Books, New York, 2nd edition.

[Raymer et al., 1996] Raymer, M. L., Punch, W. F., Goodman, E. D., and Kuhn, L. A. (1996). Genetic programming for improved data mining: An application to the biochemistry of protein interactions. In Koza, J. R., Goldberg, D. E., Fogel, D. B., and Riolo, R. L., editors, *Genetic Programming 1996: Proceedings of the First Annual Conference*, pages 375–380, Stanford University, CA. MIT Press, Cambridge, MA.

[Rechenberg, 1994] Rechenberg, I. (1994). *Evolutionsstrategie '93.* Frommann Verlag, Stuttgart, Germany.

[Reynolds, 1992] Reynolds, C. W. (1992). An evolved, vision-based behavioral model of coordinated group motion. In Meyer, J.-A. and Wilson, S. W., editors, *From Animals to Animats (Proceedings of Simulation of Adaptive Behaviour).* MIT Press, Cambridge, MA.

[Reynolds, 1994a] Reynolds, C. W. (1994a). Evolution of corridor following behavior in a noisy world. In *Simulation of Adaptive Behaviour (SAB-94).*

[Reynolds, 1994b] Reynolds, C. W. (1994b). Evolution of obstacle avoidance behaviour: Using noise to promote robust solutions. In Kinnear, Jr., K. E., editor, *Advances in Genetic Programming*, chapter 10, pages 221–241. MIT Press, Cambridge, MA.

[Rissanen, 1984] Rissanen, J. (1984). Universal coding, information, prediction and estimation. *IEEE Transactions on Information Theory*, 30:629–636.

[Robinson and McIlroy, 1995a] Robinson, G. and McIlroy, P. (1995a). Exploring some commercial applications of genetic programming. In Fogarty, T. C., editor, *Evolutionary Computing*, volume 993 of *Lecture Notes in Computer Science.* Springer-Verlag, Berlin, Germany.

[Robinson and McIlroy, 1995b] Robinson, G. and McIlroy, P. (1995b). Exploring some commercial applications of genetic programming. Project 4487, British Telecom, Systems Research Division, Martelsham, Ipswich, UK.

[Ronge, 1996] Ronge, A. (1996). Genetic programs and co-evolution. Technical Report TRITA-NA-E9625, Stockholm University, Dept. of Numerical Analysis and Computer Science.

[Rosca, 1995a] Rosca, J. (1995a). Towards automatic discovery of building blocks in genetic programming. In Siegel, E. V. and Koza, J. R., editors, *Working Notes for the AAAI Symposium on Genetic Programming*, pages 78–85, MIT, Cambridge, MA. AAAI, Menlo Park, CA.

[Rosca, 1995b] Rosca, J. P. (1995b). Entropy-driven adaptive representation. In Rosca, J. P., editor, *Proceedings of the Workshop on Genetic Programming: From Theory to Real-World Applications*, pages 23–32, Tahoe City, CA.

[Rosca, 1997] Rosca, J. P. (1997). Analysis of complexity drift in genetic programming. In Koza, J. R., Deb, K., Dorigo, M., Fogel, D. B., Garzon, M., Iba, H., and Riolo, R. L., editors, *Genetic Programming 1997: Proceedings of the Second Annual Conference*, pages 286–294, Stanford University, CA. Morgan Kaufmann, San Francisco, CA.

[Rosca and Ballard, 1994a] Rosca, J. P. and Ballard, D. H. (1994a). Genetic programming with adaptive representations. Technical Report TR 489, University of Rochester, Computer Science Department, Rochester, NY.

[Rosca and Ballard, 1994b] Rosca, J. P. and Ballard, D. H. (1994b). Learning by adapting representations in genetic programming. In *Proceedings of the 1994 IEEE World Congress on Computational Intelligence, Orlando, FL*, Orlando, FL. IEEE Press, New York.

[Rose, 1994] Rose, A. (1994). Determining the intron sequences required to enhance gene expression. Summary of Presentation 1-25, University of California at Davis, Molecular and Cellular Biology, University of California at Davis, CA.

[Ross et al., 1996] Ross, S. J., Daida, J. M., Doan, C. M., Bersano-Begey, T. F., and McClain, J. J. (1996). Variations in evolution of subsumption architectures using genetic programming: The wall following robot revisited. In Koza, J. R., Goldberg, D. E., Fogel, D. B., and Riolo, R. L., editors, *Genetic Programming 1996: Proceedings of the First Annual Conference*, pages 191–199, Stanford University, CA. MIT Press, Cambridge, MA.

[Rost and Sander, 1993] Rost, B. and Sander, C. (1993). Prediction of protein secondary structure at better than 70% accuracy. *J. Mol. Biol.*, 232:584–599.

[Rush et al., 1994] Rush, J. R., Fraser, A. P., and Barnes, D. P. (1994). Evolving co-operation in autonomous robotic systems. In *Proceedings of the IEE International Conference on Control,* March 21-24, 1994, London. IEE, London, UK.

[Ryan, 1994] Ryan, C. (1994). Pygmies and civil servants. In Kinnear, Jr., K. E., editor, *Advances in Genetic Programming*, chapter 11, pages 243–263. MIT Press, Cambridge, MA.

[Ryan and Walsh, 1997] Ryan, C. and Walsh, P. (1997). The evolution of provable parallel programs. In Koza, J. R., Deb, K., Dorigo, M., Fogel, D. B., Garzon, M., Iba, H., and Riolo, R. L., editors, *Genetic Programming 1997: Proceedings of the Second Annual Conference*, pages 295–302, Stanford University, CA. Morgan Kaufmann, San Francisco, CA.

[Ryu and Eick, 1996] Ryu, T.-W. and Eick, C. F. (1996). MASSON: discovering commonalties in collection of objects using genetic programming. In Koza, J. R., Goldberg, D. E., Fogel, D. B., and Riolo, R. L., editors, *Genetic Programming 1996: Proceedings of the First Annual Conference*, pages 200–208, Stanford University, CA. MIT Press, Cambridge, MA.

[Samuel, 1963] Samuel, A. (1963). Some studies in machine learning using the game of checkers. In Feigenbaum, E. and Feldman, J., editors, *Computers and Thought*. McGraw-Hill, New York.

[Sankoff and Kruskal, 1983] Sankoff, S. and Kruskal, J. (1983). *Time Warps, String Edits and Macromolecules: The Theory and Practice of Sequence Comparison*. Addison-Wesley, Reading, MA.

[Schoenauer et al., 1995] Schoenauer, M., Lamy, B., and Jouve, F. (1995). Identification of mechanical behaviour by genetic programming part II: Energy formulation. Technical report, Ecole Polytechnique, 91128 Palaiseau, France.

[Schoenauer et al., 1996] Schoenauer, M., Sebag, M., Jouve, F., Lamy, B., and Maitournam, H. (1996). Evolutionary identification of macro-mechanical models. In Angeline, P. J. and Kinnear, Jr., K. E., editors, *Advances in Genetic Programming 2*, chapter 23, pages 467–488. MIT Press, Cambridge, MA.

[Schwefel, 1995] Schwefel, H.-P. (1995). *Evolution and Optimum Seeking*. Sixth-Generation Computer Technology Series. John Wiley & Sons, New York.

[Schwefel and Rudolph, 1995] Schwefel, H.-P. and Rudolph, G. (1995). Contemporary evolution strategies. In *Advances in Artificial Life*, pages 893–907. Springer-Verlag, Berlin.

[Sedgewick, 1992] Sedgewick, R. (1992). *Algorithms in Pascal*. Addison-Wesley, Reading, MA.

[Segovia and Isasi, 1997] Segovia, J. and Isasi, P. (1997). Genetic programming for designing ad hoc neural network learning rules. In Koza, J. R., Deb, K., Dorigo, M., Fogel, D. B., Garzon, M., Iba, H., and Riolo, R. L., editors, *Genetic Programming 1997:*

Proceedings of the Second Annual Conference, page 303, Stanford University, CA. Morgan Kaufmann, San Francisco, CA.

[Sharman and Esparcia-Alcazar, 1993] Sharman, K. C. and Esparcia-Alcazar, A. I. (1993). Genetic evolution of symbolic signal models. In *Proceedings of the Second International Conference on Natural Algorithms in Signal Processing, NASP'93*, Essex University.

[Sherrah et al., 1997] Sherrah, J. R., Bogner, R. E., and Bouzerdoum, A. (1997). The evolutionary pre-processor: Automatic feature extraction for supervised classification using genetic programming. In Koza, J. R., Deb, K., Dorigo, M., Fogel, D. B., Garzon, M., Iba, H., and Riolo, R. L., editors, *Genetic Programming 1997: Proceedings of the Second Annual Conference*, pages 304–312, Stanford University, CA. Morgan Kaufmann, San Francisco, CA.

[Siegel, 1994] Siegel, E. V. (1994). Competitively evolving decision trees against fixed training cases for natural language processing. In Kinnear, Jr., K. E., editor, *Advances in Genetic Programming*, chapter 19, pages 409–423. MIT Press, Cambridge, MA.

[Sims, 1991a] Sims, K. (1991a). Artificial evolution for computer graphics. Technical Report TR-185, Thinking Machines Corporation, CA.

[Sims, 1991b] Sims, K. (1991b). Artificial evolution for computer graphics. *ACM Computer Graphics*, 25(4):319–328. SIGGRAPH '91 Proceedings.

[Sims, 1992a] Sims, K. (1992a). Interactive evolution of dynamical systems. In Varela, F. J. and Bourgine, P., editors, *Toward a Practice of Autonomous Systems: Proceedings of the First European Conference on Artificial Life*, pages 171–178, Paris, France. MIT Press, Cambridge, MA.

[Sims, 1992b] Sims, K. (1992b). Interactive evolution of equations for procedural models. In *Proceedings of IMAGINA conference*, Monte Carlo, January 29–31, 1992.

[Sims, 1993a] Sims, K. (1993a). Evolving images. Lecture. Presented at Centre George Pompidou, Paris on March 4, 1993.

[Sims, 1993b] Sims, K. (1993b). Interactive evolution of equations for procedural models. *The Visual Computer*, 9:466–476.

[Sims, 1994] Sims, K. (1994). Evolving 3D morphology and behavior by competition. In Brooks, R. and Maes, P., editors, *Proc. Artifi-*

cial Life IV, pages 28–39, Cambridge, MA. MIT Press, Cambridge, MA.

[Smith, 1996] Smith, P. (1996). Conjugation – A bacterially inspired form of genetic recombination. In Koza, J. R., editor, *Late Breaking Papers at the Genetic Programming 1996 Conference*, pages 167–176, Stanford University, CA. Stanford Bookstore, Stanford, CA.

[Smith, 1980] Smith, S. F. (1980). *A Learning System Based on Genetic Adaptive Algorithms.* University of Pittsburgh.

[Soule and Foster, 1997a] Soule, T. and Foster, J. (1997a). Code size and depth flows in genetic programming. In Koza, J., Deb, K., Dorigo, M., Fogel, D., Garzon, M., Iba, H., and Riolo, R., editors, *Genetic Programming 1997: Proceedings of the Second Annual Conference,* July 13–16, 1997, pages 313–320, Stanford University, Stanford, CA. Morgan Kaufmann, San Francisco, CA.

[Soule and Foster, 1997b] Soule, T. and Foster, J. A. (1997b). Code size and depth flows in genetic programming. In Koza, J. R., Deb, K., Dorigo, M., Fogel, D. B., Garzon, M., Iba, H., and Riolo, R. L., editors, *Genetic Programming 1997: Proceedings of the Second Annual Conference*, pages 313–320, Stanford University, Stanford, CA. Morgan Kaufmann, San Francisco, CA.

[Soule et al., 1996] Soule, T., Foster, J. A., and Dickinson, J. (1996). Code growth in genetic programming. In Koza, J. R., Goldberg, D. E., Fogel, D. B., and Riolo, R. L., editors, *Genetic Programming 1996: Proceedings of the First Annual Conference*, pages 215–223, Stanford University, CA. MIT Press, Cambridge, MA.

[Specht, 1990] Specht, D. (1990). Probabilistic neural networks. *Neural Networks*, 3:109–118.

[Specht, 1991] Specht, D. (1991). A general regression neural network. *IEEE Transactions on Neural Networks*, 2:568–576.

[Spector, 1994] Spector, L. (1994). Genetic programming and AI planning systems. In *Proceedings of Twelfth National Conference on Artificial Intelligence*, Seattle, WA. AAAI, Menlo Park, CA/MIT Press, Cambridge, MA.

[Spector, 1996] Spector, L. (1996). Simultaneous evolution of programs and their control structures. In Angeline, P. J. and Kinnear, Jr., K. E., editors, *Advances in Genetic Programming 2*, chapter 7, pages 137–154. MIT Press, Cambridge, MA.

[Spector and Alpern, 1994] Spector, L. and Alpern, A. (1994). Criticism, culture, and the automatic generation of artworks. In *Proceedings of Twelfth National Conference on Artificial Intelligence*, pages 3–8, Seattle, WA. AAAI, Menlo Park, CA/MIT Press, Cambridge, MA.

[Spector and Alpern, 1995] Spector, L. and Alpern, A. (1995). Induction and recapitulation of deep musical structure. In *Proceedings of International Joint Conference on Artificial Intelligence, IJCAI'95 Workshop on Music and AI*, Montreal, Quebec, Canada.

[Spector and Luke, 1996a] Spector, L. and Luke, S. (1996a). Cultural transmission of information in genetic programming. In Koza, J. R., Goldberg, D. E., Fogel, D. B., and Riolo, R. L., editors, *Genetic Programming 1996: Proceedings of the First Annual Conference*, pages 209–214, Stanford University, CA. MIT Press, Cambridge, MA.

[Spector and Luke, 1996b] Spector, L. and Luke, S. (1996b). Culture enhances the evolvability of cognition. In Cottrell, G., editor, *Proceedings of the Eighteenth Annual Conference of the Cognitive Science Society*, pages 672–677. Lawrence Erlbaum Associates, Mahwah, NJ.

[Spector and Stoffel, 1996a] Spector, L. and Stoffel, K. (1996a). Automatic generation of adaptive programs. In Maes, P., Mataric, M. J., Meyer, J.-A., Pollack, J., and Wilson, S. W., editors, *Proceedings of the Fourth International Conference on Simulation of Adaptive Behavior: From Animals to Animats 4*, pages 476–483, Cape Code. MIT Press, Cambridge, MA.

[Spector and Stoffel, 1996b] Spector, L. and Stoffel, K. (1996b). Ontogenetic programming. In Koza, J. R., Goldberg, D. E., Fogel, D. B., and Riolo, R. L., editors, *Genetic Programming 1996: Proceedings of the First Annual Conference*, pages 394–399, Stanford University, CA. MIT Press, Cambridge, MA.

[Spencer, 1993] Spencer, G. F. (1993). Automatic generation of programs for crawling and walking. In Forrest, S., editor, *Proceedings of the 5th International Conference on Genetic Algorithms, ICGA-93*, page 654, University of Illinois at Urbana-Champaign. Morgan Kaufmann, San Francisco, CA.

[Spencer, 1994] Spencer, G. F. (1994). Automatic generation of programs for crawling and walking. In Kinnear, Jr., K. E., editor, *Advances in Genetic Programming*, chapter 15, pages 335–353. MIT Press, Cambridge, MA.

[Stansifer, 1995] Stansifer, R. (1995). *The Study of Programming Languages*. Prentice Hall, Englewood Cliffs, NJ.

[Stillger and Spiliopoulou, 1996] Stillger, M. and Spiliopoulou, M. (1996). Genetic programming in database query optimization. In Koza, J. R., Goldberg, D. E., Fogel, D. B., and Riolo, R. L., editors, *Genetic Programming 1996: Proceedings of the First Annual Conference*, pages 388–393, Stanford University, CA. MIT Press, Cambridge, MA.

[Stoffel and Spector, 1996] Stoffel, K. and Spector, L. (1996). High-performance, parallel, stack-based genetic programming. In Koza, J. R., Goldberg, D. E., Fogel, D. B., and Riolo, R. L., editors, *Genetic Programming 1996: Proceedings of the First Annual Conference*, pages 224–229, Stanford University, CA. MIT Press, Cambridge, MA.

[Tackett, 1993] Tackett, W. A. (1993). Genetic programming for feature discovery and image discrimination. In Forrest, S., editor, *Proceedings of the 5th International Conference on Genetic Algorithms, ICGA-93*, pages 303–309, University of Illinois at Urbana-Champaign. Morgan Kaufmann, San Francisco, CA.

[Tackett, 1994] Tackett, W. A. (1994). *Recombination, Selection, and the Genetic Construction of Computer Programs.* PhD thesis, University of Southern California, Department of Electrical Engineering Systems.

[Tackett, 1995] Tackett, W. A. (1995). Mining the genetic program. *IEEE Expert*, 10(3):28–38.

[Tackett and Carmi, 1994] Tackett, W. A. and Carmi, A. (1994). The donut problem: Scalability and generalization in genetic programming. In Kinnear, Jr., K. E., editor, *Advances in Genetic Programming*, chapter 7, pages 143–176. MIT Press, Cambridge, MA.

[Taylor, 1995] Taylor, S. N. (1995). Evolution by genetic programming of a spatial robot juggling control algorithm. In Rosca, J. P., editor, *Proceedings of the Workshop on Genetic Programming: From Theory to Real-World Applications*, pages 104–110, Tahoe City, CA.

[Teller, 1993] Teller, A. (1993). Learning mental models. In *Proceedings of the Fifth Workshop on Neural Networks: An International Conference on Computational Intelligence: Neural Networks, Fuzzy Systems, Evolutionary Programming, and Virtual Reality*, San Francisco, CA.

[Teller, 1994a] Teller, A. (1994a). The evolution of mental models. In Kinnear, Jr., K. E., editor, *Advances in Genetic Programming*, chapter 9, pages 199–219. MIT Press, Cambridge, MA.

[Teller, 1994b] Teller, A. (1994b). Genetic programming, indexed memory, the halting problem, and other curiosities. In *Proceedings of the 7th annual Florida Artificial Intelligence Research Symposium*, pages 270–274, Pensacola, FL. IEEE Press, New York.

[Teller, 1994c] Teller, A. (1994c). Turing completeness in the language of genetic programming with indexed memory. In *Proceedings of the 1994 IEEE World Congress on Computational Intelligence*, volume 1, pages 136–141, Orlando, FL. IEEE Press, New York.

[Teller, 1996] Teller, A. (1996). Evolving programmers: The co-evolution of intelligent recombination operators. In Angeline, P. J. and Kinnear, Jr., K. E., editors, *Advances in Genetic Programming 2*, chapter 3, pages 45–68. MIT Press, Cambridge, MA.

[Teller and Veloso, 1995a] Teller, A. and Veloso, M. (1995a). A controlled experiment: Evolution for learning difficult image classification. In *Seventh Portuguese Conference On Artificial Intelligence*, volume 990 of *Lecture Notes in Computer Science*, pages 165–176, Funchal, Madeira Island, Portugal. Springer-Verlag, Berlin, Germany.

[Teller and Veloso, 1995b] Teller, A. and Veloso, M. (1995b). PADO: Learning tree structured algorithms for orchestration into an object recognition system. Technical Report CMU-CS-95-101, Department of Computer Science, Carnegie Mellon University, Pittsburgh, PA.

[Teller and Veloso, 1995c] Teller, A. and Veloso, M. (1995c). Program evolution for data mining. *The International Journal of Expert Systems*, 8(3):216–236.

[Teller and Veloso, 1996] Teller, A. and Veloso, M. (1996). PADO: A new learning architecture for object recognition. In Ikeuchi, K. and Veloso, M., editors, *Symbolic Visual Learning*, pages 81–116. Oxford University Press, Oxford, UK.

[Thedens, 1994] Thedens, D. R. (1994). Detector design by genetic programming for automated border definition in cardiac magnetic resonance images. In Koza, J. R., editor, *Genetic Algorithms at Stanford 1994*, pages 170–179. Stanford Bookstore, Stanford, CA.

[Tuerk and Gold, 1990] Tuerk, C. and Gold, L. (1990). Systematic evolution of ligands by exponential enrichment. *Science*, 249:505–510.

[Turing, 1936] Turing, A. M. (1936). On computable numbers, with an application to the Entscheidungsproblem. *Proc. London Math. Soc.*, 42:230–265.

[Veach, 1996] Veach, M. S. (1996). Recognition and reconstruction of visibility graphs using a genetic algorithm. In Koza, J. R., Goldberg, D. E., Fogel, D. B., and Riolo, R. L., editors, *Genetic Programming 1996: Proceedings of the First Annual Conference*, pages 491–498, Stanford University, CA. MIT Press, Cambridge, MA.

[von Neumann and Morgenstern, 1944] von Neumann, J. and Morgenstern, O. (1944). *Theory of Games and Economic Behavior.* Princeton University Press, Princeton, NJ.

[Walker et al., 1995] Walker, R. F., Haasdijk, E. W., and Gerrets, M. C. (1995). Credit evaluation using a genetic algorithm. In Goonatilake, S. and Treleaven, P., editors, *Intelligent Systems for Finance and Business*, chapter 3, pages 39–59. John Wiley & Sons, New York.

[Walsh and Ryan, 1996] Walsh, P. and Ryan, C. (1996). Paragen: A novel technique for the autoparallelisation of sequential programs using genetic programming. In Koza, J. R., Goldberg, D. E., Fogel, D. B., and Riolo, R. L., editors, *Genetic Programming 1996: Proceedings of the First Annual Conference*, pages 406–409, Stanford University, CA. MIT Press, Cambridge, MA.

[Watson et al., 1987] Watson, J. D., Hopkins, N. H., Roberts, J. W., Steitz, J. A., and Weiner, A. M. (1987). *Molecular Biology of the Gene.* Benjamin-Cummings, Menlo Park, CA.

[Whigham, 1995a] Whigham, P. A. (1995a). Grammatically-based genetic programming. In Rosca, J. P., editor, *Proceedings of the Workshop on Genetic Programming: From Theory to Real-World Applications*, pages 33–41, Tahoe City, CA.

[Whigham, 1995b] Whigham, P. A. (1995b). Inductive bias and genetic programming. In Zalzala, A. M. S., editor, *First International Conference on Genetic Algorithms in Engineering Systems: Innovations and Applications, GALESIA*, volume 414, pages 461–466, Sheffield, UK. IEE, London, UK.

[Whigham, 1995c] Whigham, P. A. (1995c). A schema theorem for context-free grammars. In *1995 IEEE Conference on Evolutionary*

Computation, volume 1, pages 178–181, Perth, Australia. IEEE Press, New York.

[Whigham and McKay, 1995] Whigham, P. A. and McKay, R. I. (1995). Genetic approaches to learning recursive relations. In Yao, X., editor, *Progress in Evolutionary Computation*, volume 956 of *Lecture Notes in Artificial Intelligence*, pages 17–27. Springer-Verlag, Berlin, Germany.

[White and Sofge, 1992] White, D. and Sofge, D. E. (1992). *Handbook of Intelligent Control. Neural, Fuzzy and Adaptive Approaches.* Van Nostrand Reinhold, New York.

[Whitley, 1989] Whitley, D. (1989). The genitor algorithm and selection pressure: Why rank-based allocation of reproductive trials is best. In Schaffer, J. D., editor, *Proc. 3rd Int. Conference on Genetic Algorithms*, pages 116–121, San Mateo, CA. Morgan Kaufmann, San Francisco, CA.

[Wilhelms, 1990] Wilhelms, J. (1990). Dynamics for computer graphics: A tutorial. In *ACM Siggraph'90 Course Notes*, chapter 8, pages 85–115. Dallas Convention Center, TX.

[Wineberg and Oppacher, 1994] Wineberg, M. and Oppacher, F. (1994). A representation scheme to perform program induction in a canonical genetic algorithm. In Davidor, Y., Schwefel, H.-P., and M R., editors, *Parallel Problem Solving from Nature III*, volume 866 of *Lecture Notes in Computer Science.*

[Wineberg and Oppacher, 1996] Wineberg, M. and Oppacher, F. (1996). The benefits of computing with introns. In Koza, J. R., Goldberg, D. E., Fogel, D. B., and Riolo, R. L., editors, *Genetic Programming 1996: Proceedings of the First Annual Conference*, pages 410–415, Stanford University, CA. MIT Press, Cambridge, MA.

[Winkeler and Manjunath, 1997] Winkeler, J. F. and Manjunath, B. S. (1997). Genetic programming for object detection. In Koza, J. R., Deb, K., Dorigo, M., Fogel, D. B., Garzon, M., Iba, H., and Riolo, R. L., editors, *Genetic Programming 1997: Proceedings of the Second Annual Conference*, pages 330–335, Stanford University, CA. Morgan Kaufmann, San Francisco, CA.

[Wong and Leung, 1996] Wong, M. L. and Leung, K. S. (1996). Evolving recursive functions for the even-parity problem using genetic programming. In Angeline, P. J. and Kinnear, Jr., K. E., editors, *Advances in Genetic Programming 2*, chapter 11, pages 221–240. MIT Press, Cambridge, MA.

[Yourdon and Constantine, 1979] Yourdon, E. and Constantine, L. (1979). *Structured Design: Fundamentals of a Discipline of Computer Program and Systems Design.* Prentice Hall, Englewood Cliffs, NJ.

[Zannoni and Reynolds, 1996] Zannoni, E. and Reynolds, R. (1996). Extracting design knowledge from genetic programs using cultural algorithms. In Fogel, L., Angeline, P., and Bäck, T., editors, *Proceedings of the Fifth Evolutionary Programming Conference, San Diego, CA, 1996*, Cambridge, MA. MIT Press, Cambridge, MA.

[Zhang et al., 1996] Zhang, B.-T., Kwak, J.-H., and Lee, C.-H. (1996). Building software agents for information filtering on the internet: A genetic programming approach. In Koza, J. R., editor, *Late Breaking Papers at the Genetic Programming 1996 Conference*, page 196, Stanford University, CA. Stanford Bookstore, Stanford, CA.

[Zhang and Muehlenbein, 1994] Zhang, B.-T. and Muehlenbein, H. (1994). Synthesis of sigma-pi neural networks by the breeder genetic programming. In *Proceedings of IEEE International Conference on Evolutionary Computation (ICEC-94), World Congress on Computational Intelligence*, pages 318–323, Orlando, FL. IEEE Computer Society Press.

[Zhang and Muehlenbein, 1995] Zhang, B.-T. and Muehlenbein, H. (1995). Balancing accuracy and parsimony in genetic programming. *Evolutionary Computation*, 3(1):17–38.

[Zhang and Mühlenbein, 1996] Zhang, B.-T. and Mühlenbein, H. (1996). Adaptive fitness functions for dynamic growing/pruning of program trees. In Angeline, P. J. and Kinnear, Jr., K. E., editors, *Advances in Genetic Programming 2*, chapter 12, pages 241–256. MIT Press, Cambridge, MA.

[Zhao et al., 1996] Zhao, J., Kearney, G., and Soper, A. (1996). Emotional expression classification by genetic programming. In Koza, J. R., editor, *Late Breaking Papers at the Genetic Programming 1996 Conference*, pages 197–202, Stanford University, CA. Stanford Bookstore, Stanford, CA.

[Zomorodian, 1995] Zomorodian, A. (1995). Context-free language induction by evolution of deterministic push-down automata using genetic programming. In Siegel, E. V. and Koza, J. R., editors, *Working Notes for the AAAI Symposium on Genetic Programming*, pages 127–133, MIT, Cambridge, MA. AAAI, Menlo Park, CA.

Person Index

A

B

C

D

E

F

G

H

I

J

K

L

M

N

O

P

Q

R

S

T

V

W

Z

Subject Index

A

D

E

F

J

K

L

N

O

P

Q

R

S

T

U